Topics in Applied Physics Volume 50

Topics in Applied Physics Founded by Helmut K. V. Lotsch

Light Scattering in Solids II

Basic Concepts and Instrumentation

Edited by M. Cardona and G. Güntherodt

With Contributions by
M. Cardona   R. K. Chang   G. Güntherodt
M. B. Long   H. Vogt

With 88 Figures

Springer-Verlag Berlin Heidelberg GmbH 1982

Professor Dr. *Manuel Cardona*

Max-Planck-Institut für Festkörperforschung, Heisenbergstraße 1,
D-7000 Stuttgart 80, Fed. Rep. of Germany

Professor Dr. *Gernot Güntherodt*

Universität zu Köln, II. Physikalisches Institut, Zülpicher Straße 77,
D-5000 Köln 41, Fed. Rep. of Germany

ISBN 978-3-662-31173-8   ISBN 978-3-540-39075-6 (eBook)
DOI 10.1007/978-3-540-39075-6

Originally published by Springer-Verlag Berlin Heidelberg New York in 1982
Softcover reprint of the hardcover 1st edition 1982

2153/3130-543210

This book is dedicated to the memory of
S.P.S. PORTO
who contributed so much to the field of
Light Scattering in Solids

Preface

This book is the second of a series of four volumes devoted to the scattering of light by solids; the first one appeared as Vol. 8 of *Topics in Applied Physics* in 1975. Since then so much progress has occurred that the editors have found it necessary to put together three new volumes in order to give a glimpse of the field. The volumes cover inelastic scattering, in its spontaneous and stimulated forms, by phonons, magnons, and electronic excitations in crystalline and amorphous solids. Molecular phenomena are sometimes included when they help to clarify or to contrast typical solid-state phenomena and for reference purposes in light scattering from adsorbates at surfaces. Several important topics, such as the rapidly developing field of light scattering by polymers, have been left out completely. The material covered ranges from instrumentation to theory touching also problems of specific materials.

The range of information obtained with light scattering is very wide and pertains to many branches of the natural sciences, from biophysics to chemical engineering and ecology. While these books address mostly the interests of the solid-state physicists, a number of chapters, such as those concerned with instrumentation or with the general theory of light scattering, will find widespread interest among all practitioners of Raman and Brillouin spectroscopy. Most articles should also be of interest to light scattering specialists in other fields interested in looking beyond the narrow range of their immediate concerns.

The present volume is devoted to the general principles and the main experimental techniques, both linear and nonlinear, of light scattering. The Introduction describes the scope of this and the other volumes of the series. The second chapter provides an extensive discussion of the theory of light scattering with particular emphasis on resonant phenomena. An effort is made to unify the various ideas involved around the concept of differential susceptibilities. Simple rules are given to calculate scattering cross sections or efficiencies and the results are compared with experimental data. Chapter 3 discusses one of the most important recent developments in the field of instrumentation for light scattering spectroscopy, namely multichannel detectors. At the moment it is fair to say that this development, is "state of the art". Considerable amount of effort is still required in order to make these systems commercially available. It is our believe that they are the way of the future. The spectroscopist will find in this chapter valuable hints for design and purchase. The final chapter reviews *nonlinear* light scattering spectroscopy with special emphasis on hyper-Raman techniques and coherent antistokes Raman scattering (CARS). These methods are, at present, the object of very intensive research effort.

The editors would like to express their deep appreciation to a large number of scientists, graduate students and colleagues at the Max Planck Institute in Stuttgart, at Brown University, Rhode Island, at the IBM T.J.Watson Research Center, Yorktown Heights, and at various other institutions for their collaboration, help and contributions. The names of all those who helped to "push the frontiers" of the field of light scattering will be found throughout the references of the various chapters, releasing the editors from the cumbersome burden of mentioning them all here. It is a particular pleasure to thank all of the contributors for keeping their deadlines despite other commitments and for their cooperation in considering the editors' suggestions.

Especial recognition is due to the manufactureres and suppliers of equipment for Raman and Brillouin spectroscopy, in particular monochromators, interferometers, detectors, lasers, and dyes. They have helped to free the scientist from the burden of instrumentation. Without them the progress this book is about to document would not have been possible.

Finally, we would like to dedicate this book to the memory of S. P. S. Porto who so much contributed to the field of Raman spectroscopy.

Stuttgart and Köln,   *Manuel Cardona*
December 1981   *Gernot Güntherodt*

Contents

Contributors

Cardona, Manuel
   Max-Planck-Institut für Festkörperforschung, Heisenbergstraße 1,
   D-7000 Stuttgart 80, Fed. Rep. of Germany

Chang, Richard K.
   Yale University, Dept. Engineering and Applied Science,
   New Haven, CT 06520, USA

Güntherodt, Gernot
   Universität zu Köln, II. Physikalisches Institut, Zülpicher Straße 77,
   D-5000 Köln 41, Fed. Rep. of Germany

Long, Marshall B.
   Yale University, Dept. Engineering and Applied Science,
   New Haven, CT 06520, USA

Vogt, Hans
   Max-Planck-Institut für Festkörperforschung, Heisenbergstraße 1,
   D-7000 Stuttgart 80, Fed. Rep. of Germany

1. Introduction

M. Cardona and G. Güntherodt

> Piedras que se calientan y rebrotan la vida
> opacas, dolorosas y silenciosas piedras ...
> allá dentro, muy dentro de vuestros huesos duros
> hay átomos, hay música, vibra un verdor extraño
> Rafael Lorente

Recent and significant advances in the pure and applied aspects of the scattering of light by bulk solids as well as by their surfaces and adsorbates generated the need for an extended treatment. The 1975 volume [1.1] is concerned mainly with inelastic light scattering in semiconductors. During the past six years since the completion of that book, light scattering in the classical semiconductors, in particular, resonant Raman scattering, has reached a level of maturity and profound understanding. Regardless of this, the field of light scattering in solids has experienced a continuing and enhanced growth, mainly because of its applications to an enormous variety of different materials and new systems. These comprise, for instance, doped semiconductors and their surfaces, semiconductor superlattices and heterostructures, magnetic semiconductors, ionic conductors, layered charge density wave materials, graphite intercalation compounds, metals and their adsorbates, superconductors and mixed valence compounds. Another reason for the intense activity in the field arose from the commercial availability of efficient, highly reliable, low stray light and good signal-to-noise experimental set-ups for Raman scattering. On the other hand, multiple pass or synchronously scanned tandem Fabry-Perot interferometry, needed for Brillouin scattering from opaque materials and central modes, is just beginning to get out of the specialist's hands and becoming a widespread and easy-to-handle tool with commercially available instrumentation.

The rapid developments on various frontiers of the field of light scattering have led us to the conclusion that a need exists for reviewing them at the present stage. This need has been reflected by the strong response to our request for contributions; we actually had to allow for three volumes instead of the one originally planned. However, limitations to a practical size of even three volumes have forced us to select a certain number of topics which we hope are representative of the activities in the field to date and address a large readership.

1.1 Survey of 1975 Volume

The 1975 volume on "Light Scattering in Solids" [1.1] contains a rather brief general survey of light scattering by phonons (Chap. 1 by M.Cardona), including a historical survey, an introduction to the polarizability theory of light scattering, in particular the principles of scattering by two phonons which have been so profusely developed during the past few years. It also contains a brief discussion of hot luminescence, as compared with resonant Raman scattering with intermediate states in the continuum, and appropriate references to this subject still of considerable current interest.

Chapter 2, by A.Pinczuk and E.Burstein, discusses in detail the phenomenological theory of light scattering by phonons and establishes the connection with the microscopic theory. Scattering by electronic excitations in doped semiconductors is also considered. The unifying feature of the treatment is the concept of Raman susceptibilities which is extended to cover morphic effects (effects of strains and/or electric fields on the Raman spectra).

Chapter 3, by R.M.Martin and L.Falicov, discusses the phenomenon of resonant scattering with special emphasis on solids. The various types of resonant behavior are classified and discussed according to the nature of the intermediate and the final states (discrete or continua). Emphasis is placed on the types of singularities, i.e., poles of energy denominators, involved in the resonances. The theory of temporal dependence of the scattering process is also briefly touched upon.

Chapter 4, by M.V.Klein, is devoted to electronic Raman scattering with emphasis on doped semiconductors. The discussion includes scattering by charge density fluctuations (leading to plasmons) and the various mechanisms involved: spin-flip single particle scattering, LO-phonon-plasmon coupled modes, scattering by bound electrons and by bound holes, coupling of nonpolar phonons to electronic excitations and scattering by multicomponent carrier systems (acoustic plasmons).

Chapter 5, by M.H.Brodsky, deals with Raman scattering by amorphous semiconductors with emphasis on violation of wavevector conservation selection rules and the analogy to scattering by homological single crystals and/or molecules.

The theoretical and experimental aspects of Brillouin scattering by acoustic phonons in solids are treated by A.S.Pine in Chap. 6. Both classical Brillouin scattering and scattering by acousto-electric domains in polar materials are discussed. Resonant phenomena, including the prediction of resonances near excitonic polaritons, since then profusely confirmed in a number of beautiful experiments [Ref. 1.2, Chap. 7], is also touched upon.

Finally, Chap. 7 by Y.R.Shen treats stimulated Raman scattering and a number of related phenomena with emphasis on the concept of third-order nonlinear susceptibilities.

1.2 Contents of the Present Volume

The present volume emphasizes the main principles and the instrumentation pertaining to light scattering. Chapter 2 will deal with aspects of all chapters of the 1975 book in an updated, comprehensive introductory contribution by M.Cardona. Emphasis is placed on unifying the various macroscopic and microscopic treatments. This chapter will serve as a basis for the theoretical background and nomenclature used in the present edition. An effort is made to keep the treatment simple and self-contained, within the reach of the average experimental physicist. The chapter is divided into four sections. In Sect. 2.1, we treat the phenomenological polarizability theory of light scattering in a semiclassical fashion. Quantum-mechanical concepts are introduced, however, whenever a rounding-off of the classical discussion is required. Emphasis is placed on the concept of scattering cross sections (molecules) and efficiencies (solids) and in obtaining formulas which enable one to calculate these quantities in *absolute* units. Several sections are devoted to the determination of absolute scattering efficiencies in solids. Sections 2.1.15, 2.1.16 are devoted to light scattering in amorphous and disordered materials. This subject, treated in [Ref. 1.1, Chap. 5] by *Brodsky*, has received considerable interest in recent years. As no special chapter on it is planned for the current treatise, we have included in this section an extensive updating of this aspect of light scattering.

Section 2.2 describes the quantum theory of light scattering, starting with molecules (Franck-Condon treatment) and going over to solids. A few aspects of scattering by electronic excitations not covered in [Ref. 1.1, Chap. 4] are also treated. Emphasis is placed on the calculation of absolute scattering efficiencies and of resonance phenomena. The resonance profiles are evaluated for typical interband critical points of solids and illustrated for specific cases which occur in the tetrahedral semiconductors. The treatment includes first-order scattering (by one phonon) and second-order scattering by two phonons. Forbidden scattering induced by Fröhlich interaction in the case of LO-phonons is also discussed.

Finally, Sect. 2.3 gives experimental results of resonant scattering phenomena, including resonances near the E_0, $E_0 + \Delta_0$, E_1 and $E_1 + \Delta_1$ gaps of tetrahedral semiconductors, the yellow excitonic series of Cu_2O and the indirect gap of AgBr. Resonant multiphonon processes and their relationship to hot luminescence are also treated.

The expressions in Sect. 2.1 are written in SI units unless otherwise indicated [we point out that the susceptibility in SI units includes a factor of 4π, i.e., $(\chi_{SI} = 4\pi\chi_{CGS})$]. In Sect. 2.2, however, we use atomic units $(e = m = \hbar = 1)$ as is customary for microscopic calculations. We have kept, however, the factor of 4π in the susceptibility, i.e., we define the dielectric constant of a medium as $\varepsilon = \varepsilon_0(1 + \chi)$, where ε_0 is the permittivity of the vacuum.

Chapter 3 is devoted to a novel technique in light scattering, the optical multichannel detection and its applications to time-resolved and spatially-resolved measurements, particularly in the applied research areas. The article addresses mainly nonspecialists and potential users in trying to make the manufacturers' "jargon" transparent and understandable. It will also help the potential user in making up his mind about what system and components to choose from.

Chapter 4 reports on light scattering in solids due to nonlinear optical effects such as the hyper-Raman effect in its spontaneous, resonant, stimulated and coherent forms, the coherent antistokes Raman scattering (CARS), the Raman-induced Kerr effect and multiwave mixing. The unified theoretical treatment of these phenomena is given with the concept of higher-order nonlinear susceptibilities. Experimental details and instrumentation for these new forms of spectroscopy are also discussed. The aspects dealt with bear basic scientific interests as well as new concepts for technological applications.

1.3 Contents of the Following Volume

The following volume [1.2] will be devoted to specific examples of Raman and Brillouin scattering by phonons in several families of materials. The aim is to illustrate the various types of phenomena observed and investigated with specific examples.

Chapters 2–5 are concerned with material oriented aspects in light scattering. Chapter 2 describes light scattering in graphite intercalation compounds, a model-type class in the recently very active field of "synthetic metals". The substantial contributions of Raman scattering to an understanding of the lattice dynamics, the graphite-intercalant interactions, and the superlattice formation will be discussed.

Chapter 3 reviews light scattering from electronic and magnetic excitations in transition metal halides comprising cubic fluoride perovskites, tetragonal fluoride rutiles and trigonal compounds. Besides magnon scattering in pure and mixed antiferromagnets, emphasis is placed on a summary of present-day knowledge of electronic Raman scattering from transition metal ions and of electron-phonon coupled modes.

Chapter 4 summarizes the contributions of light scattering to an understanding of disorder and dynamical processes in superionic conductors. For selected representatives of the different classes of superionic conductors the prominent features of light scattering will be demonstrated, such as vibrational excitations associated with oscillatory motion of ions at a lattice site and quasi-elastic scattering arising from the diffusive motion of the ions.

Chapter 5 illustrates various aspects of Raman scattering in metals, including defect-induced first-order scattering and second-order scattering in the carbides and nitrides of transition metals. Also, scattering in the A15 compounds and the relationship to their low-temperature phase transition is treated. Experimental data are followed by a formal theory of light scattering in metals which includes in a unified way phonons and electronic excitations. This is followed by a discussion of light scattering in charge density wave materials (transition metal dichalcogenides) including recently observed phenomena in the superconducting phase of some of these materials.

In Chapter 6, a long-time sought for review, combines expertise in most recent technological developments of Fabry-Perot interferometry with a description of the physical background when applying this technique to *opaque* materials, supported films and central modes. It is shown how the technological advances in high-contrast multipass and synchronously scanned tandem Fabry-Perot interferometers have opened new research areas such as the study of elastic waves near surfaces and interfaces, of surface spin waves, and the scattering from diffusive excitations.

Chapter 7 treats resonant light scattering by excitonic polaritons. The emphasis is on the kinematics, scattering efficiencies, resonance behavior as observable in Brillouin as well as Raman scattering, and, in particular, the elusive role of the additional boundary conditions (ABC). A brief general review of the field of exciton polaritons is given. Multiphonon processes and their relationship to the resonant scattering – hot luminescence dichotomy are also treated. Finally, the phenomenon of exciton polariton mediated electronic scattering, of considerable current interest, is discussed.

1.4 Contents of the Forthcoming Volume

A forthcoming volume on Light Scattering in Solids [1.3] is being planned and should appear in the near future. It will cover light scattering by free carrier excitations in semiconductors (A.Pinczuk, M.Cardona, and G.Abstreiter), spin-dependent Raman scattering in magnetic semiconductors (G.Güntherodt and R.Zeyher), Raman scattering in rare earth chalcogenides (G.Güntherodt and R.Merlin), spin-flip Raman scattering in CdS (S.Geschwind and R.Romestain), enhanced Raman scattering by molecules on metals (mainly experimental by A.Otto, and theory by R.Zeyher and K.Arya) and, finally, pressure dependent effects in light scattering (B.A.Weinstein and R.Zallen). The complete and detailed contents of this forthcoming volume [1.3] will be given at the end of the present volume.

1.5 Recent Topics and Highlights of Light Scattering in Solids

In this section, we want to outline a number of selected developments in the pure and applied aspects of light scattering in solids which have caused a significant impact on the field during the past five years. We do not attempt a complete survey, but rather want to stress the versatility of the method and the diversity of the various phenomena that have been investigated. This brief overview is addressed to all those who want to glance at the progress in the field without going into details. Our apologies to all those whose work is not mentioned here. It may appear throughout the more detailed chapters of this volume and the forthcoming ones [1.2, 3].

1.5.1 Instrumentation, Techniques

As will be described in detail in Chap. 2, resonant Raman scattering at various interband gaps of semiconductors has reached a high degree of sophistication in recent years. A great deal of this success has been due to the steady improvements in the field of dye lasers. The present-day state of the art of commercially available, stable dyes is shown in Fig. 2.7.

There is an increasing need for laser excitation lines in the ultraviolet and vacuum ultraviolet spectral region for use in resonant Raman scattering at high-energy interband gaps [1.4] and for work on metallic systems which have high reflectivities in the visible spectral region and a plasma frequency in the near uv. For intense uv laser lines of wavelength shorter than 1200 Å, one has to await, for instance, the availability of a free-electron laser [1.5]. For longer wavelengths frequency multiplication of visible cw laser radiation appears to be a reasonable compromise for the time being. Temperature tuned (stability ±0.01 K) intracavity second-harmonic generation (SHG) in ADP and KDP crystals with true-cw output power up to 300 mW at 2572.5 Å with a maximum 32% power-conversion efficiency, has been reported [1.6]. More realistic, practical applications in light scattering experiments yielded typical cw output between 20 and 40 mW at 2572 Å (4.82 eV) by intracavity – ADP-SHG of the Ar^+ 5145 Å laser line and 5–10 mW at 2956 Å (4.19 eV) by intracavity – ADA-SHG of the rhodamine 6 G dye laser [1.4]. The power-conversion efficiency is usually severely limited by thermal instabilities (local heating) of the nonlinear material, absorption due to trace impurities and by damage to antireflection coatings of the crystal.

For the same plasma tube, mode-locking seems to increase the SHG efficiency about 20 times and may be used for extra-cavity SHG [1.7]. Frequency doubling of synchronously pumped (cavity dumped) dye lasers is superior in uv output power to that of cw (cavity dumped) ones. SHG inside ring laser cavities has yielded several mW tunable uv power using 2.5 W pump power. Tunable uv radiation may also be obtained by frequency mixing of a dye laser output (or its SH) with that of a YAG laser (or its harmonics) using KDP crystals.

Whether synchrotron radiation is going to be the "laser" of the 1980's [1.8] still depends on a detailed evaluation of the scientific case [1.9] for applying it to Raman scattering in solids. First attempts in this direction have been undertaken in x-ray resonant Raman studies in metals [1.8, 10]. The resonance enhancement near K-absorption edges may be useful for studying linewidths and other final state interaction effects.

Several attempts have recently been undertaken to cut down measuring times in light scattering experiments and even to perform time-resolved measurements by replacing the single channel photomultiplier (PM) detection system by an optical multichannel analyzer (OMA) system, somewhat anal-ogous to the previously used photographic plates but with the advantage of electronic data processing. The method has so far not been widely applied to light scattering in solids mainly because of the continuing lack of good signal-to-noise ratios and the somewhat cumbersome instrumentation problems. The basic design of the monochromators to be used with OMA systems for light scattering experiments is shown in Fig. 3.8. A two-grating assembly with subtractive dispersion serves for stray light rejection, whereas a third grating yields the dispersion, i.e., frequency resolution. Zoom lenses may be used for optimum focussing onto the two-dimensional detector array.

The molecular-iodine filter [1.11] with a resonant absorption band coinciding with the 5145 Å Ar^+ laser line, has been used successfully to remove light elastically scattered by sample imperfections (rejection greater than 10^7). Thus, it became feasible to study dynamic central peaks (width $\delta, 0.1 < \delta < 10 \, cm^{-1}$) in the quasi-elastic light-scattering spectrum which arise from entropy and phonon density fluctuations [1.12]. However, as will be described in [1.2] by J. R. Sandercock, multiple-pass tandem Fabry-Perot interferometers with their large free spectral range represent the ultimate and most elegant way of studying central peaks. In particular, no computer-assisted data handling is required.

A new technique, interference enhanced Raman scattering (IERS), has recently been introduced to observe Raman scattering from very thin evaporated films of highly absorbing materials ($\alpha > 10^5 \, cm^{-1}$) [1.13]. The three-layer sandwich structure consists of a highly reflecting (bottom) layer and a nonabsorbing dielectric (middle) layer. Their thickness is adjusted so that no reflected light comes from the front side of the thin sample (top) layer because of destructive interference. On the other hand, the light scattered from the interior of the sample layer is enhanced via constructive interference with that coming from the reflector at the bottom. Depending on the optical constants of the material under investigation, one expects theoretically, a gain in scattering intensity of $10 - 10^3$ over that obtained from a thick bulk sample using a conventional Raman backscattering configuration. A gain of a factor of 20 has been obtained for Te films. Applications of IERS to films of metallic and oxidized Ti are aimed at a study of their structural and vibrational properties, in particular, the amorphous versus crystalline state of the interface.

Surface-enhanced Raman scattering (see below) at solid-electrolyte and more recently at solid-vacuum interfaces of special adsorbate-substrate systems has yielded enhancements of up to $10^5 - 10^6$, as compared to the free molecule scattering cross section. Therefore, the more involved technique of surface-electromagnetic-wave-enhanced Raman scattering by overlayers on metals [1.14] has, so far, received little attention. The theoretical estimate of the scattering intensity by a thin overlayer on a Ag film using an attenuated total-reflection prism scattering configuration showed an enhancement by two orders of magnitude over that in a standard backscattering configuration.

Pulsed laser annealing has become a widely used tool for repairing the damage introduced by ion-implantation doping of semiconductors. Raman scattering allows the direct determination of the lattice temperature within 10 ns after a heating pulse by measuring the Stokes/antistokes ratio of phonon

Raman scattering of a probe laser pulse [1.15]. This technique allows the strict testing of thermal melting models versus nonthermal-equilibrium models of laser annealing in a μm-thick layer on a time scale of tens of ns.

The modern picosecond laser pulse technology has recently been applied to time-resolved spontaneous Raman scattering from nonequilibrium LO-phonons [1.16]. The dynamics of nonthermal LO-phonons in GaAs has been studied by an excite-and-probe scheme using 2.5 ps pulses, resulting in a relaxation time of phonons populated at 77 K of 7 ± 1 ps. This information about phonon relaxation processes is obviously more direct than an analysis of the broadening of Raman or infrared spectra.

While light scattering under uniaxial stress on large single crystals is rather straightforward [1.17], the application of hydrostatic pressure imposes more serious experimental difficulties, the use of the diamond-anvil-type pressure cell [1.18] being restricted to strong scatterers only (minimum about 10^3 counts per second for ~ 200 mW incident power). These aspects will be dealt with in more detail in [1.3] by *Weinstein* and *Zallen*. A comprehensive review concerning morphic effects in Raman scattering, including stress-induced and pressure-induced Raman scattering, has already been given by *Anastassakis* [1.19].

With the continuing development of high-power pulsed lasers, the development of coherent and nonlinear scattering techniques (such as coherent antistokes Raman scattering (CARS), hyper-Raman, etc.) has been quite impressive. The reader is referred to Chap. 4.

1.5.2 Semiconductors

Resonant Raman scattering in semiconductors is described at length in Chap. 2. The direct interrelationship between resonant Raman scattering and the electronic band structure has been demonstrated most clearly in resonant Raman scattering by overtones of TO-phonons (2 TO) in GaAs [1.20]. From the selective resonance enhancement near the indirect $\Gamma_8^v - L_6^c$ and $\Gamma_8^v - X_6^c$ valence (v)-conduction (c) band gaps, it follows directly that the L point conduction band minima lie below those at the X points. This experimentally determined ordering of the conduction band minima in GaAs is of importance for the understanding of the "Gunn effect". Of considerable interest is also the resonant two-phonon scattering as it yields reliable values for electron two-phonon interaction constants.

Light scattering by free electrons and electron-phonon coupled modes in heavily doped semiconductors has been the subject of intense research during recent years [1.21, 22]. In these investigations, the wave vector and frequency-dependent dielectric function of the charge carriers could be studied via the wave vector dependence of the coupled plasmon-LO phonon modes in n-GaAs. Moreover, their resonance enhancement as well as spin-flip single particle excitations have been investigated. In nonpolar materials and for TO-phonons in polar materials, small self-energies result from the deformation potential-type electron-phonon interaction [1.23]. Very recently, the self-energy of phonons due to holes in *p*-type Ge has been shown to be *q*-dependent by means of Raman scattering [1.24]. A comprehensive account of these results will be presented in more detail by *Pinczuk* et al. in [1.3].

Raman scattering by coupled plasmon-LO phonon modes has also been shown to be useful for determining the carrier concentration, for instance, in GaAs or $Al_xGa_{1-x}As$ without the need of applying electrical contacts [1.25]. This method may be used for routine-type in situ characterization of molecular beam epitaxy (MBE) layers.

Raman scattering from plasmon-phonon coupled modes has also been used to study the electron-hole pair density of a pulse-photoexcited plasma in undoped GaP, prior to the formation of the electron-hole liquid [1.26a]. This method appears to be more sensitive than the conventional line-shape fitting of the electron-hole plasma radiative recombination spectrum. Of course, these investigations are actually aiming at studying plasmon modes of the electron-hole liquid and finding easier means of determining its density. We also mention the recent observation of acoustic plasmons in GaAs [2.26b].

The prediction that Brillouin scattering in semiconductors should undergo, like Raman scattering, a resonance enhancement when the incoming photon energy approaches an exciton energy, or in the case of strong exciton-photon interaction a polariton branch, has recently been verified in GaAs [1.27] and CdS [1.28]. The resonant interaction of acoustic phonons with excitonic polaritons allows a determination of the dispersion and other parameters of the polaritons and of the phonon-polariton interaction. It is hoped that it will help to solve the delicate problem of additional boundary conditions (ABC) for polaritons at the sample-vacuum interface [Ref. 1.2, Chap. 7].

1.5.3 Semiconductor Surfaces

Besides the above-mentioned inelastic light scattering by charge carriers in semiconductor surface layers with thicknesses of about 100 to 3000 Å, there has been strong interest in performing these experiments on quantized two-dimensional plasmas in inversion and accumulation surface space charge layers of semiconductors [1.29]. Recently, the resonance enhancement of inelastic light scattering from a quasi-two-dimensional electron gas confined to the interface of abruptly doped $GaAs/n-Al_xGa_{1-x}As$ heterojunctions has been reported [1.30]. These investigations constitute the first evidence for single-particle intersubband excitations in a two-dimensional electron system by means of Raman scattering.

Oxidation of semiconductor surfaces as well as oxide-semiconductor interfaces can also be studied by Raman scattering. The two anomalous peaks in the Raman spectra of oxidized GaAs, InAs and GaSb, InSb [1.31] have been attributed to excess As and Sb, respectively, in the interface between the oxides and the underlying semiconductor. The reduction in scattering intensity of the

TO, LO-modes of the oxidized III–V semiconductors due to the crystalline, semimetallic As or Sb layers can be used to estimate the thickness of the interface layers.

Resonantly excited bulk LO-phonon Raman spectroscopy on (110)-GaAs has recently been shown to be a sensitive technique for measuring the band bending and Fermi level position at the surface as a function of cleavage conditions, oxygen exposure and doping [1.32]. Differences in the LO scattering intensity of UHV-cleaved and air-cleaved samples arise due to the surface electric field present in the latter. The minimum detectable surface electric field of 1.5×10^4 V/cm and the associated change in barrier height of 0.05 eV make this method comparable in sensitivity with that of photoemission spectroscopy and ellipsometry.

1.5.4 Semiconductor Superlattices

During the past five years, Raman scattering in multiple $GaAs\text{-}Ga_{1-x}Al_xAs$ heterostructures has received much attention. One major issue has been the search for Brillouin zone folding effects in Raman scattering. The effect reported for the scattering by polar phonons in $GaAs\text{-}Ga_{1-x}Al_xAs$ superlattices could not be confirmed recently [1.33]. Instead, the results are explained within a phenomenological theory based on optical anisotropy induced by layering [1.33].

On the other hand, however, Raman scattering from folded longitudinal acoustic phonons in GaAs-AlAs superlattices has been observed [1.34]. In this case, the new periodicity of the superlattice along the direction perpendicular to the layers is considered as giving rise to Brillouin zone folding and to gaps in the phonon spectrum.

1.5.5 Amorphous Semiconductors, Laser Annealing

An updated account of the characterization of amorphous semiconductors and of the determination of their local order by means of Raman scattering is given in Chap. 2. Vibrational spectroscopy of amorphous semiconductors using Raman scattering has provided a broad scale of information ranging from vibrational excitations of large atomic clusters [1.35] and Si-H bonds in hydrogenated *a*-Si [1.36] to one-fold coordination in low-temperature deposited thin-film *a*-I [1.37].

Laser annealing is nowadays in common use to "heal" either laser-induced or ion-implantation-induced damage to crystal lattices (see also Sect. 1.5.1). Raman scattering has been shown, among other techniques, to be a valuable tool in identifying microcrystallites in the annealed region or to examine the reordered state of the annealed material, like, for instance, in Si [1.38] or GaAs [1.39].

1.5.6 Brillouin Scattering from Opaque Materials

The study of opaque materials by means of Brillouin scattering has revealed, besides the usual elasto-optic contribution in transparent materials, a new dominant scattering mechanism – the acoustic surface ripple [1.40]. The latter basically acts as a "phase grating" and is produced in the surface by both bulk and surface phonons. In general, phonon modes in the presence of surfaces and interfaces, acting as mechanical boundaries, give rise to new excitations specific to the surface (Rayleigh, Lamb, Segawa, Love modes). These new excitations, localized typically within a wavelength from the surface, are only seen in opaque materials where the scattering volume is near the surface. A comprehensive review on this subject is given by *Sandercock* in [1.2]. The investigations of opaque materials have only been possible through multipass or even tandem-multipass Fabry-Perot interferometers. Highlights in this field have been the studies of solid metal surfaces [1.41], Rayleigh waves [1.42] and their lifetime in amorphous silicon [1.43], frequency shifts and linewidths in vitreous silica down to $0.3\,K$ [1.44], layer compounds [1.45] and the metallic glass $Fe_{0.80}B_{0.20}$ [1.46]. The latter material also showed scattering by surface spin waves which had been first discovered in the magnetic semiconductor EuO [1.47]. Light scattering by thermal acoustic spin waves exhibits, besides modified bulk spin waves, a spin wave propagating along the surface of the crystal. The former exhibit the usual Stokes/antistokes anomaly [1.48] typical for magnons, whereas the surface spin wave is observed either in the Stokes or in the antistokes scattering, depending on the direction of the magnetic field (nonreciprocal propagation properties [1.49]).

1.5.7 Layer Materials

Since 1975, extensive studies of Raman scattering have been performed in the layered transition-metal dichalcogenides in which electronic instabilities give rise to charge density waves accompanied by periodic lattice distortions. Such distortions can be either commensurate or incommensurate with the undistorted lattice. Raman scattering is an extremely valuable tool in identifying the amplitude and phase modes of commensurate or incommensurate charge density waves accompanying superlattice formation [1.50–54].

It has been shown that the strength of second-order Raman scattering in charge density wave materials is directly related to the magnitude of the Kohn anomaly because of the singularity of the q-dependent electron-phonon interaction at $q = 2k_F$, where k_F is the Fermi wave vector [1.55].

Studies of the Raman-active phonons in the highly anisotropic charge density wave material TaS_3 have provided insight into the anisotropy of the bonding, the temperature dependence of the order parameter of the charge density wave and the absolute magnitude of the electronic gap which opens up at the Fermi energy due to the charge density wave [1.56]. The relaxation of phonons via the excitation of carriers across the electronic gap shows up in the temperature dependence of the phonon line widths.

More details on Raman studies of phonon anomalies and particularly on two-phonon scattering in transition metal compounds will be given by *Klein* in [1.2].

The outstanding contributions of Raman scattering to the study of the lattice dynamics of graphite intercalation compounds will be reviewed by *M.S.Dresselhaus* and *G.Dresselhaus* in [1.2]. Of particular interest have been, and still are, the coupling between the intercalants and the graphite bounding layers and the dynamics of the charged intercalants [1.57]. The question of whether the superlattice formation leads to zone-folding effects or to a disorder-induced phonon density of states [1.58] in the Raman scattering is still of current interest.

1.5.8 Superconductors

The first observation of superconducting-gap excitations by means of Raman scattering has been reported recently [1.59]. The layered transition metal dichalcogenide 2H-NbSe$_2$ has been shown to exhibit in the superconducting state new Raman-active A and E modes close in energy to the BCS gap 2Δ. These modes have been attributed to a coupling of the superconducting-gap excitations to charge density waves [1.59] or, in turn, to a self-energy effect on Raman-active phonons of frequency $\omega \gtrsim 2\Delta$, leading to bound excitations ($\omega' < 2\Delta$) induced by the electron-phonon coupling [1.60]. The advantage of the method for studying superconducting phenomena obviously lies in the energy resolution, the symmetry dependence of the scattering and the possible determination of gap anisotropies from an A, E mode splitting.

Raman studies of the superconducting A 15 compound V$_3$Si exhibiting a strongly temperature-dependent asymmetric line shape of the E_g optical phonon, have provided evidence for a direct Gorkov-type coupling between the $q = 0$ optic phonons and the electronic instability associated with the martensitic transformation [1.61].

For superconductors with NaCl structure like TiN or YS, it has been shown recently that their phonon anomalies contribute dominantly to the first-order scattering intensity, which primarily is defect-induced [1.62]. The scattering intensity can be described quantitatively by a one-phonon density of states multiplied by electron-phonon matrix elements which can be expressed in terms of phonon-induced local intra-ionic charge deformabilities also describing the phonon anomalies. The application of the concept of local cluster deformabilities to the calculation of Raman intensities will be described in more detail by *Güntherodt* and *Merlin* in [1.3].

1.5.9 Spin-Dependent Effects

Spin-flip Raman scattering has been extensively used to study electron dynamics and spin diffusion in n-CdS [1.63, 64]. The novel technique of "Raman echoes" (in analogy to spin echoes) has been applied to study the metal-insulator transition in CdS [1.65]. The review by *Geschwind* and *Romestain* in [1.3] will give more detailed insight into this subject.

New Raman scattering mechanisms corresponding to simultaneous spin-phonon excitations have been discovered recently in rare-earth and transition-metal compounds [1.66, 67]. In this case, the phonon Raman scattering contains information about the order and the dynamics of the spin system. In the model type class of rare-earth magnetic semiconductors, the europium chalcogenides, spin-phonon excitations occur in Raman scattering because of the large spin-orbit coupling in the excited, intermediate $4f$-hole state [1.66]. This mechanism leads, in Raman scattering from phonons, to the observation of spin-disorder-induced scattering in the paramagnetic phase [1.66], phonon-magnon scattering in the ferromagnetic phase [1.68], and to "magnetic Bragg" scattering from spin superstructures in the antiferromagnetic phase [1.69]. On the other hand, in the layered, antiferromagnetic transition metal compound VI_2, the phonon modulation of the exchange interaction has been found to be responsible for the Raman scattering from zone-folded phonons induced by the spin superstructure ("magnetic Bragg" scattering) [1.67]. The interested reader is referred to the chapter by *Güntherodt* and *Zeyher* in [1.3].

Besides the profuse literature on resonant Raman scattering in semiconductors, there have been only a few examples of resonant light scattering from magnetic excitations. However, resonant two-magnon Raman scattering as compared with resonant two-phonon Raman scattering in α-Fe_2O_3 [1.70] and NiO [1.71] has been shown to be a useful tool for identifying the localized magnetic (initial) states in these materials, providing information comparable to that from spin-polarized photoemission. The otherwise inaccessible excited-state exchange constants can be determined by comparing the two-magnon scattering cross sections with the theoretically predictable one-magnon cross sections.

Raman scattering in paramagnetic rare-earth trifluorides and trichlorides with axial symmetry has revealed the interaction of phonons with localized electrons in unfilled $4f$ shells in terms of "magnetic phonon splitting" [1.72]. The latter is a splitting of doubly-degenerate (E_g) optical phonon states due to an external magnetic field (parallel to the crystal axis) and arises from the interaction between the lattice deformation due to optic phonons and the multipole moments of second and higher-order of the rare-earth charge distribution. The Raman scattering features have been interpreted theoretically in terms of magneto-elastic interactions [1.73].

Phonon frequency shifts observed by Raman scattering in heavily doped n-Si have been attributed to self-energy effects of the phonon-deformation-potential interaction [1.74]. Analogously, phonon self-energies due to the $4f$-

electron-phonon interaction are made responsible for frequency shifts of optic phonons observed by Raman scattering in $Ce_xLa_{1-x}F_3$ [1.75]. The data are interpreted in terms of phonon-induced virtual quadrupole transitions to excited crystal-field states of Ce^{3+}, i.e., as dynamic quadrupolar deformations of the $4f$ shell by phonons. We would like to point out that the very similar concept of phonon-induced local charge deformabilities mentioned in Sect. 1.5.8 for superconductors is also applicable to the description of lattice dynamics and Raman intensities in mixed valence rare-earth compounds (see Chap. 5 of [1.3]).

1.5.10 Surface-Enhanced Raman Scattering

The investigations of the enhancement mechanisms of Raman scattering from monolayers of adsorbed molecules (by a factor of $10^5 - 10^6$ over the same quantity of molecules in the gas or liquid phase) probably constitute the most active field in light scattering at present. A step forward from initial Raman experiments on pyridine monolayers adsorbed to a Ag-electrolyte interface [1.76] was undertaken by looking at cyanide molecules on a Ag-air interface [1.77]. Present-day studies are concerned with *in situ* surface preparation under ultrahigh vacuum conditions and characterization under controlled surface coverage [1.78, 79]. Surface roughness inherent in the evaporated Ag substrate films has been shown to play an essential role in the enhanced Raman scattering from CO on Ag [1.78] or pyridine on Ag [1.79], the latter exhibiting an enhancement factor of 10^4.

Solid-solid interfaces in the form of tunnel junction structures evaporated onto optical diffraction gratings permit not only the direct comparison of inelastic electron tunneling spectra with the surface-enhanced Raman spectra, but also the direct excitation of the surface plasmon modes of the Ag substrate [1.80]. The latter are considered as an intermediate state in the surface-enhanced Raman process on Ag.

An updated review of the current experimental state of the field will be given by *Otto* in [1.3]. The various theoretical models that have appeared recently, going beyond earlier proposals [1.81], will be summarized by *Zeyher* and *Arya* in [1.3].

1.5.11 Miscellaneous

In the investigations of homogeneously mixed valence compounds exhibiting valence fluctuations of the rare-earth ions, Raman scattering has provided information about the symmetry of the electron-phonon coupling [1.82] and of the charge fluctuations [1.83–85]. Furthermore, electronic Raman scattering by $4f$ multiplets permitted a study of the configuration interaction between localised f electrons and conduction electrons [1.86]. A more detailed account of this work will be given by *Güntherodt* and *Merlin* in [1.3]. Theoretical predictions concerning the power spectrum of temporal fluctuations in the occupation number of an ionic configuration hybridized with conduction-electron states [1.87] have not yet been confirmed by means of light scattering.

Light scattering has also been applied to rather exotic solids, including one-dimensional organic conductors like TTF-TCNQ [1.88–91] or metallic polymers like $(SN)_x$ [1.92, 93]. The investigations in the TTF-TCNQ charge transfer salts have focused mainly on an identification of lattice and intramolecular modes [1.88, 90] and on the relationship between the Raman frequencies and the average charge transfer [1.88, 91]. Questions still concern the frequency change of Raman modes associated with the Peierls distortion [1.89].

Light scattering by solitons, like, for instance, by spin fluctuations in quasi-one-dimensional ferromagnets in a regime of temperature and magnetic fields where solitons exist as thermal excitations [1.94], may become a challenging experimental task. The dispute [1.95] over the previously observed central peak in neutron scattering of $CsNiF_3$ [1.96], whether arising from scattering by solitons [1.96] or by spin-wave density fluctuations [1.95], calls for precise absolute intensity data normalized to single-spin-wave intensities.

So far, not much attention has been paid to the effects predicted for scattering of electromagnetic radiation in the x-ray spectral range [1.97]. With the continuing boom in dedicated synchrotron radiation, interesting relationships between the x-ray Raman edge and the x-ray absorption edge [1.98] have to be investigated in more detail. The possibility of obtaining structural information from the extended modulation of the x-ray Raman scattering edge, similar to that from extended x-ray absorption fine structure (EXAFS), appears to be very intriguing.

For a review of recent developments in light scattering in the Soviet Union, we would like to refer the reader to the Proceedings of the Second Joint USA-USSR Symposium on Light Scattering in Solids [1.99]. Another review about Brillouin scattering from bulk magnons in $CrBr_3$, $FeBO_3$, and $CoCO_3$ by *Borovik-Romanov* and *Kreines* [1.100] should be mentioned here.

A detailed account of the various light scattering activities in Japan cannot be given here. As an example, we would just like to mention the work on spin-dependent Raman scattering from phonons in magnetic semiconductors, especially the theoretical work based on that by *Moriya* [1.101]. A detailed description by *Güntherodt* and *Zeyher* will be found in [1.3].

References

1.1   M.Cardona (ed.): *Light Scattering in Solids*, Topics Appl. Phys., Vol. 8 (Springer, Berlin, Heidelberg, New York 1975)

1.2   M.Cardona, G.Güntherodt (eds.): *Light Scattering in Solids* III, Topics Appl. Phys., Vol. 51 (Springer, Berlin, Heidelberg, New York 1982)

1.3   M.Cardona, G.Güntherodt (eds.): *Light Scattering in Solids* IV, Topics Appl. Phys., Vol. 54 (Springer, Berlin, Heidelberg, New York 1982)

1.4   J.M.Calleja, J.Kuhl, M.Cardona: Phys. Rev. B**17**, 876 (1978)

1.5   D.A.G.Deacon, L.R.Elias, J.M.J.Madey, G.J.Raman, H.A.Schwettman, T.I.Smith: Phys. Rev. Lett. **38**, 892 (1977)

1.6   P.Huber: Opt. Commun. **15**, 196 (1975)

1.7   Information courtesy of Spectra-Physics and CASS Corp.

1.8   P.Eisenberger: In *Light Scattering in Solids*, ed. by Balkanski, Leite, Porto (Flammarion, Paris 1975) p. 959

1.9   European Synchrotron Radiation Facility Supplement, The Scientific Case, ed. by Y.Farge, P.J.Duke (European Science Foundation, Strasbourg 1979)

1.10   C.J.Sparks, Jr.: Phys. Rev. Lett. **33**, 262 (1974)

1.11   G.E.Devlin, J.L.Davis, L.Chase, S.Geschwind: Appl. Phys. Lett. **19**, 138 (1971)

1.12   K.B.Lyons, P.A.Fleury: Phys. Rev. Lett. **37**, 161, 1088 (1976)
       P.A.Fleury, K.B.Lyons: *Lattice Dynamics*, ed. by M.Balkanski (Flammarion, Paris 1978) p. 731
       P.A.Fleury, K.B.Lyons: In *Structural Phase Transitions* I, ed. by K.A.Müller, H.Thomas, Topics Current Phys., Vol. 23 (Springer, Berlin, Heidelberg, New York 1981) p. 9

1.13   R.J.Nemanich, C.C.Tsai, G.A.N.Connell: Phys. Rev. Lett. **44**, 273 (1980)
       G.A.N.Connell, R.J.Nemanich, C.C.Tsai: Appl. Phys. Lett. **36**, 31 (1980)

1.14   Y.J.Chen, W.P.Chen, E.Burstein: Phys. Rev. Lett. **36**, 1207 (1976)

1.15   H.W.Lo, A.Compaan: Phys. Rev. Lett. **44**, 1604 (1980)
       A.Aydinli, M.C.Lee, H.W.Lo, A.Compaan: Phys. Rev. Letters **47**, 1676 (1981)
       B.Strizker, A.Popieszczyk, J.A.Tagle: Phys. Rev. Letters **47**, 1677 (1981)

1.16   D. von der Linde, J.Kuhl, H.Klingenberg: Phys. Rev. Lett. **44**, 1505 (1980)

1.17   M.Chandrasekhar, H.R.Chandrasekhar, M.Grimsditch, M.Cardona: Phys. Rev. B**22**, 4825 (1980)

1.18   R.S.Hawke, K.Syassen, W.B.Holzapfel: Rev. Sci. Instrum. **45**, 1598 (1974)
       B.A.Weinstein, G.J.Piermarini: Phys. Rev. **12**, 1172 (1975)

1.19   E.M.Anastassakis: In *Dynamical Properties of Solids*, Vol. 4, ed. by G.K.Horton, A.A.Maradudin (North Holland, Amsterdam 1980) p. 157

1.20   R.Trommer, M.Cardona: Solid State Commun. **21**, 153 (1977)

1.21   For the most recent review see: M.Cardona: J. Phys. Soc. Jpn., Vol. **49**A, Suppl. (1980) p. 23

1.22   G.Abstreiter, R.Trommer, M.Cardona, A.Pinczuk: Solid State Commun. **30**, 703 (1979), and preceding papers by the same group of authors

1.23   M.Chandrasekhar, J.B.Renucci, M.Cardona: Phys. Rev. B**17**, 1623 (1978)

1.24   D.Olego, M.Cardona: Phys. Rev. **23**, 6592 (1981)

1.25   G.Abstreiter, E.Bauser, A.Fischer, K.Ploog: Appl. Phys. **16**, 345 (1978)

1.26a J.E.Kardontchik, E.Cohen: Phys. Rev. Lett. **42**, 669 (1979)

1.26b A.Pinczuk, J.Shah, P.A.Wolff: Phys. Rev. Letters **47**, 1487 (1981)

1.27   R.G.Ulbrich, C.Weisbuch: Phys. Rev. Lett. **38**, 865 (1977); **39**, 654 (1977)

1.28   G.Winterling, E.S.Koteles, M.Cardona: Phys. Rev. Lett. **39**, 1286 (1977)

1.29   E.Burstein, A.Pinczuk, S.Buchner: *Physics of Semiconductors*, Institute of Physics Conference Series, Vol. 43 (Institute of Physics, Bristol 1979) p. 1231

1.30   G.Abstreiter, K.Ploog: Phys. Rev. Lett. **42**, 1308 (1979)

1.31   R.L.Farrow, R.K.Chang, S.Mroczkowski, F.H.Pollak: Appl. Phys. Lett. **31**, 768 (1977)

1.32   H.J.Stolz, G.Abstreiter: Solid State Commun. **36**, 857 (1980)

1.33   R.Merlin, C.Colvard, M.V.Klein, H.Morkoc, A.Y.Cho, A.C.Grossard: Appl. Phys. Lett. **36**, 43 (1980)

1.34   C.Colvard, R.Merlin, M.V.Klein, A.C.Gossard: Phys. Rev. Lett. **45**, 298 (1980)

1.35   R.J.Nemanich, S.A.Solin, G.Lucovsky: Solid State Commun. **21**, 273 (1977)

1.36   M.H.Brodsky, M.Cardona, J.J.Cuomo: Phys. Rev. B**16**, 3556 (1977)

1.37   B.V.Shanabrook, J.S.Lannin, I.C.Hisatsune: Phys. Rev. Lett. **46**, 130 (1981)

1.38   J.F.Morhange, G.Kanellis, M.Balkanski: Solid State Commun. **31**, 805 (1979)

1.39   R.Tsu, J.E.Baglin, G.J.Lasher, J.C.Tsang: Appl. Phys. Lett. **34**, 153 (1979)

1.40   S. Mishra, R. Bray: Phys. Rev. Lett. **39**, 222 (1977)
1.41   J. R. Sandercock: Solid State Commun. **26**, 547 (1978)
1.42   W. Senn, G. Winterling, M. Grimsditch, M. H. Brodsky: *Physics of Semiconductors*, Institute of Physics Conference Series, Vol. 43 (Institute of Physics, Bristol 1979) p. 1231
1.43   R. Vacher, H. Sussner, M. Schmidt: Solid State Commun. **34**, 279 (1980)
1.44   R. Vacher, H. Sussner, S. Hunklinger: Phys. Rev. B**21**, 5950 (1980)
1.45   R. T. Harley, P. A. Fleury: J. Phys. C**12**, 2863 (1979)
1.46   P. H. Chang, A. P. Malozemoff, M. Grimsditch, W. Senn, G. Winterling: Solid State Commun. **27**, 617 (1978)
1.47   P. Grünberg, F. Metawe: Phys. Rev. Lett. **39**, 1561 (1977)
1.48   W. Wettling, M. G. Cottam, J. R. Sandercock: J. Phys. C**8**, 211 (1975)
1.49   R. W. Damon, J. R. Eshbach: J. Phys. Chem. Sol. **19**, 308 (1961)
1.50   J. E. Smith, Jr., J. C. Tsang, M. W. Shafer: Solid State Commun. **19**, 283 (1976)
1.51   J. C. Tsang, J. E. Smith, Jr., M. W. Shafer: Phys. Rev. Lett. **37**, 1407 (1976)
1.52   J. A. Holy, M. V. Klein, W. L. McMillan, S. F. Meyer: Phys. Rev. Lett. **37**, 1145 (1976)
1.53   E. F. Steigmeier, G. Harbeke, H. Auderset, F. J. DiSalvo: Solid State Commun. **20**, 667 (1976)
1.54   J. R. Duffey, R. D. Kirby, R. V. Coleman: Solid State Commun. **20**, 617 (1976)
1.55   P. F. Maldague, J. C. Tsang: *Lattice Dynamics*, ed. by M. Balkanski (Flammarion, Paris 1978) p. 602
1.56   J. C. Tsang, C. Hermann, M. W. Shafer: Inst. of Phys. Conf. Ser. **43**, 457 (1979)
1.57   J. J. Song, D. D. L. Chung, P. C. Eklund, M. S. Dresselhaus: Solid State Commun. **20**, 1111 (1976); M. S. Dresselhaus, G. Dresselhaus: In *Physics and Chemistry of Materials with Layered Structures*, Vol. 6, ed. by F. Lévy (D. Reidel, Dordrecht 1979) p. 423
1.58   N. Caswell, S. A. Solin: Phys. Rev. B**20**, 2551 (1979)
1.59   R. Sooryakumar, M. V. Klein: Phys. Rev. Lett. **45**, 660 (1980)
1.60   C. A. Balseiro, L. M. Falicov: Phys. Rev. Lett. **45**, 662 (1980)
1.61   H. Wipf, M. V. Klein, B. S. Chandrasekhar, T. H. Geballe, J. H. Wernick: Phys. Rev. Lett. **41**, 1752 (1978)
1.62   G. Güntherodt, A. Jayaraman, W. Kress, H. Bilz: Phys. Lett. **82A**, 26 (1981)
1.63   R. Romestain, S. Geschwind, G. E. Devlin: Phys. Rev. Lett. **39**, 1583 (1977)
1.64   S. Geschwind, R. E. Walstedt, R. Romestain, V. Narayanamurti, R. B. Kümmer, G. E. Devlin: J. de Phys., Coll. C6, **41**, C5–105 (1980)
1.65   P. Hu, S. Geschwind, T. M. Jedju: Phys. Rev. Lett. **37**, 1357 (1976)
1.66   R. Merlin, R. Zeyher, G. Güntherodt: Phys. Rev. Lett. **39**, 1215 (1977)
1.67   G. Güntherodt, W. Bauhofer, G. Benedek: Phys. Rev. Lett. **43**, 1428 (1979)
1.68   G. Güntherodt, R. Merlin, P. Grünberg: Phys. Rev. B**20**, 2834 (1979)
1.69   R. P. Silberstein, L. E. Schmutz, V. J. Tekippe, M. S. Dresselhaus, R. L. Aggarwal: Solid State Commun. **18**, 1173 (1976)
1.70   T. P. Martin, R. Merlin, D. Huffmann, M. Cardona: Solid State Commun. **22**, 566 (1977)
1.71   R. Merlin: J. de Phys., Coll. C5, **41**, C5–233 (1980)
1.72   G. Schaack: Solid State Commun. **17**, 505 (1975); Physica **89**B, 195 (1977)
1.73   P. Thalmeier, P. Fulde: Z. Phys. B**26**, 323 (1977)
1.74   M. Chandrasekhar, J. B. Renucci, M. Cardona: Phys. Rev. B**17**, 1623 (1978)
1.75   K. Ahrens, G. Schaack: Phys. Rev. Lett. **42**, 1488 (1979)
1.76   M. Fleischmann, P. J. Hendra, A. J. McQuillan: Chem. Phys. Lett. **26**, 163 (1974)
1.77   A. Otto: Surface Science **75**, L392 (1978)
1.78   T. A. Wood, M. V. Klein: Solid State Commun. **35**, 263 (1980)
1.79   J. E. Rowe, C. V. Shank, D. A. Zwemer, C. A. Murray: Phys. Rev. Lett. **44**, 1770 (1980)
1.80   J. C. Tsang, J. R. Kirtley, J. A. Bradley: Phys. Rev. Lett. **43**, 771 (1979)
1.81   E. Burstein, Y. J. Chen, C. Y. Chen, S. Lundqvist, E. Tosatti: Solid State Commun. **29**, 567 (1979)
1.82   G. Güntherodt, R. Merlin, A. Frey, M. Cardona: Solid State Commun. **27**, 551 (1978)
1.83   G. Güntherodt, A. Jayaraman, H. Bilz, W. Kress: In *Valence Fluctuations in Solids*, ed. by L. Falicov, W. Hanke, M. B. Maple (North-Holland, Amsterdam, New York, Oxford 1981) p. 121

1.84   N. Stüsser, M. Barth, G. Güntherodt, A. Jayaraman: Solid State Commun. **39**, 965 (1981)
1.85   W. Kress, H. Bilz, G. Güntherodt, A. Jayaraman: Proc. Intern. Conf. on Phonons, Bloomington, **Ind.** (1981) to be published
1.86   G. Güntherodt, A. Jayaraman, E. Anastassakis, E. Bucher, H. Bach: Phys. Rev. Lett. **46**, 855 (1981)
1.87   A. Lopez, C. Balseiro: Phys. Rev. B**17**, 99 (1978)
1.88   H. Kuzmany, B. Kundu, H. J. Stolz: *Lattice Dynamics*, ed. by M. Balkanski (Flammarion, Paris 1978) p. 584
1.89   H. Temkin, D. B. Fitchen: *Lattice Dynamics*, ed. by M. Balkanski (Flammarion, Paris 1978) p. 587
1.90   K. D. Truong, C. Carlone: Phys. Rev. B**20**, 2238 (1979)
1.91   S. Matsuzaki, T. Moriyama, K. Toyoda: Solid State Commun. **34**, 857 (1980)
1.92   H. J. Stolz, B. Wendel, A. Otto: Phys. Stat. Sol. (b) **78**, 277 (1976)
1.93   H. Temkin, D. B. Fitchen: Solid State Commun. **19**, 1181 (1976)
1.94   D. L. Mills, P. S. Riseborough, S. E. Trullinger: Bull. Am. Phys. Soc. **25**, 239 (1980)
1.95   G. Reiter: Phys. Rev. Lett. **46**, 202 (1981)
1.96   J. K. Kjems, M. Steiner: Phys. Rev. Lett. **41**, 1137 (1978)
1.97   C. V. Raman: Indian J. **2**, 387 (1928)
1.98   M. Popescu: Phys. Lett. **73A**, 260 (1979)
1.99   J. L. Birman, H. Z. Cummins, K. K. Rebane (eds.): *Light Scattering in Solids*, Proc. Second Joint USA-USSR Symp. (Plenum Press, New York 1979)
1.100   A. S. Borovik-Romanov, N. M. Kreines: J. Magnetism and Magnet. Mat. **15–18**, 760 (1980)
1.101   T. Moriya: J. Phys. Soc. Jpn. **23**, 490 (1967)

Additional References

1. V. V. Artamonov, M. Ya. Valakh, V. A. Korneichuk: Resonant interaction between plasmons and two-phonon states in ZnSe crystals. Solid State Commun. **39**, 703 (1981)
2. J. Briggs, P. R. Findley, Z. L. Wu, W. C. Walker, D. F. Nicoli: Quasielastic light scattering from superionic PbF_2. Phys. Rev. Lett. **47**, 1011 (1981)
3. J. P. Ipatova, A. V. Subashiev, V. A. Voitenko: Electron light scattering from doped Si. Solid State Commun. **37**, 893 (1981)
4. A. Kueny, M. Grimsditch, K. Miyano, I. Banerjee, Charles M. Falco, Ivan K. Schuller: Anomalous behavior of surface acoustic waves in Cu/Nb superlattices. Phys. Rev. Lett. **48**, 166 (1982)
5. K. B. Lyons, P. A. Fleury: Magnetic energy fluctuations: observation by light scattering. Phys. Rev. Lett. **48**, 202 (1982)

2. Resonance Phenomena

M. Cardona

With 60 Figures

The first volume of this series [2.1] was written in 1973–74. At that time, reliable cw dye lasers had become commercially available and their range of tunability was being rapidly extended. It was clear that these lasers were going to become the standard source for light scattering experiments and there was going to be increasing interest in this type of work. A few experiments involving first and second-order scattering had already been performed with dye lasers for simple solids, in particular, for germanium-zincblende-type semiconductors [2.2–4] and for Cu_2O [2.5, 6]. With few exceptions, it was conventional to give the scattering efficiencies in relative units: the efficiencies for second-order scattering were compared with the first-order efficiencies.

It had become clear at that time that the band structure and lattice dynamics of those semiconductors were sufficiently well known to allow the quantitative interpretation of resonance phenomena in light scattering. From such an interpretation, very detailed information on electron-phonon interaction should be obtained. The main ideas concerning the forbidden q-dependent and E-dependent scattering by LO-phonons [2.7] were known although there was still considerable confusion concerning the interpretation of LO-multiphonon processes [2.8, 9]. Brillouin scattering resonant with polaritons had been predicted [2.10] but not yet observed [2.11]. Stimulated Raman phenomena were beginning to be studied in solids and the potential of the CARS-technique (Coherent Antistokes Raman Scattering) had been recognized [Ref. 2.1, Chap. 7]. The main principles underlying scattering by electronic excitations were known [Ref. 2.1, Chap. 4] but experimental results, especially concerning resonance phenomena, were scarce. The potential of Raman scattering as a tool for the characterization of amorphous materials had become clear [Ref. 2.1, Chap. 5] but the present boom in amorphous materials research had not yet started. Since that time, all of these fields have experienced an enormous development.

For the past ten years, the editors of these volumes have been heavily engaged in research on light scattering in solids. In the course of this work, they have extensively used the theoretical underpinning given in [2.1]. While doing so, they have become aware of the shortcomings of that volume and they have accumulated a great deal of practical information for the theoretical analysis of experimental data. In particular, as resonant cross sections have become measurable in absolute units, the need for theoretical expressions relating them to electron-phonon coupling constants, with the correct numerical factors, has become acute. These expressions are now scattered throughout the literature and the numerical constants entering in them are not always given correctly. The necessity to unify the various theoretical treatments, and to point out their equivalence or difference, also appeared. The concepts of cross section, efficiency, Raman tensor, susceptibility and polarizability had to be used consistently and with precision.

This, and the fact that the introduction [Ref. 2.1, Chap. 1] had been kept to a minimum, induced us to write the present chapter and to make it as complete as possible within space limitations. In it, we treat mainly the theory of light scattering by phonons and by electronic excitations in crystalline and amorphous solids and we illustrate it with a number of typical experimental examples. In Sect. 2.1, we handle this theory within the macroscopic framework of the so-called polarizability theory, which is valid provided the laser frequency does not get "too close" to singularities in the electronic polarizability. This approximation, also referred to as quasistatic or adiabatic, is based on the assumption that while a solid is vibrating, we can define a susceptibility or polarizability with a time dependent component related to the vibration [Ref. 2.1, Chaps. 1 and 2]. Contact with this type of approach is kept throughout the chapter: most microscopic expressions are at some point transformed into others involving a susceptibility and, vice versa, it is shown, whenever possible, how to transform a susceptibility-type expression into the more general one in which the quasistatic restriction has been lifted.

In Sect. 2.1.1, we start with a treatment appropriate to molecules and we then generalize it to solids (Sect. 2.1.3). We first treat *elastic* scattering and carry over the treatment to the inelastic case (Sect. 2.1.4). We give a description of the Raman and Brillouin selection rules and the methods of obtaining the type of activity (Raman, ir) of the phonons (Sects. 2.1.9 and 10). We also cover in detail the modifications in the cross section introduced by the electro-optic effect for LO ir-active phonons (Sect. 2.1.12). The macroscopic theory of Brillouin scattering is given in Sect. 2.1.14.

A long section (2.1.15) is devoted to amorphous materials. This topic was covered in [Ref. 2.1, Chap. 5] and no additional chapter on it is planned for the rest of the series. Along with the enormous development in the field of amorphous semiconductors, in particular amorphous silicon, since 1975, Raman scattering has become a standard tool for the characterization of these materials. We have, therefore, felt compelled to include a great deal of recent experimental results together with the theory of light scattering by amorphous materials in Sect. 2.1.15. A related field, that of disorder-induced light scattering in crystals, is treated in Sect. 2.1.16.

For the sake of completeness and in order to make contact between the *third-order* Raman susceptibility and the cross section for spontaneous scattering, we briefly treat the phenomenon of stimulated Raman scattering in Sect. 2.1.17. We then devote some length to the various methods used for the determination of absolute scattering cross sections and the sign of the Raman susceptibility (Sect. 2.1.18). Although we try to keep these sections mainly within the spirit of the "macroscopic" theories, we introduce here and there quantum concepts in an *ad hoc* way as they are needed to round off the discussion. In Sects. 2.2.1–12, we discuss various aspects of the quantum theory of light scattering, trying to unify the various points of view and keeping in mind all the time the connection with the macroscopic theory: we try in all cases to reduce the exact expressions to derivatives or finite differences of static susceptibilities. In Sect. 2.1 we used SI units. In Sect. 2.2 we write all expressions in *atomic units* ($e = \hbar = m = 1$), unless otherwise indicated. We lump into the susceptibilities χ, however, the factor of 4π characteristic of the susceptibility in rationalized (SI) units [a factor of $(4\pi)^{-1}$ is usually written in front of χ so that the reader can remove it if he so desires, in order to obtain the usual cgs-type susceptibility]. Section 2.2.2 treats succinctly the electronic Raman scattering as a complement to [Ref. 2.1, Chap. 4]. Section 2.2.3 gives the Frank-Condon treatment of scattering efficiencies and establishes the connection with the polarizability theories. Sections 2.2.4–6 present perturbation theory treatments of scattering efficiencies using as intermediate states uncorrelated electron-hole pairs. Section 2.2.5 gives a review of the electronic contribution to the optical properties of germanium-zincblende materials and uses the resulting break-up into E_0, $E_0 + \Delta_0$, E_1, $E_1 + \Delta_1$ critical points for the calculation of Raman efficiencies or Raman polarizabilities. The Fröhlich-interaction-induced forbidden LO-scattering is treated in Sect. 2.2.8. Emphasis is placed on the simplicity of the treatment and in establishing contact with the polarizability theory (the latter is not usually done in the literature). The various theories which have been given for multi-phonon LO-scattering are discussed and compared in Sect. 2.2.11. Section 2.2.10 gives expressions for second-order scattering via standard deformation potential interaction and Sect. 2.2.12 treats the elasto-optic constants, i.e., Brillouin scattering.

In the remaining sections we present experimental results (some had already been given as an illustration to the theoretical parts). Emphasis is on germanium and zincblende-type materials. We give examples of first-order allowed and forbidden scattering near E_0 and $E_0 + \Delta_0$, and near E_1 and $E_1 + \Delta_1$. From these data, several deformation potentials are obtained. We also show examples of resonances in the two-phonon scattering efficiencies and their application to the determination of electron two-photon coupling constants. Multiphonon LO-processes are also illustrated.

Cu_2O has almost become single handedly a laboratory for the study of a number of resonant Raman phenomena which are not found in the zincblende-type materials. This is mainly because of the forbidden nature of the $n = 1$ yellow exciton: this exciton is extremely sharp. Some examples of these resonant scattering phenomena are given in Sect. 2.3.6. Experimental data on elasto-optic constants are presented in Sect. 2.3.4.

Before closing this section, we would like to mention a few general references which the reader will find useful. They are the proceedings of the conferences on light scattering in solids [2.12–14] and of the Raman Scattering Conferences [2.15, 16], the recent books authored by *Hayes* and *Loudon* [2.17], that edited by *Weber* [2.18], and the two books edited by *Anderson* [2.19]. Information on lattice vibrations spectra will be found in a recent atlas by *Bilz* and *Kress* [2.20] and general information on the interaction of solids with electromagnetic radiation with emphasis on group-theoretical aspects, is given in a monumental treatise by *Birman* [2.21]. Considerable information will be also found in the proceedings of the recent International Conferences on the Physics of Semiconductors and on Amorphous and Liquid Semiconductors. They are listed in [2.22]. Also of interest are the proceedings of the conference on lattice dynamics [2.23] and those of the Soviet-American symposia on the theory of light scattering [2.24, 25]. Part of the material presented here is discussed, from a somewhat different point of view, in [2.26a].

Last but not least, we mention the recent S.P.S.Porto commemorative issue of the Journal of Raman Spectroscopy which contains research articles by some of the most prominent practitioners of the field [2.26b].

2.1 Classical Theory: Elastic Scattering by Molecules, Liquids, and Solids

2.1.1 Scattering Cross Section: Thompson Scattering

In order to introduce the theory of light scattering and the various parameters involved as much as to get a feeling for the order of magnitude of scattering efficiencies, we discuss briefly the classical theory of elastic scattering of light. The basic formula is that for the radiation energy emitted per unit time by an electric dipole moment M vibrating at the frequency ω [2.27]:

$$\frac{dW_s}{d\Omega} = \frac{\omega^4}{(4\pi)^2 \varepsilon_0 c^3} |\hat{e}_s \cdot M|^2 ,\tag{2.1}$$

where $d\Omega$ is the element of solid angle, ε_0 the permittivity (dielectric constant) of the medium (considered isotropic and nonmagnetic; if vacuum, the permittivity of vacuum), c the speed of light in the medium and $\hat{e}_s$ the unit vector representing the polarization of the scattered light selected by the measuring system at the point of observation. If this detector is unpolarized, one must average (2.1) for all *possible* polarizations $\hat{e}_s$ *perpendicular* to the direction of propagation. The same thing applies to $\hat{e}_L$ (see below) if the incident radiation is unpolarized.

Let us assume that the radiating dipole is an atom, a molecule, or any other complex whose dimensions are small compared with the wavelength of light and let us call $\underset{\sim}{\alpha}$ the polarizability tensor of this complex: the dipole moment induced in the complex by an incident electric field $E_L = \hat{e}_L E_L$ is

$$M = \underset{\sim}{\alpha} \cdot \hat{e}_L E_L .\tag{2.2}$$

Replacing (2.2) into (2.1), we find

$$\frac{dW_s}{d\Omega} = \frac{\omega^4}{(4\pi)^2 \varepsilon_0 c^3} |\hat{e}_s \cdot \underline{\alpha} \cdot \hat{e}_L|^2 E_L^2 . \tag{2.3}$$

The differential scattering cross section $d\sigma/d\Omega$ is obtained by dividing (2.3) by the energy incident per unit area and unit time $W_L = \varepsilon_0 c E_L^2$:

$$\frac{d\sigma}{d\Omega} = \frac{\omega^4}{(4\pi\varepsilon_0)^2 c^4} |\hat{e}_s \cdot \underline{\alpha} \cdot \hat{e}_L|^2 . \tag{2.4}$$

Let us note that $\varepsilon_0^2 c^4$ is independent of ε_0 in an isotropic medium. The modifications of (2.4) necessary for an anisotropic medium (ε_0 a tensor) can be found in [2.28a, b].

In general, the polarizability tensor $\underline{\alpha}$ will not be isotropic. In this case, the plane of polarization of the scattered light will not be the same as that of the incident radiation: the scattering process changes the plane of polarization. If $\underline{\alpha}$ is isotropic, i.e., if $\underline{\alpha} = \mathbb{1}\alpha$, where $\mathbb{1}$ is the unit matrix (such is the case for all atoms and also for isotropic spherical small macroscopic particles), (2.4) becomes:

$$\frac{d\sigma}{d\Omega} = \frac{\omega^4 \alpha^2}{(4\pi\varepsilon_0)^2 c^4} |\hat{e}_s \cdot \hat{e}_L|^2 \tag{2.5}$$

and the scattering is completely "polarized". Integrated over all directions of space, (2.5) yields

$$\sigma = \frac{4\pi\omega^4\alpha^2}{(4\pi\varepsilon_0)^2 c^4} \langle |\hat{e}_s \cdot \hat{e}_L|^2 \rangle = \frac{4\pi}{3} \frac{\omega^4\alpha^2}{(4\pi\varepsilon_0)^2 c^4} . \tag{2.6}$$

In order to estimate the magnitude of σ, we must obtain a typical estimate of α. In the visible, α is of electronic origin. Let us assume that α is produced by an electronic charge e tied to atomic cores by a restoring force $k = \omega_0^2 m$, where m is the electron mass and ω_0 the vibrating frequency of the harmonic oscillator so defined. In this case, solving the appropriate equation of motion we find

$$\alpha = \frac{e^2/m}{\omega_0^2 - \omega^2 - i\omega\gamma} , \tag{2.7}$$

where γ is the damping constant. Equation (2.7) can be easily modified to yield an anisotropic $\underline{\alpha}$ by introducing different values of the parameters e^2/m, ω_0 and γ for each of the principal directions. Confining ourselves to the isotropic case and replacing (2.7) into (2.5), we obtain

$$\frac{d\sigma}{d\Omega} = \frac{r_e^2 \omega^4}{(\omega_0^2 - \omega^2)^2 + \omega^2\gamma^2} |\hat{e}_L \cdot \hat{e}_s|^2 , \tag{2.8}$$

where $r_e = e^2/4\pi\varepsilon_0 mc^2 = 2.8 \times 10^{-15}$ m is the so-called classical radius of the electron (the radius at which the Coulomb energy equals the relativistic rest energy). The square of r_e gives the order of magnitude of the differential cross section at frequencies ω of the order of, but not to close to, ω_0. This is a usual case in the visible since, for many atoms, ω_0 lies in the blue or near ultraviolet. The scattering cross sections are thus in the region $\omega \lesssim \omega_0$ (but not too close to ω_0) of the order of $r_e^2 \simeq 10^{-29}$ m$^2 = 0.1$ barn. This cross section is quite small. It can be, however, substantially enhanced for ω very close to the resonance frequency ω_0. If we assume a "quality factor" $Q = \omega_0/\gamma = 10^4$ for the harmonic oscillator, rather modest for atomic and molecular cases, (2.8) yields a quite considerable cross section of 10 Mbarns for $\omega = \omega_0$. The phenomenon is then referred to as "resonance fluorescence".

At low frequencies $\omega \ll \omega_0$, (2.8) becomes

$$\frac{d\sigma}{d\Omega} = r_e^2 \frac{\omega^4}{\omega_0^4} |\hat{e}_L \cdot \hat{e}_s|^2 . \tag{2.9}$$

Equation (2.9) represents the famous ω^4-law which also applies to Raman and Brillouin scattering at frequencies *well below* all resonance frequencies [2.29]. In the opposite case, $\omega \gg \omega_0$, we again encounter the classical scattering cross section of free electrons (Thompson scattering). The *total* Thompson cross section is:

$$\sigma_T = \frac{8\pi}{3} r_e^2 . \tag{2.10}$$

2.1.2 Depolarized Scattering

The simplest case of an anisotropic polarizability is that of a body with axial symmetry. In this case, $\underset{\sim}{\alpha}$ has two independent components $\alpha_{||}$ and $\alpha_\perp$:

$$\underset{\sim}{\alpha} = \begin{pmatrix} \alpha_\perp & & \\ & \alpha_\perp & \\ & & \alpha_{||} \end{pmatrix} = \langle \alpha \rangle \mathbb{1} + \beta \begin{pmatrix} -1 & & \\ & -1 & \\ & & 2 \end{pmatrix} = \langle \underset{\sim}{\alpha} \rangle + \underset{\sim}{\beta}$$

with $\qquad\qquad\qquad\qquad\qquad\qquad\qquad\qquad\qquad\qquad\qquad\qquad$ (2.11)

$$\langle \alpha \rangle = \tfrac{1}{3}(\alpha_{||} + 2\alpha_\perp) \quad \text{and} \quad \beta = \tfrac{1}{3}(\alpha_{||} - \alpha_\perp).$$

In (2.11) we have decomposed $\underset{\sim}{\alpha}$ into the sum of a diagonal tensor $\langle \underset{\sim}{\alpha} \rangle$ and a traceless tensor $\underset{\sim}{\beta}$. The isotropic tensor $\langle \underset{\sim}{\alpha} \rangle$ produces polarized scattering while $\underset{\sim}{\beta}$ changes the polarization of the incident field. While cross terms proportional to $\langle \alpha \rangle \beta$ may in principle occur, they often vanish since $\langle \alpha \rangle$ and $\underset{\sim}{\beta}$, belonging to different irreducible representations of the symmetry group, are usually connected by an arbitrarily fluctuating phase.

Let us now consider fixed polarizations $\hat{e}_L$ and $\hat{e}_s$ and an anisotropic scattering complex which rotates rapidly and occupies with equal probability all possible positions. We assume, however, that the frequency of rotation is smaller than the line width of the incident radiation; otherwise, *inelastic rotational structure* results. The corresponding scattering efficiency is obtained by replacing (2.11) into (2.4) and averaging over all possible directions for the z-axis of the $\underset{\sim}{\alpha}$-tensor. The cross section has two components: a *polarized* one, given by $\langle \alpha \rangle$, and a partly depolarized one given by β. The average of the depolarized component is most readily obtained if one considers that the traceless tensor $\underset{\sim}{\beta}$ transforms upon rotation into a linear combination of the following "orthogonal" components:

$$
\begin{pmatrix} \beta & & \\ & -\beta & \\ & & 0 \end{pmatrix}, 3^{-1/2} \begin{pmatrix} -\beta & & \\ & -\beta & \\ & & 2\beta \end{pmatrix}, \begin{pmatrix} 0 & \beta & 0 \\ \beta & 0 & 0 \\ 0 & 0 & 0 \end{pmatrix}, \begin{pmatrix} 0 & 0 & \beta \\ 0 & 0 & 0 \\ \beta & 0 & 0 \end{pmatrix}, \begin{pmatrix} 0 & 0 & 0 \\ 0 & 0 & \beta \\ 0 & \beta & 0 \end{pmatrix}. \quad (2.12)
$$

The five tensors (2.12) are basis functions of angular momentum $J=2$. The depolarized scattering can be readily found by averaging the contributions of the five matrices (2.12) to (2.4). We obtain:

$$
\hat{e}_s \| \hat{e}_L; \frac{d\sigma_p}{d\Omega} = \frac{\omega^4}{(4\pi\varepsilon_0)^2 c^4} \left(\langle \alpha \rangle^2 + \frac{4}{15} \beta^2 \right)
$$

$$
\hat{e}_s \perp \hat{e}_L; \frac{d\sigma_d}{d\Omega} = \frac{\omega^4}{(4\pi\varepsilon_0)^2 c^4} \frac{3}{15} \beta^2.
$$

$$\quad (2.13)$$

The ratio of the depolarized cross section σ_d to the polarized one σ_p (depolarization ratio) thus becomes:

$$
D = \frac{\sigma_d}{\sigma_p} = \frac{3\beta^2}{15\langle \alpha^2 \rangle + 4\beta^2} \leqq \frac{3}{4}. \quad (2.14)
$$

Hence, the maximum value possible for the depolarization ratio in a medium in which the scattering centers are oriented at random is 3/4. This result also holds for Raman and Brillouin scattering (off resonance!) and is particularly useful for interpreting polarized spectra of amorphous materials [2.30]. We point out, however, that this result only holds true if $\underset{\sim}{\alpha}$ is a symmetric tensor. If antisymmetric components of $\underset{\sim}{\alpha}$ are allowed, such as if a magnetic field is present, D has, in principle, no upper boundary [2.3, 31, 32].

We have explicitly assumed that the frequency of rotation which produces the averaging of the tensor $\underset{\sim}{\beta}$ is negligibly small. This is often not the case for molecules. Under these conditions, the β-scattering changes the state of rotation of the molecules and the scattering becomes inelastic: the so-called pure rotational bands result. As gaseous molecules condense to form a liquid, these rotational modes become hindered or heavily damped. The scattering becomes quasi-elastic [Ref. 1.2, Chap. 6].

2.1.3 Elastic Scattering in Solids

In the case of a solid with N scattering particles, we must add the N-contributions to the E_s field with the appropriate phase (Fig. 2.1):

$$E_s \propto \sum_{R_i}^{N} e^{i(k_L - k_s)R_i} \alpha(R_i). \tag{2.15}$$

At this point, we introduce the susceptibility $\chi(r)$ defined as the dipole moment *per unit volume* induced by a unit field

$$\varepsilon_0 \chi(r) = \sum_{R_i}^{N} \alpha(R_i) \delta(r - R_i), \tag{2.16}$$

where δ represents the Dirac function.

Multiplying the summands of (2.15) by $\delta(r - R_i)$ and integrating over the volume, we find

$$E_s \propto \int_V e^{i(k_L - k_s)r} \chi(r) dV_r = V\chi(k_L - k_s), \tag{2.17}$$

where V is the radiating volume and $\chi(k_L - k_s)$ the Fourier component of the susceptibility corresponding to the *scattering vector* $q = k_L - k_s$. Thus, scattering along a given direction k is only possible if there is a nonvanishing suscepti-bility for the corresponding q. In a perfect crystalline solid, the susceptibility $\chi(k_L, k_s)$ vanishes unless $k_L - k_s = q$ equals either zero or a reciprocal lattice vector G (Bloch's theorem). Since $|q| \simeq 2\pi/\lambda \ll |G|$ ($|G| \geq 2\pi/a_0$, $a_0 =$ lattice con-stant), we find that $q = G$ is impossible and only forward scattering ($q = 0$, geometrical optics) occurs.

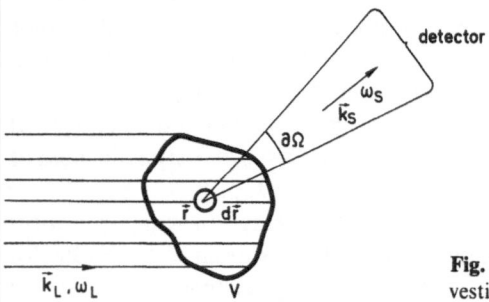

Fig. 2.1. Experimental configuration for the in-vestigation of light scattering

This result is an expression of the law of conservation of crystal momentum found in crystalline solids; crystal momentum (i.e., wave vector) must be conserved *modulo* a reciprocal lattice vector, hence for allowed scattering, either $q = G_i$ or $q = 0$. Because of the small magnitude of q in the case of light, the only possibility is $q = 0$ (forward scattering). In the case of x-rays, $|G_i| \neq 0$ becomes possible (Bragg scattering). In an inhomogeneous or amorphous solid, however, $\chi(q) \neq 0$ for $|q| \simeq \dfrac{2\pi}{\lambda}$ and scattering can take place. The corresponding cross section is then

$$\frac{d\sigma(q)}{d\Omega} = \frac{\omega^4 V^2}{(4\pi)^2 c^4} |\hat{e}_s \cdot \underset{\sim}{\chi}(q) \cdot \hat{e}_L|^2 . \tag{2.18}$$

Note that this scattering "cross section" is proportional to the *square* of the scattering volume V, a result of having assumed that all atoms scatter *coherently*. This is not the case for inelastic (Raman and Brillouin) scattering: the various elements of volume scatter incoherently and the differential cross section is proportional to V.

This result, however, should not be taken too seriously. For forward elastic scattering, for instance, there is an uncertainty in the angle α that the scattered beam forms with the incident beam which is equal to $\Delta\alpha \propto \lambda^{3/2} V^{-1/2}$, where λ is the wavelength of the light (the same phenomenon which produces the finite size of x-ray spots for finite samples). Hence, the radiation is confined to a solid angle $\Omega \propto \lambda^3 / V_0$. This fact, which cancels one of the factors of V in (2.18), leaves the *total* cross section $\sigma(q)$ simply proportional to V as in the case of Raman scattering. The only effect of coherence in the elastic scattering is to concentrate the radiation in a cone of solid angle $\Omega \propto \lambda^3 / V_0$ around the scattering direction.

The scattering cross section of (2.18) is proportional to the square of the magnitude of the fluctuations in χ of wave vector q:

$$\sigma \propto \chi^*(\omega, q) \cdot \chi(\omega, q), \tag{2.19}$$

a result, which, appropriately generalized, we shall find holds rather widely in light scattering.

It is also of interest to discuss the resonance of the scattering cross section in a solid near a characteristic transition frequency ω_0, where $\chi(\omega)$ can have a Lorentzian structure such as that of (2.7) or a somewhat weaker singularity at the so-called Van Hove critical points [2.48]. Let us assume, for instance, that the fluctuations in $\chi(\omega)$ are produced by fluctuations $\varrho(q)$ in the particle density ϱ and that these fluctuations, in turn, produce fluctuations in the critical energy ω_0 according to

$$\Delta\omega_0 = \frac{d\omega_0}{d\ln\varrho} \frac{\Delta\varrho}{\varrho_0} = -D \frac{\Delta\varrho}{\varrho_0},$$

where D is the so-called "deformation potential". The cross section becomes

$$\sigma(\omega, q) \propto \left|\frac{d\chi}{d\omega_0}\right|^2 \cdot |D\varrho(q)|^2 , \tag{2.19a}$$

where $\varrho(q)$ represents the q-component of the spectrum of spatial fluctuations in the density. Thus, in principle, a measurement of $\sigma(\omega, q)$ should yield the spectrum of $|d\chi/d\omega_0|^2$ as a function of ω. Structure in the susceptibility should appear in $|d\chi/d\omega_0|^2$ much more sharply than in $\chi(\omega)$, thus enabling us to uncover weak Van Hove singularities or critical points. This method, which has hardly been used, is similar to the so-called modulation spectroscopy techniques [2.48] where the modulation in $\chi(\omega)$ produced by an external parameter (temperature, electric field, pressure...) is measured.

We have discussed above the elastic scattering by fluctuations in χ. Time-dependent fluctuations in χ produce, as will be seen in Sect. 2.1.8, inelastic scattering. The time dependence of the fluctuations, however, may be non-sinusoidal or overdamped. In this case, a more or less broad quasi-elastic line appears, centered around the incident frequency ω_L. Such is the case of scattering by "entropy density fluctuations" first treated by *Einstein* [2.33a]. In this case, the $\chi(q)$ of (2.18) is obtained from (for cubic solids)

$$\chi(q) = \left(\frac{\partial \chi}{\partial S}\right)_p (\overline{[\delta S]_q^2})^{1/2} , \tag{2.19b}$$

where $\overline{[\delta S]_q^2}$, the average entropy density fluctuation, is given by [2.34]

$$\overline{[\delta S]_q^2} = V C_p , \qquad \left(\frac{\partial \chi}{\partial S}\right)_p = T\left(\frac{\partial \chi}{\partial T}\right)_p C_p^{-1} V^{-1} \tag{2.19c}$$

where C_p is the specific heat per unit volume at constant pressure. Using (2.19c), (2.19b) can be rewritten as:

$$\chi(q) = \frac{T}{V^{1/2} C_p^{1/2}} \left(\frac{\partial \chi}{\partial T}\right)_P = \frac{T}{V^{1/2} C_p^{1/2}} \left[\left(\frac{\partial \chi}{\partial T}\right)_V + \left(\frac{\partial \chi}{\partial V}\right)_T \left(\frac{\partial V}{\partial T}\right)_P\right] . \tag{2.19d}$$

[In 2.19d the units of C_p are those of a (volume)$^{-1}$.] Hence, the quasi-elastic scattering due to entropy density fluctuations is determined by the "thermo-optic" constant $(\partial\chi/\partial T)_p$. In (2.19d), we have split the $\chi(q)$ for entropy density fluctuations into two terms, one depending on $(\partial\chi/\partial T)_V$ and the other one $(\partial\chi/\partial V)_T$, an elasto-optic constant. We shall see in Sect. 2.1.14 that the elasto-optic constants determine the Brillouin cross section. If we neglect $(\partial\chi/\partial T)_V$, the total cross section produced by (2.19d) is simply proportional to the Brillouin cross section. For a gas, the proportionality constant, the so-called Landau Placzek [2.33b] relation, becomes

$$\frac{\sigma_{\text{entropy}}}{\sigma_{\text{Brillouin}}} = \frac{C_p}{C_v} - 1 , \tag{2.19e}$$

where $\sigma_{\text{Brillouin}}$ includes Stokes and antistokes components. The approximation $(\partial\chi/\partial T)_V = 0$ is valid for gases but not for solids. The quasi-elastic scattering by entropy density fluctuations in solids has been treated by *Wehner* and *Klein* [2.35].

2.1.4 Inelastic Scattering by Molecules

Let us first discuss briefly (and dispose of) the scattering by rotational modes of molecules. The symmetric part of the polarizability tensor $\langle\underline{\alpha}\rangle$ is invariant under rotations of the molecule and thus, as seen in Sect. 2.1.2, only produces elastic scattering. The asymmetric part $\underline{\beta}$, however, when referred to coordinates at rest is changed profoundly by rotations, oscillating between various linear combinations of the basis tensors (2.12). Each one of the elements of $\underline{\beta}$ averages to zero as time elapses. Its time dependence is $\exp(i\omega_r t) + \exp(-i\omega_r t)$, where ω_r is the rotational frequency. When multiplying these elements by the incident field $E_0\exp(-i\omega_L t)$, components of the polarization proportional to $\exp[-i(\omega_L\pm\omega_r)t]$ appear. Hence a scattered field at frequencies $\omega_s = \omega_L\pm\omega_r$ results (inelastic scattering). The frequencies ω_r, approximately proportional to the angular momentum J, are actually quantized ($\omega_r\propto nJ$, $n = 1, 2, 3 \ldots$). Hence the rotational structure consists of series of equally spaced lines below ω_L (Stokes) and other series above ω_L (antistokes). The total scattered intensity (all lines) will, of course, be equal to that calculated in Sect. 2.1.2 under the assumption $\omega_r = 0$. For a detailed treatment of pure rotational scattering, the reader is referred to [2.18] (see also Fig. 2.2).

We shall now treat vibrational scattering under the assumption that the rotational state of the molecule does not change ($\omega_r = 0$). In general, actually, rotational-vibrational bands are observed in which the rotational, as much as the vibrational state of the molecule, changes in the scattering process. The

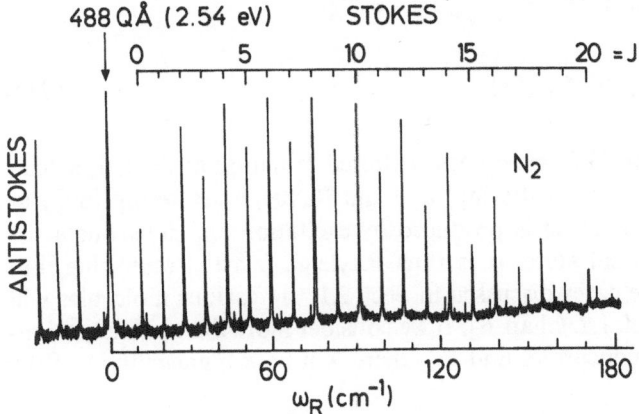

Fig. 2.2. Rotational Raman spectrum of nitrogen molecules obtained with a photographic plate as detector (response nonlinear in intensity) [Ref. 18, Fig. 3.5]

cross section obtained under the assumption $\omega_r = 0$ will apply to the sum of the cross sections of all rotational structures associated with one given vibrational jump.

Let us consider a vibrational mode of the molecule of frequency ω_v. This mode is characterized by displacements of the N atoms with time dependences $\exp(\pm i\omega_v t)$ and amplitudes u_i ($i = 1, 2 \ldots N$). The amplitudes u_i are related to the normal mode coordinate ξ through the normal mode transformation [2.36]:

$$M_i^{1/2} u_i(\pm \omega_v t) = (e_i \xi e^{-i\omega_v t} + e_i^* \xi^* e^{i\omega_v t}),\qquad (2.20)$$

where

$$\sum_{i=1}^{N} |e_i|^2 = 1$$ and M_i is the mass of atom i. The set of vectors e_i represents the eigenvector of the vibrational mode.

We shall now discuss the so-called quasistatic, adiabatic or polarizability theory of Raman scattering [2.31]. This "classical" theory is based on (2.1) (dipole radiation) and the assumption that ω_v is small compared with the "electronic energies" which determine the polarizability α. The latter condition can be expressed for an α of the form given in (2.7):

$$\omega_v^2 \ll |\omega_0^2 - \omega^2 - i\omega\gamma| .\qquad (2.21)$$

Equation (2.21) simply expresses the fact that the laser is many ω_v's away from the resonance. Under these conditions, we can treat the phonon as a static deformation of the molecule and define at each instant of time a polarizability $\alpha(\omega_L, \xi)$ which depends on the normal mode coordinate ξ. By expanding $\alpha(\omega_L, \xi)$ in powers of ξ, we find

$$
\begin{aligned}
\alpha(\omega_L, \xi) = \alpha(\omega_L) &+ \frac{\partial \alpha}{\partial \xi} \xi e^{-i\omega_v t} + \frac{\partial \alpha}{\partial \xi^*} \xi^* e^{i\omega_v t} \\
&+ \frac{1}{2} \frac{\partial^2 \alpha}{\partial \xi^2} \xi^2 e^{-2i\omega_v t} + \frac{1}{2} \frac{\partial^2 \alpha}{\partial \xi^{*2}} \xi^{*2} e^{2i\omega_v t} \\
&+ \frac{1}{2} \frac{\partial^2 \alpha}{\partial \xi \partial \xi^*} (\xi \xi^* + \xi^* \xi) + \ldots .
\end{aligned}
\qquad (2.21a)
$$

By replacing (2.21a) into (2.2), we obtain scattered radiation at the frequencies $\omega_L \pm \omega_v$, $\omega_L \pm 2\omega_v$ (overtone scattering) and the Rayleigh scattering (bilinear term in $\xi \xi^*$)$\omega_s = \omega_L$. The latter is produced by excitation and subsequent de-excitation of a vibrational state. It is thus Rayleigh (elastic) scattering, but contrary to some of the cases discussed in Sect. 2.1, the various molecules will scatter *incoherently* [Ref. 1.2, Chap. 6]. If we consider more than one vibrational mode, e.g., two frequencies ω_v and $\omega_{v'}$, there will be a bilinear term of the form:

$$\frac{1}{2} \frac{\partial^2 \alpha}{\partial \xi_v \partial \xi_{v'}^*} (\xi_v \xi_{v'}^* + \xi_v^* \xi_{v'}).\qquad (2.22)$$

This term will give rise to scattered light at the frequencies $\omega_L + \omega_v - \omega_{v'}$ and $\omega_L - \omega_v + \omega_{v'}$ (*difference scattering*). In this case, we shall also obtain *combination scattering* at the frequencies $\omega_L - \omega_v - \omega_{v'}$ (Stokes) and $\omega_L + \omega_v + \omega_{v'}$ (antistokes). The higher-order terms omitted in (2.21a) will yield scattering by three phonons, four phonons, etc.

Let us first discuss the one-phonon scattering cross section which ensues from replacing (2.21a) into (2.4):

$$\frac{d\sigma_s}{d\Omega} = \frac{\omega_s^4}{(4\pi\varepsilon_0)^2 c^4} \left| \hat{e}_s \cdot \frac{\partial \boldsymbol{\alpha}}{\partial \xi} \cdot \hat{e}_L \right|^2 \langle \xi\xi^* \rangle \text{ (Stokes)}$$

$$\frac{d\sigma_a}{d\Omega} = \frac{\omega_s^4}{(4\pi\varepsilon_0)^2 c^4} \left| \hat{e}_s \cdot \frac{\partial \boldsymbol{\alpha}}{\partial \xi^*} \cdot \hat{e}_L \right|^2 \langle \xi^*\xi \rangle \text{ (antistokes)},$$

(2.23)

where $\langle \ \rangle$ represents the thermodynamical average over the ground *state of the molecule.*

We have so far assumed that the frequency ω_v is infinitely sharp. Under these conditions, the observed Raman line will be a δ-function $\delta(\omega_R - \omega_v)$ as a function of the "*Raman shift*" ω_R. Hence, (2.23) can also be written as (we only give the Stokes component)

$$\frac{\partial \sigma_s}{\partial \Omega \partial \omega_R} = \frac{\omega_s^4}{(4\pi\varepsilon_0)^2 c^4} \left| \hat{e}_s \cdot \frac{\partial \boldsymbol{\alpha}}{\partial \xi} \cdot \hat{e}_L \right|^2 \langle \xi\xi^* \rangle \delta(\omega_R - \omega_v).$$

(2.24)

In general, ω_v will be broadened; this fact can be taken into account by replacing $\delta(\omega_R - \omega_v)$ by a Lorentzian or another appropriate lineshape function.

It is easy to estimate semiclassically the average phonon amplitude $\langle \xi^*\xi + \xi\xi^* \rangle^{1/2}$. In terms of the normal mode coordinate ξ, the potential energy at the point of maximum displacement ξ is simply $|\xi + \xi^*|^2 \omega_0^2/2 = 2\xi^2\omega_0^2$. By equating this to the total vibrational energy when in the nth harmonic oscillator level $\hbar\omega_v(n + 1/2)$, we find

$$\langle \xi^*\xi + \xi\xi^* \rangle_n = \frac{\hbar}{2\omega_v}(2n+1); n = \frac{1}{\exp(\hbar\omega_v/kT) - 1}$$

(2.25)

(n is the Bose-Einstein statistical factor). Equation (2.25) yields, when replaced into (2.23), the sum of Stokes and antistokes cross sections. In order to separate the Stokes from the antistokes contribution, we second-quantize the displacements ξ and ξ^*, i.e., we replace them by the operators ξ and $\xi^\dagger$, respectively. The Stokes and antistokes contributions to (2.25) are then [2.37]:

Stokes: $\langle \xi\xi^\dagger \rangle = \langle n|\xi|n+1 \rangle \langle n+1|\xi^\dagger|n \rangle = \frac{\hbar}{2\omega_v}(n+1)$

(2.26)

antistokes: $\langle \xi^\dagger\xi \rangle = \langle n|\xi^\dagger|n-1 \rangle \langle n-1|\xi|n \rangle = \frac{\hbar}{2\omega_v}n$.

The standard phonon creation and annihilation operators b^+ and b are related to $\xi^\dagger$ and ξ by

$$b = \sqrt{\frac{2\omega_v}{\hbar}}\, \xi, \; b^\dagger = \sqrt{\frac{2\omega_v}{\hbar}}\, \xi^\dagger. \tag{2.26'}$$

In order to further discuss the cross sections (2.24) we must consider the structure of the "Raman polarizability" $\partial\boldsymbol{\alpha}/\partial\xi$. The normal coordinate ξ corresponds to an ensemble of static atomic displacements given by (2.20) with $t = 0$. Hence $\partial\boldsymbol{\alpha}/\partial\xi$ can be written as

$$\frac{\partial\boldsymbol{\alpha}}{\partial\xi} = \sum_{i=1}^{N} \frac{\partial\boldsymbol{\alpha}}{\partial\boldsymbol{u}_i} \times \frac{\partial\boldsymbol{u}_i}{\partial\xi} = \sum_{i=1}^{N} \frac{\partial\boldsymbol{\alpha}}{\partial\boldsymbol{u}_i} M_i^{-1/2} \hat{e}_i, \tag{2.27}$$

and likewise,

$$\frac{\partial\boldsymbol{\alpha}}{\partial\xi^*} = \sum_{i=1}^{N} \frac{\partial\boldsymbol{\alpha}}{\partial\boldsymbol{u}_i} M_i^{-1/2} \hat{e}_i^*.$$

We note, in passing, that (2.26) when replaced into (2.23) leads to the conclusion that

$$\frac{d\sigma_s}{d\Omega} = \frac{n+1}{n} \frac{d\sigma_a}{d\Omega} = \frac{d\sigma_a}{d\Omega} \exp(\hbar\omega_v/kT). \tag{2.28}$$

As we shall see below, (2.28) breaks down near a resonance. It is also not valid for scattering by magnetic excitations ([2.38], see also [Ref. 1.2, Chap. 6]). For phonons away from resonance, (2.28) can be used to determine the temperature of the scattering volume of the sample from the ratio $(d\sigma_s/d\Omega)/(d\sigma_a/d\Omega)$. In spite of the widespread use of cylindrical lenses to focus the laser on the sample, the temperature of the focussed spot is usually higher than that of the rest of the sample and thus difficult to determine in any other way.

Equation (2.27) enables us to calculate $\partial\boldsymbol{\alpha}/\partial\xi$ if the effect on $\boldsymbol{\alpha}$ of all independent atomic displacements $\boldsymbol{u}_i$ is known. These $\partial\boldsymbol{\alpha}/\partial\boldsymbol{u}_i$ can, in principle, be obtained by performing calculations of the polarizability of the molecule with all possible small deformations $\boldsymbol{u}_i$. Such a first principles calculation is difficult and not very accurate. Hence, a number of semi-empirical methods have been developed to estimate $\partial\boldsymbol{\alpha}/\partial\boldsymbol{u}_i$. Among them we mention *Wolkenstein*'s bond polarizability method [2.39, 40] which assigns to each molecular bond a polarizability which is a function of the bond length l only. We thus have two differential polarizabilities, one parallel to the bond direction $\partial\alpha_\parallel/\partial l$ and a perpendicular one $\partial\alpha_\perp/\partial l$. Attempts have been made to use in this connection the concept of *transferability*, i.e., to use the bond polarizabilities obtained by fitting Raman data of one or several molecules to interpret the Raman spectra of other molecules containing the same bonds [2.41].

2.1.5 Resonant First-Order Raman Scattering

We shall now discuss the phenomenon of resonant Raman scattering. For this purpose, we shall replace the polarizability α by a sum of terms of the form of (2.7), where the various ω_0 will be the electronic excitation energies. In order to preserve the tensorial character of α, we multiply (2.7) by the "oscillator strength tensor" F. Thus, for ω near a given ω_0, we approximate α by

$$\alpha = \frac{(e^2/m)F}{\omega_0^2 - \omega^2 - i\omega\gamma} + \text{constant}. \tag{2.29}$$

Contributions of (2.29) to $\partial\alpha/\partial\xi$ arise in two different ways: through the dependence of the electronic energy ω_0 on ξ and through that of F. We can thus write by differentiating (2.29):

$$\frac{d\alpha}{d\xi} = -\frac{2\omega_0(e^2/m)F}{[\omega_0^2 - \omega^2 - i\omega\gamma]^2}\frac{d\omega_0}{d\xi} + \frac{(e^2/m)}{(\omega_0^2 - \omega^2 - i\omega\gamma)}\frac{dF}{d\xi}. \tag{2.30}$$

If one so desires, one can add several contributions of the form of (2.30) for different transition energies ω_0. For simplicity, we discuss in what follows only one of these terms.

Equation (2.30), together with (2.24), determines the shape of the resonance in σ if one replaces ω by $(\omega_L + \omega_s)/2 = \omega_L \pm \omega_v/2$. Although we have assumed $\omega_v \simeq 0$, and thus either ω_L or ω_s can be used for ω in (2.30), the use of their average yields a somewhat better approximation to the resonance behavior (it may, otherwise, be shifted from ω_0 by $\pm\omega_v/2$). Equation (2.30) contains two contributions: that of the change in excitation energy ω_0 with ξ, a diagonal matrix element of the *electron-phonon interaction*, and that of the change in oscillator strength with ξ. The former is more strongly resonant than the latter since it contains an additional power of $\omega_0^2 - \omega^2 - i\omega\gamma$ in the denominator. The resonance due to this term is also expected to be stronger than that of pure rotational spectra which is simply $\propto |\chi|^2$. Thus, for ω_L close to ω_0, the former term will be dominant unless $d\omega_0/d\xi$ vanishes, possibly as a result of a symmetry selection rule (a situation which often arises as symmetry restrictions on $d\omega_0/d\xi$ are stronger than those on $dF/d\xi$). So as to discuss the second term in (2.30), we recall the quantum-mechanical expression for F:

$$F = \frac{2\omega_0 m}{\hbar}\langle o|r|i\rangle \cdot \langle i|r|o\rangle, \tag{2.31}$$

where the "dipole" matrix elements $\langle i|r|o\rangle$ are taken between the initial (o) and the final (i) states of the electronic transitions under consideration. The dot in (2.31) represents the diadic product of the two vectors. The derivative of F with respect to ξ contains two contributions, a trivial one related to $d\omega_0/d\xi$

and another one related to the changes in matrix elements with ξ. *Quantum mechanically*, such changes arise from a mixture of $|o\rangle$ and $|i\rangle$ with other wave functions through the perturbation Hamiltonian $\partial H/\partial\xi$. Using perturbation theory we can write

$$\frac{d}{d\xi}\langle i|r|o\rangle = \sum_{j\neq i}\frac{\langle i|dH/d\xi|j\rangle\,\langle j|r|o\rangle}{\hbar(\omega_i-\omega_j)}. \tag{2.32}$$

Hence, these terms referred to sometimes as three-band terms because of the presence of the three states o, j, and i, are determined by the off-diagonal matrix elements of the electron-phonon interaction. The terms containing $d\omega_0/d\xi$ are then called two-band terms. From a fit of the cross section measured as a function of ω_L with (2.30), we can obtain information about diagonal and nondiagonal matrix elements of the electron-phonon interaction. It should now be clear why symmetry restrictions are stiffer on the diagonal $\langle i|\partial H/\partial\xi|i\rangle$ than on the nondiagonal elements $\langle i|\partial H/\partial\xi|i\rangle$ of the electron-phonon interaction: the product of states $\langle i|,|j\rangle$ generates many more symmetry types than simply $\langle i|,|i\rangle$. The contribution (2.32) is particularly large whenever small frequency differences $\omega_i-\omega_j$ are involved. Such is often the case for states separated by spin-orbit coupling (e.g., p-like atomic states).

Most simple molecules have their electronic excitations in the violet and near uv. Hence, their Raman cross sections do not show much of a resonance in the visible, a region to which most Raman measurements have been confined (because of the availability of cw laser lines). As an example of incipient resonant behaviour, we show in Fig. 2.3 the relative cross sections of vibrational Raman lines of CCl_4 and C_2H_3Cl in the visible [Ref. 2.18, p. 155]. The absorption edge ω_0 of these gases lies around 6.4 eV [2.42].

The scattering cross section corresponding to the second term of (2.30) has the form:

$$\frac{d\sigma}{d\Omega} \propto \frac{1}{(\omega^2-\omega_0^2)^2+\omega^2\gamma^2} \propto \mathrm{Im}\{\alpha\}. \tag{2.33}$$

The imaginary part of the polarizability α is, to a good approximation within the resonance region, proportional to the absorption coefficient of the molecules in gaseous form or in solution. As an example, we show in Fig. 2.4 the resonant behavior of the stretching mode (211 cm^{-1}) of I_2 molecules dissolved in hexane as compared with the corresponding optical density (proportional to the absorption coefficient) [2.43]. The shapes of both "resonance" spectra are very similar although a shift between them of about 400 cm^{-1}, larger than $\omega_v/2$, exists. These shifts, not totally understood, are common in resonance Raman spectroscopy [2.44a].

We would also like to point out that as a result of an interference of a resonant (2.30) and a nonresonant ($\sim$constant) contribution to $d\alpha/d\xi$ of

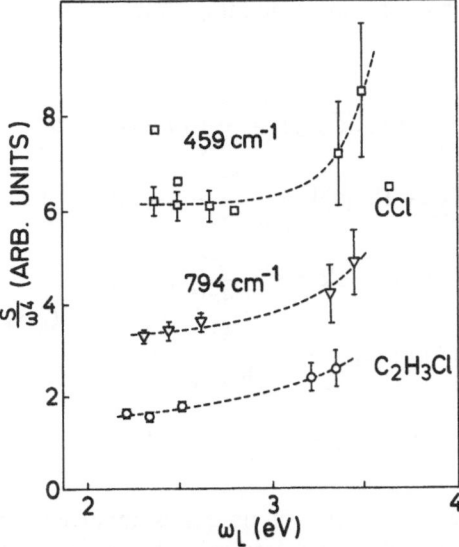

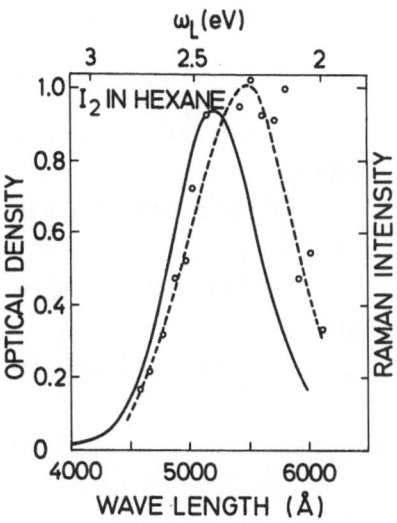

Fig. 2.3. Raman scattering efficiencies (in arbitrary units) measured for vibrational modes of CCl_4 and vinyl chloride showing resonance at high photon energies [2.18, Fig. 4.13]

Fig. 2.4. Raman scattering efficiencies measured for vibrational modes of I_2 (211 cm^{-1}) dissolved in hexane showing resonant behavior (points and dashed line). The solid line represents the optical density (proportional to the absorption coefficient) in arbitrary units [2.43]

opposite signs, a decrease in σ when approaching ω_0 may result. The cross section thus goes through zero before ω_0 and increases above this point. This behavior is sometimes referred to as an *antiresonance*. Such antiresonances, if they arise from the first term in (2.30), can usually be fitted with an expression of the type

$$\frac{\sigma}{\omega^4} = \left(A - \frac{B}{(\omega_0^2 - \omega^2)^2} \right)^2 . \tag{2.33a}$$

where A and B are adjustable constants.

2.1.6 Resonant Second-Order Raman Scattering

The cross section for scattering by two phonons is also obtained by replacing (2.21a) into (2.4) and keeping only quadratic and bilinear terms in ξ and ξ^*. The cross section is then proportional to the second derivative of α with respect to ξ. One must distinguish between scattering by two of the same phonons (overtones) and by two different phonons (combinations). In the latter case, sum and difference processes are possible. The corresponding statistical factors are:

$$
\text{Stokes} \begin{cases} \text{overtones} & |\langle n+2|\xi^\dagger|n+1\rangle \langle n+1|\xi^\dagger|n\rangle|^2 = \dfrac{\hbar^2}{4\omega_v^2}(n+1)(n+2) \\[2ex] \text{combinations} & |\langle n'+1|\xi'^\dagger|n'\rangle \langle n+1|\xi^\dagger|n\rangle|^2 = \dfrac{\hbar^2}{4\omega_v\omega_{v'}}(n+1)(n'+1) \end{cases}
$$

$$
\text{Antistokes} \begin{cases} \text{overtones} & |\langle n-2|\xi|n-1\rangle \langle n-1|\xi|n\rangle|^2 = \dfrac{\hbar^2}{4\omega_v^2}n(n-1) \\[2ex] \text{combinations} & |\langle n'-1|\xi'|n'\rangle\langle n-1|\xi|n\rangle|^2 = \dfrac{\hbar^2}{4\omega_v\omega_{v'}}nn' \end{cases}
$$

(2.34)

$$
\text{Difference} \qquad \langle n'-1|\xi|n'\rangle \langle n+1|\xi^\dagger|n\rangle = \dfrac{\hbar^2}{4\omega_v\omega_{v'}}(n+1)n'.
$$

It is possible, in principle, to elucidate the type of processes involved by measuring the temperature dependence of the cross section and comparing it with the predictions of (2.34). We should mention here that in solids, the number of phonons N is so large that combinations $(\sim N^2)$ overwhelm the *true* overtones $(\sim N)$. The latter are never observed. Nevertheless, one talks about "overtones" when two phonons of the same *branch* are involved in the process [2.44a, c].

Let us now focus on the cross section for combination Stokes scattering (the treatment of all other cases is similar):

$$
\frac{\partial^2\sigma(\omega_R, \omega_L)}{\partial\omega_R\partial\Omega} = \frac{\omega_s^4}{(4\pi\varepsilon_0)^2 c^4}\left|\hat{e}_s \cdot \frac{\partial^2\underline{\alpha}}{\partial\xi\partial\xi'}\cdot\hat{e}_L\right|^2 \frac{\hbar^2}{4\omega_v\omega_{v'}}(n+1)(n'+1)
$$
$$
\cdot \delta[\omega_R - (\omega_v + \omega_{v'})],
$$

(2.35)

where, as already mentioned, the δ-function can be replaced by a Lorentzian or another line-shape function.

The shape of the resonance spectrum $d\sigma(\omega_L)/d\Omega$ as a function of ω_L is determined by the second-order Raman polarizability $\partial^2\underline{\alpha}/\partial\xi\partial\xi'$ which can be obtained and discussed in a manner similar to its first-order counterpart. We can thus write

$$
\frac{\partial^2\underline{\alpha}}{\partial\xi\partial\xi'} = \frac{\partial^2\underline{\alpha}}{\partial\omega_0^2}\frac{\partial\omega_0}{\partial\xi}\frac{\partial\omega_0}{\partial\xi'} + \frac{\partial\underline{\alpha}}{\partial\omega_0}\frac{\partial^2\omega_0}{\partial\xi\partial\xi'} + \frac{\partial^2\underline{\alpha}}{\partial\underline{F}^2}\frac{\partial\underline{F}}{\partial\xi'}\frac{\partial\underline{F}}{\partial\xi'} + \frac{\partial\underline{\alpha}}{\partial\underline{F}}\frac{\partial^2\underline{F}}{\partial\xi\partial\xi'}. \quad (2.36)
$$

The first term in the rhs of (2.36) is the most strongly resonant one. It involves the diagonal matrix elements $\partial\omega_0/\partial\xi$ of the first-order electron-phonon interaction taken to second order (bilinear terms). However, because of the symmetry restrictions on $\partial\omega_0/\partial\xi$, this term will often vanish. Symmetry restrictions are weaker for $\partial^2\omega_0/\partial\xi\partial\xi'$ and this term often dominates. It has the same resonant behavior as the first-order scattering (2.30). Whenever this is the case, one can obtain from the ratio of second to first-order cross sections the ratio of second to first-order diagonal elements of the electron-phonon interaction. Thus, taking only the first term in (2.30) and the second in (2.36),

$$\frac{\dfrac{d\sigma^{(2)}_{vv'}}{d\Omega}}{\dfrac{d\sigma^{(1)}_{v''}}{d\Omega}} = \frac{\left|\dfrac{\partial^2\omega_0}{\partial\xi\partial\xi'}\right|^2}{\left|\dfrac{\partial\omega_0}{\partial\xi''}\right|^2}\frac{\hbar\omega_{v''}}{2\omega_v\omega_{v'}}\frac{(n'+1)(n+1)}{(n''+1)} \tag{2.37}$$

for Stokes scattering. The terms involving derivatives of $\underline{F}$ in (2.36) contain information about the off-diagonal matrix elements of the first and second-order electron-phonon interaction.

2.1.7 Absolute Raman Scattering Cross Sections for Molecules

We should point out at the start that the cross sections given above are *power* cross sections. They give the ratio of scattered to incident power. In photon counting systems, *quantum* cross sections are usually measured. The power cross sections are converted into quantum ones by multiplication by ω_L/ω_s; instead of ω_s^4, a factor $\omega_s^3\omega_L$ appears in the quantum cross sections.

A compilation of work on absolute Raman cross sections of molecules has recently appeared [Ref. 2.18, p. 144]. These cross sections are not easy to determine, the literature containing a number of erroneous results.

The most successful method to date seems to be the comparison of the area under the vibrational Raman line with that of the Rayleigh line [2.44b] or the pure rotational Raman bands [2.45]. The latter correspond to cross sections which can be exactly calculated from the rather well-known static polarizability tensor of the molecule. With these methods, the cross section of the Q branch (no rotational excitation) stretching mode of the N_2 molecule ($\omega_v = 2331\ \text{cm}^{-1}$) has been determined rather accurately to be [2.46a]

$$\left(\frac{d\sigma}{d\Omega}\right) = (5.05 \pm 0.1) \times 10^{-48}\,\omega_s^4\,\text{cm}^6\,\text{sr}^{-1}, \tag{2.38}$$

where ω_s is given in wave numbers $[\text{cm}^{-1}]$. This cross section is six orders of magnitude smaller than the one for elastic scattering given in Sect. 2.1.1.

It is easy to perform an approximate calculation of this cross section with the theory discussed above. In fact, for $\omega \ll \omega_0$ and Stokes scattering, (2.23) becomes for the N_2 molecule [we use for $d\alpha/d\xi$ (2.30) with $\underline{F} = 0.5$]

$$\frac{d\sigma}{d\Omega} = r_e^2 \frac{\omega_s^4}{\omega_0^6}\left|\frac{d\omega_0}{du}\right|^2\frac{\hbar}{4M\omega_v}(n+1), \tag{2.39}$$

where M is the atomic mass and u the atomic displacement. Replacing into (2.39) $\omega_0 = 8$ eV (64,500 cm^{-1}, the average electronic transition energy of N_2) and $d\omega_0/du = 20$ eV/Å, a typical value of the electron-phonon coupling constant, we find for the stretching vibration of N_2

$$\frac{d\sigma}{d\Omega} = 6.4 \times 10^{-48} \, \omega_s^4 (n+1) \, \text{cm}^6 \, \text{sr}^{-1} \tag{2.40}$$

which, since at room temperature $n \simeq 0$, represents the measured value (2.38) rather well.

It is also possible to use the same method to estimate the second-order cross section which, for the stretching modes of N_2 is

$$\frac{d\sigma}{d\Omega} = r_e^2 \frac{\omega_s^4}{\omega_0^6} \left| \frac{d^2\omega_0}{du^2} \right|^2 \frac{\hbar^2}{16M^2\omega_v^2} (n+1)^2. \tag{2.41}$$

Typical values of $d^2\omega_0/du^2$ are between 10^2 and 10^3 eV/Å^2 (Sect. 2.3.3). These numbers yield for the two-phonon cross sections a value one to two orders of magnitude smaller than that of the one-phonon case [2.44a].

2.1.8 First-Order Raman Scattering in Crystals

The classical theory of Raman scattering in solids can be obtained from the corresponding one for molecules in the same manner as done in Sect. 2.1.3 for the elastic scattering. The solid possessing within a scattering volume V a number N of unit cells, can be considered as one big molecule. The vibrational eigenvectors of a given mode are particularly simple as a result of the translational invariance. Using Bloch's theorem, the eigenvectors of the lth unit cell, labelled by the position vector R_l, can be related to those of the unit cell at the origin e_i through

$$e_{il}(q) = e_i e^{iq \cdot R_l}, \tag{2.42}$$

where q is the so-called crystal momentum, pseudomomentum, or wave vector of the mode. Instead of normalizing e_{il} to the whole crystal [as done in (2.20) for the molecule], it is convenient to normalize it to the unit cell:

$$\sum_i^{\text{unit cell}} |e_i(q)|^2 = 1. \tag{2.43}$$

The vibrational amplitude of a given atom of mass M_i of cell R_l in the mode of frequency $\omega_v(q)$, thus becomes (2.20)

$$u_{il}(q) = \frac{M_i^{-1/2}}{\sqrt{N}} [e_i(q)\xi_i(q)e^{i[q \cdot R_l - \omega_v(q)t]}$$
$$+ e_i^*(q)\xi_i^*(q)e^{-i[q \cdot R_l - \omega_v(q)t]}]. \tag{2.44}$$

The scattered field contains the phase factors

$$e^{-i[\omega_L \pm \omega_v(q)]t} e^{i(k_L \pm q)\cdot r}, \tag{2.45}$$

where the $+$ sign holds for antistokes and the $-$ sign for Stokes scattering. Equation (2.45) embodies the laws of conservation of energy and wave vector:

$$\begin{aligned} \omega_s &= \omega_L \pm \omega_v \\ k_s &= k_L \pm q. \end{aligned} \tag{2.46}$$

For the same reasons as discussed in Sect. 2.1.2, (2.46) implies that only scattering for $q \simeq 0$ is possible, the magnitude of the allowed q's ranging between 0 (forward scattering) and $4\pi/\lambda$ (back scattering). The scattering cross section can be obtained from (2.23) by introducing the susceptibility $\chi(\omega_L)$ in the same manner as done for elastic scattering (Sect. 2.1.3). We find for the Stokes component

$$\frac{d\sigma_s}{d\Omega} = \frac{\omega_s^4 V^2}{(4\pi)^2 c^4} \left| e_s \cdot \frac{d\chi}{d\xi} \cdot e_L \right|^2 \langle \xi \xi^\dagger \rangle, \tag{2.47}$$

and likewise for the antistokes component with the statistical factor $\langle \xi \xi^\dagger \rangle$ replaced by $\langle \xi^\dagger \xi \rangle$. The normal coordinates ξ are, of course, only those of optical modes with $q \simeq 0$ (for acoustical modes see below under *Brillouin scattering*, Sect. 2.1.14). These statistical factors have the values given in (2.26). The expression equivalent to (2.27) for the "Raman susceptibility" is

$$\frac{d\chi}{d\xi} = \sum_i^{\text{unit cell}} \frac{\partial \chi}{\partial u_i} \frac{M_i^{-1/2}}{N^{1/2}} e_i. \tag{2.48}$$

Note the presence of the factor $N^{-1/2}$ in (2.48)! When replacing (2.48) into (2.47), this factor cancels a power of V in the numerator of (2.47) (the number of unit cells N is proportional to the volume). It is, of course, somewhat awkward to use a Raman susceptibility which depends on the volume, as shown in (2.48). Hence, we redefine it to be independent of the volume:

$$\frac{d\chi'}{d\xi} = V^{1/2} \frac{d\chi}{d\xi} = V_c^{1/2} \sum_i^{\text{unit cell}} \frac{\partial \chi}{\partial u_i} M_i^{-1/2} e_i, \tag{2.49}$$

where V_c is the volume of the primitive cell. In terms of this renormalized, volume independent Raman susceptibility, (2.47) becomes

$$\frac{d\sigma_s}{d\Omega} = \frac{\omega_s^4 V}{(4\pi)^2 c^4} \left| e_s \cdot \frac{d\chi'}{d\xi} e_L \right|^2 (n+1) \frac{\hbar}{2\omega_v} \tag{2.50}$$

or, equivalently, introducing the lineshape function $\Delta(\omega_v - \omega_R)$ (usually Lorentzian):

$$\frac{\partial^2 \sigma_s}{\partial \Omega \partial \omega_R} = \frac{\omega_s^4 V}{(4\pi)^2 c^4} \left| e_s \cdot \frac{d\chi'}{d\xi} \cdot e_L \right|^2 (n+1)\Delta(\omega_v - \omega_R)\frac{\hbar}{2\omega_v} \tag{2.51}$$

with

$$\int_0^\infty \Delta(\omega_v - \omega_R)d\omega_R = 1 .$$

The cross sections (2.50, 51) are proportional to the scattering volume V (keep in mind that the incident beam has been assumed to cover all of V and its power flux to be independent of V). For the purpose of comparison with molecular cross sections, it is convenient to refer this cross section to the volume of either the unit cell, a formula unit, or an atom. Thus, different values for the cross section of a solid may be given by different authors and the reader has to find at the outset what the volume of the sample is for which the cross section has been defined. This problem is circumvented by using, instead of σ, the scattering efficiency S defined by dropping V in (2.50, 51). This quantity, with dimensions of an inverse length, represents the ratio between scattered and incident power for a unit path length within the solid.

The scattering efficiency S just defined, multiplied by the scattering length and corrected for reflection losses in entering and leaving the sample, gives the total observable scattered intensity. The scattering length may be limited by absorption in the solid. This happens particularly near resonances ($\omega_L \simeq \omega_0$). In this case, it is customary to use the backscattering geometry. The measured effective scattering efficiency S^* (dimensionless) is then given by [2.46b]

$$S^* = S\frac{1 - \exp[-(S + \alpha_L + \alpha_s)L]}{S + \alpha_L + \alpha_s}(1 - R_L)(1 - R_s), \tag{2.52}$$

where L is the plane parallel sample thickness (assumed large so that no back-reflection occurs), α_L and α_s the absorption coefficients and R_L and R_s the reflectivities of the incident and the scattered radiation, respectively. Since we have made in (2.52) a distinction between incident and scattered radiation, it may be worth pointing out that all expressions above also require a factor n_s/n_L, where n_s and n_L are the refractive indices of scattered and incident radiation, respectively. These factors dropped out consistently when it was assumed that $\omega_L \simeq \omega_s$ and that $R_L \simeq R_s$. While this is not true near sharp absorption lines (e.g., excitons), the semiclassical theory just presented breaks down anyhow in such cases. Another experimental consideration concerns the fact that the scattered light is focussed onto the entrance slit of the spectrometer by a collecting lens. The angle of collection Ω is constant outside the sample (Ω_0). The solid angle inside the sample, that to which the differential cross sections given above applies, is given by

$$\Omega_i = 2\pi(1 - \cos\Theta_i) \tag{2.53}$$

and

$$\Omega_0 = 2\pi(1 - \cos\Theta_0),$$

where the angle Θ_i is related to that of the refracted ray outside through Snell's law. In the usual experiments, the f-number of the collecting lens is between 1.5 and 2. Hence the *maximum* values of the angles Θ_0 lie around 15°. The corresponding angles Θ_i are much smaller because of the usually large refractive indices (for Ge $n = 4$). Hence, we can expand (2.53) in power series of the angles and find

$$\frac{\Omega_i}{\Omega_0} = \left(\frac{n_0}{n_i}\right)^2. \tag{2.54}$$

The cross sections and efficiencies given above must be multiplied by this factor to relate theory to the experimental data. Because of the frequency dependence of n_i, this factor can significantly modify the dependence of S on ω_L when measured over a wide frequency range.

For the sake of completeness we introduce another useful notation for representing scattering efficiencies and cross sections. The formulas given so far contain squares of second rank tensors contracted with vectors. They can, therefore, be written as fourth rank tensors contracted with four vectors so as to form a scalar (e.g., the scattering efficiency). The scattering efficiency obtained from (2.51) can thus be written [2.17]

$$\frac{\partial^2 S}{\partial\Omega\partial\omega_R} = \left(\frac{\omega_s}{c}\right)^4 \hat{e}_{L\alpha}\hat{e}_{L\gamma}\hat{e}_{s\beta}\hat{e}_{s\delta}I_{\alpha\beta\gamma\delta}, \tag{2.55}$$

where

$$I_{\alpha\beta\gamma\delta} = \frac{1}{(4\pi)^2}\frac{\partial\chi'_{\alpha\beta}}{\partial\xi}\frac{\partial\chi'^{*}_{\gamma\delta}}{\partial\xi}(n+1)\Delta(\omega_v - \omega_R)\frac{\hbar}{2\omega_v}.$$

Equation (2.55) can also be rewritten in terms of a time-dependent polarizability operator $\boldsymbol{P}$ fluctuating at the frequency ω_v [2.46c]:

$$I_{\alpha\beta\gamma\delta} = \frac{1}{(4\pi\varepsilon_0)^2 2\pi V}\int_{-\infty}^{+\infty}\langle P_{\alpha\beta}(t)P^{\dagger}_{\gamma\delta}(0)\rangle e^{i\omega_R t}\,dt \tag{2.56}$$

with

$$P^{\dagger}_{\alpha\beta}(t) = V^{1/2}\varepsilon_0\sum_{\omega_v}\left(\frac{\partial\chi'_{\alpha\beta}}{\partial\xi^*}\right)\xi^{\dagger}(t),$$

where the operator

$$\xi^\dagger(t) = \exp(-iHt)\xi^\dagger(0)\exp(iHt)$$

has been written in the Heisenberg representation. The Raman susceptibility $(\partial\chi'_{\alpha\beta}/\partial\xi)$ is treated as a number. The angular brackets $\langle\ldots\rangle$ represent a thermodynamical average over the equilibrium configuration at temperature T.

Equation (2.56) describes the efficiency as the frequency spectrum of the average fluctuations of the polarizability P. This result is, of course, not confined to scattering by phonons. Any other elementary excitations which produce a fluctuation in P, i.e., whose amplitude parameter (equivalent to ξ) affects P, will also lead to inelastic light scattering through the equivalent equation (2.56). Among these excitations, we mention here scattering by magnons (through the dependence of χ on the magnetization) and by plasmons (χ depends on the electric field which accompanies the plasma waves).

It is interesting to discuss the symmetry properties of the fourth rank tensor $I_{\alpha\beta\gamma\delta}$. Since S is invariant upon the group of symmetry operations of the crystal, $I_{\alpha\beta\gamma\delta}$ must also be invariant, i.e., it must belong to the identity representation of the group. From the symmetry of $d\chi'_{\alpha\beta}/d\xi$, which only applies within the framework of the theory so far developed, $I_{\alpha\beta\gamma\delta}$ is invariant upon the substitutions $\alpha\rightarrow\beta$, $\gamma\rightarrow\delta$. More generally, only the sums $(I_{\alpha\beta\gamma\delta}+I_{\gamma\beta\alpha\delta}+I_{\alpha\delta\gamma\beta}+I_{\gamma\delta\alpha\beta})$ appear as independent components. There can thus be no more than 36 independent components of I (corresponding to the 6×6 independent components of $(\hat{e}_{L\beta}\hat{e}_{L\delta})\times(\hat{e}_{s\alpha}\hat{e}_{s\gamma})$. This would include antisymmetric components of $d\chi/d\xi$. If these are ruled out, the maximum number of independent components of $\tilde{I}_{\alpha\beta\gamma\delta}$ becomes 21, also the maximum possible number of elastic constants [2.47]! Hence, in a cubic material, $I_{\alpha\beta\gamma\delta}$ has the same symmetry properties as the elastic constants. It has, therefore, three independent components equivalent to the three elastic constants (if we rule out antisymmetric components of $d\chi'_{\alpha\beta}/d\xi$). These are $I_{\alpha\alpha\alpha\alpha}$, $I_{\alpha\alpha\beta\beta}$, and $I_{\alpha\beta\alpha\beta}$.

The theory of resonant Raman scattering in solids parallels that given in Sect. 2.1.5 for molecules. The only distinction is that the discrete electronic states of the molecules must be replaced by electronic energy bands and excitons. The resonance then follows a behavior similar to that given by (2.30) but integrated over a continuum of ω_0 frequencies. The corresponding results for simple and useful cases will be given in Sects. 2.2.4–7. The measurement of resonances in Raman scattering cross sections of simple solids near absorption edges or other critical points of the interband transitions (Van Hove singularities [2.49]) has occupied the attention of many workers within the past ten years. Such measurements yield information about diagonal and nondiagonal components of the electron-phonon interaction (deformation potentials in solids). They may also yield information about the electronic states producing the resonance. The experimental results usually obtained in back scattering must always be corrected for absorption and reflection in the manner indicated in (2.52). As an example, we show in Fig. 2.5 the resonance observed for the TO Raman phonons of ZnTe [2.49] near the fundamental absorption edge $\omega_0 \simeq 2.25$ eV.

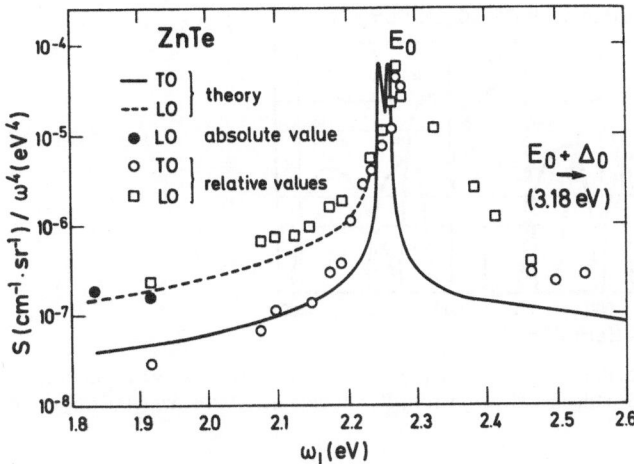

Fig. 2.5. Resonance observed for the TO-phonon Raman scattering of ZnTe near the fundamental absorption edge. The solid line is a theoretical fit [2.172]. The absolute scattering efficiency of the vertical scale was obtained by comparison with Brillouin scattering [2.50] (Sect. 2.2.18 c)

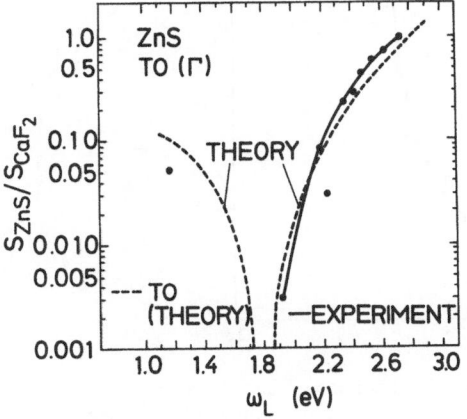

Fig. 2.6. Antiresonance observed for TO Raman phonons below the fundamental edge of Wurtzite-type ZnS. The solid line is a theoretical fit [2.51]

The data were measured by comparison with CaF_2, a material for which $\omega_0 = 12\ eV \gg \omega_L$ and thus its scattering efficiency is simply proportional to ω_s^4 in the region of measurement [2.50]. Hence, the factor ω^4 which appears in all expressions for S given so far is eliminated; one obtains a curve proportional to $|d\chi'/du|^2$. The fitted curve has been calculated in this manner (Sect. 2.2.6). A recent determination [2.49] has enabled us to give absolute values of the cross section in Fig. 2.5. Absolute cross sections and scattering efficiencies of solids will be discussed in Sect. 2.1.18.

It is also possible to find in solids antiresonant behaviour of the type predicted by (2.33a). As an example, we show in Fig. 2.6 the antiresonance observed for TO-phonons in wurtzite-type ZnS [2.51]. It was fitted with an

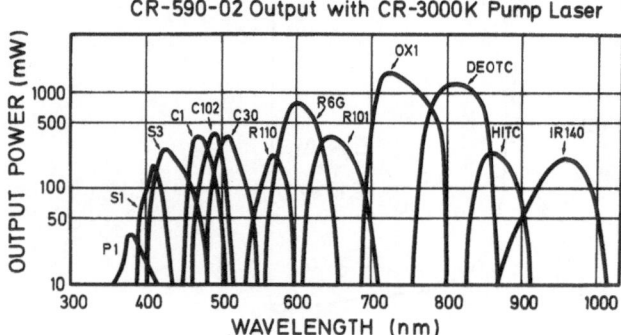

Fig. 2.7. Power spectrum obtained for various dyes as a function of dye laser wavelength in cw operation. (P: polyphenyl, S: stilbene, C: coumarin, R: rhodamine, Ox: oxazine) (Courtesy of Coherent Radiation)

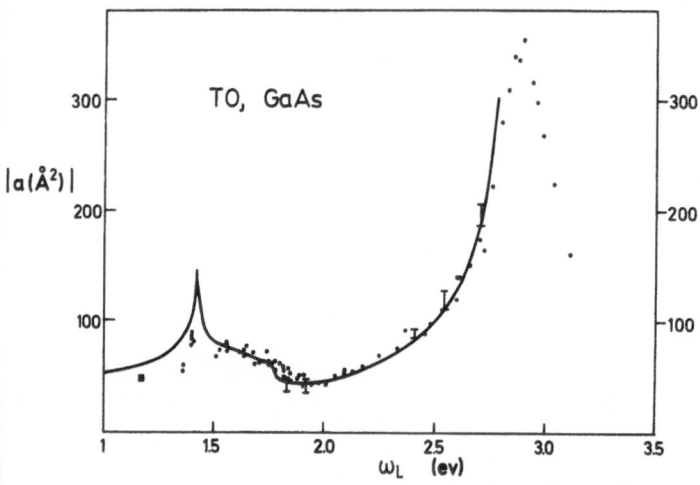

Fig. 2.8. Resonances in the Raman scattering by TO-phonons in GaAs as a function of photon energy (in terms of the Raman polarizability a). The resonant structures belong to the so-called E_0, $E_0 + \Delta_0$, E_1 and $E_1 + \Delta_1$ gaps [2.53]. The points are experimental, the solid line was obtained with (2.194, 201)

expression of the type of (2.33a). The antiresonant behavior does not occur for the corresponding phonons of zincblende-type ZnS [2.52].

The measurements of resonant Raman scattering require a continuously tunable laser or a series of closely-spaced discrete laser lines. Considerable progress has been made in extending the range of cw tunable dye lasers into the uv and near infrared in the past five years. Figure 1.1 of [2.1] shows the state of the art in 1975. The present state of the art is shown in Fig. 2.7.

Another problem in the study of resonant Raman scattering is the variation of the spectrometer throughput and detector sensitivity with photon energy.

This is particularly serious when the measurements are performed over a wide frequency range (see, for instance, Fig. 2.8 for the resonance scattering by optical phonons in GaAs between 1.35 and 3.1 eV). We have mentioned one possible way to circumvent the problem: the use of CaF_2 (optical phonons at 320 cm^{-1}) as a comparison standard for which $S\omega_L^{-4}$ is constant below ~ 5 eV. Another possibility is to calibrate the spectral response of the spectrometer with a tungsten source [2.54]. Accuracies of the relative calibration are seldom better than 20% over the whole calibration range.

2.1.9 The Raman Tensor

We have defined above a fourth rank tensor $I_{\alpha\beta\gamma\delta}$ which transforms like the identity representation (Γ_1 in Bethe's notation, A or A_g in Mulliken's). The possible independent components of this tensor (three in the case of a cubic material) can be obtained by inspection of the corresponding table for the elastic constants while keeping in mind the $\alpha \to \gamma$, $\beta \to \delta$ invariance. They are given in Table 2.1 for the 32 crystallographic point groups. Once these components are known it is possible to obtain selection rules for light scattering in crystals.

It is more convenient, however, to use instead of $I_{\alpha\beta\gamma\delta}$ the so-called second rank Raman tensor R_{ij}, an entity proportional to $(\partial\chi_{ij}/\partial\xi)$ which is defined in the literature usually to within a numerical constant [i.e., it may or may not include ω_s^2, c^2, $(n+1)^{1/2}$... and all other numerical factors of (2.50)]. This tensor is usually used to calculate selection rules and there is no need to set these factors straight unless we want to use it to obtain absolute values of S (Sect. 2.1.8).

The symmetry properties of R_{ij} are derived simply from the fact that the scattering efficiency for polarizations $\hat{e}_L$ and $\hat{e}_s$ is given by

$$dS \propto |\hat{e}_s \cdot \underline{R} \cdot e_L|^2 . \tag{2.57}$$

The polar vectors $\hat{e}_s$, $\hat{e}_L$ will belong to one (or more) irreducible representations of the point group of the crystal. Let us label these representations Γ_V. In order to make dS invariant, $\underline{R}$ must contain components which vary like irreducible components Γ_{R_i} of the product representations:

$$\Gamma_R = \Gamma_{Vs}^* \times \Gamma_{VL} = \Gamma_{R_1} + \Gamma_{R_2} + \cdots . \tag{2.58}$$

Each one of the components Γ_{R_i} will determine one set of reduced tensors which will represent scattering by phonons of a given symmetry. The irreducible symmetry of a given set of Raman active phonon Γ_p must be contained in the product ($\Gamma_{Vs}^* \times \Gamma_{VL}$) as it is produced by the annihilation of the photon Γ_{VL} and creation of Γ_{Vs}. Hence the scattering by a Γ_p phonon can be represented by

$$dS \propto |\hat{e}_s \cdot \underline{R}_{\Gamma_p} \cdot \hat{e}_L|^2 , \tag{2.59}$$

where $\underset{\sim}{R}_{\Gamma_p}$ is a component of R which belongs to the same irreducible representation as the phonon p. These principles are not confined to phonons; they can be used for scattering by other types of elementary excitations (e.g., magnons, plasmons...), whereby one has to keep in mind that in some cases, (e.g. magnons) the Raman tensor can have antisymmetric components.

We should point out that the discussion above is based on having assumed that the scattering vector is *exactly* zero ($q = 0$) or conversely, that it does not depend on q (dipole approximation). This may not be true near resonances (see, e.g., the Γ_{25} phonon of Cu_2O, Sect. 2.3.6), in particular, for polar phonons involving the Fröhlich interaction (Sect. 2.2.8). It is also obviously not true for Brillouin scattering; $q = 0$ implies a rigid translation of the system for which $d\chi/d\xi = 0$. In this case, S is determined by a third (or higher) rank tensor:

$$dS \propto |e_{s,k} R_{klm} e_{L,l} q_m|^2 . \tag{2.60}$$

Table 2.1. Raman tensors and their symmetries (labelled in Mulliken's and Bethe's notations) for the 32 crystallographic point groups (both Schoenflies and Hermann-Maughin notation given). The corresponding $I_{\alpha\beta\gamma\delta}$ tensors (2.55) are also given for the uniaxial and cubic crystals. The Raman tensors include a possible antisymmetric component which cancels for phonon scattering away from resonance. See [2.17]

Biaxial crystals

Triclinic
$$\begin{bmatrix} a & d & f \\ e & b & h \\ g & i & c \end{bmatrix}$$

1	C_1	A	Γ_1
$\bar{1}$	C_i	A_g	Γ_1^+

Monoclinic
$$\begin{bmatrix} a & d & \\ e & b & \\ & & c \end{bmatrix} \begin{bmatrix} & & f \\ & & h \\ g & i & \end{bmatrix}$$

2	C_2	A	Γ_1	B	Γ_2
m	C_s	A'	Γ_1	A''	Γ_2
2/m	C_{2h}	A_g	Γ_1^+	B_g	Γ_2^+

Orthorhombic
$$\begin{bmatrix} a & & \\ & b & \\ & & c \end{bmatrix} \begin{bmatrix} & d & \\ e & & \\ & & \end{bmatrix} \begin{bmatrix} & & f \\ & & \\ g & & \end{bmatrix} \begin{bmatrix} & & \\ & & h \\ & i & \end{bmatrix}$$

222	D_2	A	Γ_1	B_1	Γ_3	B_2	Γ_2	B_3	Γ_4
mm2	C_{2v}	A_1	Γ_1	A_2	Γ_3	B_1	Γ_2	B_2	Γ_4
mmm	D_{2h}	A_g	Γ_1^+	B_{1g}	Γ_3^+	B_{2g}	Γ_2^+	B_{3g}	Γ_4^+

Table 2.1 (continued)

Tetragonal

$$\begin{bmatrix} a & c & \\ -c & a & \\ & & b \end{bmatrix} \begin{bmatrix} d & e & \\ e & -d & \\ & & \end{bmatrix} \underbrace{\begin{bmatrix} & & f \\ & & h \\ g & i & \end{bmatrix} \begin{bmatrix} & & -h \\ & & f \\ -i & g & \end{bmatrix}}$$

4	C_4						
$\bar{4}$	S_4	A	Γ_1	B	Γ_2	E	$\Gamma_3 + \Gamma_4$
4/m	C_{4h}	A_g	Γ_1^+	B_g	Γ_2^+	E_g	$\Gamma_3^+ + \Gamma_4^+$

$I_{1111} = a^2 + d^2;$ $I_{3333} = b^2;$ $I_{1122} = \frac{1}{2}(a^2 + e^2 - c^2 - d^2)$
$I_{1133} = \frac{1}{2}(ab + fg + hi);$ $I_{1212} = c^2 + e^2;$ $I_{3232} = h^2 + f^2;$ $I_{2323} = g^2 + i^2$

$$\begin{bmatrix} a & & \\ & a & \\ & & b \end{bmatrix} \begin{bmatrix} & c & \\ -c & & \\ & & \end{bmatrix} \begin{bmatrix} d & & \\ & -d & \\ & & \end{bmatrix} \begin{bmatrix} & e & \\ e & & \\ & & \end{bmatrix} \underbrace{\begin{bmatrix} & & f \\ & & \\ g & & \end{bmatrix} \begin{bmatrix} & & \\ & & f \\ & g & \end{bmatrix}}$$

422	D_4										
4mm	C_{4v}	A_1	Γ_1	A_2	Γ_2	B_1	Γ_3	B_2	Γ_4	E	Γ_5

$I_{1111} = a^2 + d^2;$ $I_{3333} = b^2;$ $I_{1122} = \frac{1}{2}(a^2 - d^2 - c^2 + e^2)$
$I_{1133} = \frac{1}{2}(ab + fg);$ $I_{1212} = c^2 + e^2;$ $I_{3232} = f^2;$ $I_{2323} = g^2$

Trigonal

$$\begin{bmatrix} a & & \\ & a & \\ & & b \end{bmatrix} \begin{bmatrix} & c & \\ -c & & \\ & & \end{bmatrix} \begin{bmatrix} d & & \\ & -d & \\ & & \end{bmatrix} \begin{bmatrix} & e & \\ e & & \\ & & \end{bmatrix} \underbrace{\begin{bmatrix} & & f \\ & & \\ g & & \end{bmatrix} \begin{bmatrix} & & \\ & & f \\ & g & \end{bmatrix}}$$

422	D_4										
4mm	C_{4c}	A_1	Γ_1	A_2	Γ_2	B_1	Γ_3	B_2	Γ_4	E	Γ_5
$\bar{4}2m$	D_{2d}										
4/mmm	D_{4h}	A_{1g}	Γ_1^+	A_{2g}	Γ_2^+	B_{1g}	Γ_3^+	B_{2g}	Γ_4^+	E_g	Γ_5^+

$I_{1111} = a^2 + d^2;$ $I_{3333} = b^2;$ $I_{1122} = \frac{1}{2}(a^2 - c^2 + e^2 - d^2)$
$I_{1133} = \frac{1}{2}(ab + fg);$ $I_{1212} = e^2 + c^2;$ $I_{3232} = f^2;$ $I_{2323} = g^2$

$$\begin{bmatrix} a & c & \\ -c & a & \\ & & b \end{bmatrix} \underbrace{\begin{bmatrix} d & e & f \\ e & -d & h \\ g & i & \end{bmatrix} \begin{bmatrix} e & -d & -h \\ -d & -e & f \\ -i & g & \end{bmatrix}}$$

3	C_3	A	Γ_1	E	$\Gamma_2 + \Gamma_3$
$\bar{3}$	C_{3i}	A_g	Γ_1^+	E_g	$\Gamma_2^+ + \Gamma_3^+$

$I_{1111} = a^2 + d^2 + e^2;$ $I_{3333} = b^2;$ $I_{1122} = a^2 - c^2$
$I_{1133} = ab + fg + hi;$ $I_{1212} = e^2 + d^2 - c^2;$ $I_{3232} = f^2 + h^2;$ $I_{2323} = g^2 + i^2$

Table 2.1 (continued)

Trigonal

$$
\begin{bmatrix} a & & \\ & a & \\ & & b \end{bmatrix}
\begin{bmatrix} & & c \\ & -c & \\ & & \end{bmatrix}
\underbrace{\begin{bmatrix} d & & \\ d & e & \\ & & f \end{bmatrix}\begin{bmatrix} d & & -e \\ & -d & \\ -f & & \end{bmatrix}}
$$

32	D_3	A_1	Γ_1	A_2	Γ_2		E	Γ_3
3m	C_{3v}							
$\bar{3}m$	D_{3d}	A_{1g}	Γ_1^+	A_{2g}	Γ_2^+		E_g	Γ_3^+

$I_{1111} = a^2 + d^2$; $I_{3333} = b^2$; $I_{1122} = \frac{1}{2}(a^2 - c^2)$
$I_{1133} = \frac{1}{2}(ab + ef)$; $I_{1212} = c^2 + d^2$; $I_{3232} = e^2$; $I_{2323} = f^2$

Hexagonal

$$
\begin{bmatrix} a & c & \\ -c & a & \\ & & b \end{bmatrix}
\underbrace{\begin{bmatrix} & & d \\ & & f \\ e & g & \end{bmatrix}\begin{bmatrix} & & -f \\ & & d \\ -g & e & \end{bmatrix}}
\underbrace{\begin{bmatrix} i & h & \\ h & -i & \\ & & \end{bmatrix}\begin{bmatrix} h & -i & \\ -i & -h & \\ & & \end{bmatrix}}
$$

6	C_6	A	Γ_1	E_1	$\Gamma_5 + \Gamma_6$	E_2	$\Gamma_2 + \Gamma_3$
$\bar{6}$	C_{3h}	A'	Γ_1	E''	$\Gamma_5 + \Gamma_6$	E'	$\Gamma_2 + \Gamma_3$
6/m	C_{6h}	A_g	Γ_1^+	E_{1g}	$\Gamma_5^+ + \Gamma_6^+$	E_{2g}	$\Gamma_2^+ + \Gamma_3^+$

$I_{1111} = a^2 + i^2 + h^2$; $I_{3333} = b^2$; $I_{1122} = \frac{1}{2}(a^2 - c^2)$
$I_{1133} = \frac{1}{2}(ab + de + fg)$; $I_{1212} = h^2 + i^2 + c^2$; $I_{3232} = d^2 + f^2$; $I_{2323} = e^2 + g^2$

$$
\begin{bmatrix} a & & \\ & a & \\ & & b \end{bmatrix}
\begin{bmatrix} & & c \\ & -c & \\ & & \end{bmatrix}
\underbrace{\begin{bmatrix} & & \\ & & d \\ e & & \end{bmatrix}\begin{bmatrix} & & -d \\ & & \\ & -e & \end{bmatrix}}
\underbrace{\begin{bmatrix} f & & \\ & f & \\ & & \end{bmatrix}\begin{bmatrix} f & & \\ & -f & \\ & & \end{bmatrix}}
$$

622	D_6	A_1	Γ_1	A_2	Γ_2	E_1	Γ_5	E_2	Γ_6
6mm	C_{6v}								
$\bar{6}m2$	D_{3h}	A_1'	Γ_1	A_2'	Γ_2	E''	Γ_5	E'	Γ_6
6/mmm	D_{6h}	A_{1g}	Γ_1^+	A_{2g}	Γ_2^+	E_{1g}	Γ_5^+	E_{2g}	Γ_6^+

$I_{1111} = a^2 + f^2$; $I_{3333} = b^2$; $I_{1122} = \frac{1}{2}(a^2 - c^2)$
$I_{1133} = \frac{1}{2}(ab + de)$; $I_{1212} = f^2 + c^2$; $I_{2323} = e^2$; $I_{3232} = d^2$

Table 2.1 (continued)

Cubic

$$
\begin{bmatrix} a & & \\ & a & \\ & & a \end{bmatrix}
\begin{bmatrix} b & & \\ & b & \\ & & -2b \end{bmatrix}
\begin{bmatrix} & & -3^{1/2}b \\ & & 3^{1/2}b \\ & & \end{bmatrix}
\begin{bmatrix} & & \\ & & c \\ & d & \end{bmatrix}
\begin{bmatrix} & & c \\ & & \\ d & & \end{bmatrix}
$$

| 23 | T | A | Γ_1 | | E | $\Gamma_2+\Gamma_3$ | | T | Γ_4 |
| m3 | T_h | A_g | Γ_1^+ | | E_g | $\Gamma_2^+ + \Gamma_3^+$ | | T_g | Γ_4^+ |

$I_{1111} = a^2 + 4b^2$; $I_{1122} = \frac{1}{2}(a^2 - 2b^2 + dc)$; $I_{2121} = c^2$
$I_{1212} = d^2$

$$
\begin{bmatrix} a & & \\ & a & \\ & & a \end{bmatrix}
\begin{bmatrix} b & & \\ & b & \\ & & -2b \end{bmatrix}
\begin{bmatrix} & & -3^{1/2}b \\ & & 3^{1/2}b \\ & & \end{bmatrix}
\begin{bmatrix} & & \\ & & c \\ & -c & \end{bmatrix}
\begin{bmatrix} & & c \\ & & \\ -c & & \end{bmatrix}
\begin{bmatrix} & & \\ & & d \\ & d & \end{bmatrix}
\begin{bmatrix} & & d \\ & & \\ d & & \end{bmatrix}
$$

432	$\left.\begin{matrix} O \\ T_d \end{matrix}\right\}$	A_1	Γ_1	E	Γ_3	Γ_{12}	T_1	Γ_4	Γ_{25}	T_2	Γ_4	Γ_{15}
43m												
m3m	O_h	A_{1g}	Γ_1^+	E_g	Γ_3^+	$\Gamma_{12'}$	T_{1g}	Γ_4^+	$\Gamma_{15'}$	T_{2g}	Γ_5^+	$\Gamma_{25'}$

$I_{1111} = a^2 + 4b^2$; $I_{1122} = \frac{1}{2}(a^2 - 2b^2 + d^2 - c^2)$; $I_{1212} = d^2 + c^2$

Table 2.2. Selection rules for Raman scattering by Γ_{25}, (Γ_{15}) phonons in germanium and zinc-blende-type materials for the three principal surfaces [001], [111], and [110] in backscattering. The efficiencies are given in terms of the irreducible components a, b, c, d of Table 2.1

Surface	Incident polarization $\hat{e}_L$	Scattered polarization $\hat{e}_s$	Raman efficiency
[1$\bar{1}$0]	[110]	[110]	$a^2 + b^2 + d^2$ (TO)
[1$\bar{1}$0]	[001]	[001]	$a^2 + 4b^2$
[1$\bar{1}$0]	[001]	[110]	d^2 (TO) $+ c^2$
[1$\bar{1}$0]	[111]	[111]	$a^2 + \frac{4}{3}d^2$ (TO)
[100]	[01$\bar{1}$]	[01$\bar{1}$]	$a^2 + b^2 + d^2$ (LO)
[100]	[011]	[01$\bar{1}$]	$3b^2 + c^2$
[100]	[010]	[001]	d^2 (LO) $+ c^2$
[100]	[010]	[010]	$a^2 + 4b^2$
[111]	[1$\bar{1}$0]	[1$\bar{1}$0]	$a^2 + \frac{1}{3}d^2$ (LO) $+ \frac{2}{3}d^2$ (TO)
[111]	[1$\bar{1}$0]	[11$\bar{2}$)	$a^2 + \frac{2}{3}d^2$ (TO)
[111]	[11$\bar{2}$]	[11$\bar{2}$]	$a^2 + \frac{1}{3}d^2$ (LO) $+ \frac{2}{3}d^2$ (TO)

The symmetry properties of the tensor R_{klm} (2.60) are discussed in [2.17, 55]. They are equivalent to the properties of the electric field induced Raman tensor [2.56a].

In Table 2.1, we present the possible Raman tensors and their symmetries for the 32 crystallographic point groups including possible antisymmetric components. We also include the relationship between the possible R_{ij} and the tensor components $I_{\alpha\beta\gamma\delta}$ (to an arbitrary coefficient) symmetrized with respect to the four possible equivalent combinations of indices. The antisymmetric components may be easily eliminated from the table for scattering by phonons whenever the assumption (2.21) of the quasistatic treatment applies. The selection rules for Raman scattering are obtained by contracting the tensors corresponding to the phonons under consideration and given in Table 2.1 with possible incident and scattered polarization vectors $\hat{e}_L$ and $\hat{e}_s$. As an example, we present in Table 2.2 the selection rules calculated for scattering by the $\Gamma_{25'}$ (i.e., T_{2g}) phonons of materials with the O_h point group (e.g., diamond, CaF_2) on the three principal faces ([001], [110], [111]). These rules also apply to the Γ_{15} (i.e., T_1) phonons of the zincblende structure.

2.1.10 Factor Group Analysis of Phonon Symmetries

The decomposition of the Raman tensor (a second rank tensor) into irreducible symmetry components for the various crystallographic point groups is found in any standard group theory textbook. If that group contains the inversion, the representations are either odd or even while the *second rank* Raman tensor yields only even representations. Likewise, the vector-like dipole operator relevant to ir-absorption is *odd*. Hence, the well-known selection rule ir-allowed $\equiv$ Raman-forbidden and vice versa which, however, is only correct in the presence of *inversion symmetry*. In this case, the usually forbidden Raman lines can become allowed through the mechanism of (2.60); the *third rank* tensor R_{klm} decomposes into *odd* representations.

We discuss next the determination of the irreducible representations which correspond to the various phonons of a given crystal. We first define the factor group as a group of crystal transformations which leave the crystal invariant (belong to the space group) *and* map the primitive cell onto itself, i.e., a given atom within this cell is not taken out of it. To the factor group belong the following operations: all rotations and reflections of the point group plus the basis operations of a screw axis and glide planes accompanied by a lattice translation so as to leave the primitive cell atoms within the primitive cell (PC). The operations of glide planes and screw axis, since they contain translations, move atoms from the inside to the outside of the PC. In this case, a lattice translation is added to return the atom to the PC. This translation will depend on the atom under consideration.

It is easy to see that the transformations so defined form a group which is isomorphic to the crystal point group (the group obtained by removing *all* translations from the factor group). Its irreducible representations can be found by assigning to each element of the factor group P a matrix of the form

$$D(P_f)=\begin{array}{c} \\ \begin{array}{c}A\\B\\C\\ \\M\\ \\\end{array}\end{array}\begin{array}{cccccc}A & B & C & \dots & M & \dots\\ \begin{pmatrix}0 & 1 & 0 & \dots & 0 & \dots\\ 0 & 0 & 0 & \dots & 1 & \dots\\ 1 & 0 & 0 & & & \dots\\ \dots & \dots & \dots & \dots & \dots & \dots\\ \dots & \dots & \dots & 1 & \dots & \dots\\ \dots & \dots & \dots & \dots & \dots & \dots\end{pmatrix} \end{array} \tag{2.61}$$

where $ABC\dots M\dots$ represent all atoms in the PC. The element $D_{MN}(P_f)$ is set equal to 1 if atom N is transformed into M by P_f. In this manner, a set of matrices is constructed which constitute a representation of the factor group. A first and important reduction of this representation is easily effected by grouping together all equal atoms in the sequence $ABC\dots M\dots$ which then becomes $A_1A_2\dots B_1B_2\dots M_1M_2\dots$. Since P_f cannot transform an atom into a different one, a reduction of the representation has occurred. We must now reduce each one of the block representations corresponding to atoms $A_1B_1\dots M_1\dots$ [labelled $D^M(P_f)$]. The reduction of each $D^M(P_f)$ into irreducible components is then performed in the standard way by calculating the traces (i.e., characters) of $D^M(P_f)$ [labelled $\chi^M(P_f)$] and projecting them onto those of the irreducible representations of the point group $\chi_i(P)$ with the use of standard character tables

$$\sum_P \chi_i^*(P)\chi^M(P_f)=gs_i^M, \tag{2.62}$$

where g is the number of elements of the point group and s_i^M the number of times χ_i is contained in χ^M. We thus write

$$D^M(P_f)=\sum_i s_i^M D_i(P). \tag{2.63}$$

The factor group analysis is facilitated by using the tables in [2.56b, c, 65, 66a]. We are now in a position to obtain the symmetries of the phonons at the Γ-point ($q=0$) of the Brillouin zone. We consider a set of equal atoms A_i and attach to each atom a displacement vector u. Let $D^v(P)$ be the representation of the point group defined by the vector u. The corresponding "phonons" will transform under the factor group like the representation

$$D^M(P_f)\times D^v(P). \tag{2.64}$$

Table 2.3. Representation of the factor group of the diamond structure corresponding to atomic permutations. The notation for the operations of the point group is that of [2.66 b]. We list for each class only one representative matrix. Below the matrices we give the corresponding character

	E	$8C_3$	$3C_2$	$6C_4$	$6C_{2'}$	I	$8S_6$	$3\sigma_h$	$6S_4$	$6\sigma_d$
D	$\begin{pmatrix} 1 & 0 \\ 0 & 1 \end{pmatrix}$	$\begin{pmatrix} 1 & 0 \\ 0 & 1 \end{pmatrix}$	$\begin{pmatrix} 1 & 0 \\ 0 & 1 \end{pmatrix}$	$\begin{pmatrix} 0 & 1 \\ 1 & 0 \end{pmatrix}$	$\begin{pmatrix} 0 & 1 \\ 1 & 0 \end{pmatrix}$	$\begin{pmatrix} 0 & 1 \\ 1 & 0 \end{pmatrix}$	$\begin{pmatrix} 0 & 1 \\ 1 & 0 \end{pmatrix}$	$\begin{pmatrix} 0 & 1 \\ 1 & 0 \end{pmatrix}$	$\begin{pmatrix} 1 & 0 \\ 0 & 1 \end{pmatrix}$	$\begin{pmatrix} 1 & 0 \\ 0 & 1 \end{pmatrix}$
χ	2	2	2	0	0	0	0	0	2	2

The irreducible symmetries of the corresponding phonons are obtained by reducing the vector representation D^v and multiplying its components by the various $D_i^*(P)$ for which $s_i^M \neq 0$. If, after exhausting all basis atoms a given representation appears only *once*, the corresponding phonon eigenvectors are determined exclusively by symmetry. If it appears more than once, the eigenvectors must be determined by solving the dynamical matrix.

We proceed now to discuss several examples of increasing degree of difficulty of the algorithm presented above. We treat here the diamond, zincblende, CaF$_2$, wurtzite, and chalcopyrite structures.

The diamond structure has only two equal atoms per unit cell while the point group (O_h) has 48 symmetry operations, all proper and improper rotations which bring a cube onto itself. The corresponding matrices $D^A(P_f)$ are given in Table 2.3.

Using (2.62), we find from the characters in Table 2.3

$$D^M(P_f) = \Gamma_1 + \Gamma_{2'}. \tag{2.65}$$

The vector representation is, in the O_h group, Γ_{15} [2.65]. Hence, the Γ-phonons of diamond have symmetries

$$\begin{aligned} &\Gamma_1 \times \Gamma_{15} = \Gamma_{15} \text{ (acoustic phonons)} \\ &\Gamma_{2'} \times \Gamma_{15} = \Gamma_{25'} \text{ (optic phonons)}. \end{aligned} \tag{2.66}$$

The corresponding eigenvectors are thus determined by symmetry. They are shown in Fig. 2.9. The optic phonon $\Gamma_{25'}$ is *even* with respect to the center of inversion midway between the two atoms of the PC while the acoustic one (actually a uniform translation for q *strictly* equal to zero) is odd. Hence, the optic phonon is Raman active and the corresponding Raman tensors have only one off-diagonal independent component d (Table 2.1). It is customary in the literature [2.40] to give the numerical value of this component in terms of the parameter a [Å^2]:

$$4\pi a = V_c \frac{d\chi_{12}}{2du_3}, \tag{2.67}$$

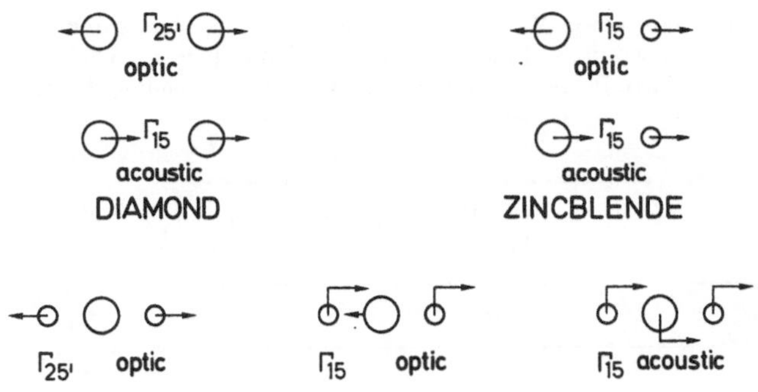

DIAMOND

ZINCBLENDE

Fig. 2.9. Symmetry determined eigenvectors of the diamond, zincblende and CaF_2 structure

where $V_c = a_0^3/4$ and $2u_3$ represents the relative displacement of one sublattice with respect to the other. The factor of 4π has been included in (2.67) so as to obtain numerical values of a in accordance with those in the literature, usually quoted in cgs units. Let us recall that $\chi_{SI} = 4\pi\chi_{cgs}$. The parameter a can be called a "Raman polarizability". The definition of a just given can also be applied to the $\Gamma_{25'}$ phonons of the fluorite structure and the Γ_{15} of zincblende. For diamond, a is nondispersive in the visible and has the value $a \simeq 4\text{Å}^2$ (probably positive according to theoretical calculations. The sign, however, is irrelevant to the Raman cross section, see Sect. 2.1.18).

The zincblende structure (e.g., GaAs) is similar to that of diamond but the two basis atoms are different. $D^A(P_f)$ contains only one-dimensional representations which, of course, must be the identity representation Γ_1 of the point group (T_d). Hence, the two sets of phonons will both simply have the symmetry of a vector in this group, i.e., Γ_{15}. Although this symmetry occurs twice, the eigenvectors are determined independent of force constants by the requirement that the center of mass does not move for the optic phonon ($M_A u_A = -M_B u_B$, see Fig. 2.9). The Γ_{15} optic phonons are Raman allowed with a Raman tensor isomorphic to that of diamond. Next in difficulty we discuss the CaF_2 (fluorite) structure, fcc with three atoms per unit cell: Ca at the origin and F at $\pm a_0/4(111)$. The point group is again O_h. A vector motion of the Ca atom generates the Γ_{15} representation.

The transformation properties of the two fluorine atoms under the operations of the factor group are the same as for diamond (Table 2.3). We thus find from these atoms the phonons $\Gamma_{25'}$ and Γ_{15}. The two Γ_{15} phonons combine to give an optical Raman inactive, ir active mode (eigenvectors determined from center of mass condition $2u_F M_F = -u_{Ca} M_{Ca}$) and an acoustical one (Fig. 2.9). The $\Gamma_{25'}$ optical phonon is Raman active; its Raman tensor has the same form as for diamond.

The hexagonal wurtzite structure (e.g. CdS) is closely related to zincblende. There are four atoms per unit cell at positions

$$Cd\begin{cases}000\\00u\end{cases} \quad S\begin{cases}\frac{1}{3}\frac{2}{3}\frac{1}{2}\\\frac{1}{3}\frac{2}{3}(\frac{1}{2}+u)\end{cases}, \tag{2.68}$$

Table 2.4. Representation of the factor group of the wurtzite structure corresponding to atomic permutations. The notation for the operations of the point group is that of [2.66 b]. We list for each class only one representative matrix. Below the matrices we give the corresponding character

	E	C_2	$2C_3$	$2C_6$	$3\sigma_v$	$3\sigma_d$
D	$\begin{pmatrix} 1 & 0 \\ 0 & 1 \end{pmatrix}$	$\begin{pmatrix} 0 & 1 \\ 1 & 0 \end{pmatrix}$	$\begin{pmatrix} 1 & 0 \\ 0 & 1 \end{pmatrix}$	$\begin{pmatrix} 0 & 1 \\ 1 & 0 \end{pmatrix}$	$\begin{pmatrix} 0 & 1 \\ 1 & 0 \end{pmatrix}$	$\begin{pmatrix} 1 & 0 \\ 0 & 1 \end{pmatrix}$
χ	2	0	2	0	0	2

where $u \simeq 0.375$. The point group is C_{6v}. The representation of the factor group corresponding to atomic permutations is given in Table 2.4. Using the character tables for C_{6v} of [2.66a], we find

$$D^A(P_f) = D^B(P_f) = \Gamma_1 + \Gamma_3$$

and for the representations of the vector displacement, we find Γ_1 (z polarization) and Γ_5 (x, y polarization). Hence the representations of the phonons at $k = 0$ are

$$2(\Gamma_1 \times \Gamma_1) + 2(\Gamma_1 \times \Gamma_3) + 2(\Gamma_5 \times \Gamma_1) + 2(\Gamma_5 \times \Gamma_3) = 2\Gamma_1 + 2\Gamma_3 + 2\Gamma_5 + 2\Gamma_6$$
$$(2.69)$$

The equivalence between the Γ- and the Mulliken notations can be seen in Table 2.1 ($\Gamma_6 \equiv E_2$, $\Gamma_5 = E_1$, $\Gamma_3 = B$, $\Gamma_1 = A_1$).

Of the phonons in (2.69), one set of Γ_5 and one Γ_1 correspond to acoustic phonons (uniform translation). All others are optical phonons, Γ_5 and Γ_1 are ir allowed and Γ_5, Γ_6, and Γ_1 Raman allowed. Note that because of the absence of inversion symmetry a phonon *can* be simultaneously Raman and ir active. The Raman tensors for E_1, E_2, and A_1 phonons are given in Table 2.1. The Γ_3 modes are both ir and Raman *silent*.

We discuss next a tetrahedral structure with a slightly higher degree of complication, that of chalcopyrite ($CuFeS_2$, generally ABC_2). This structure is characteristic of germanium-derived semiconductors such as $CuGaS_2$ and $ZnGeAs_2$ [2.57a]. There are two formula units (eight atoms) per PC. The atomic positions in the PC are: ABC_2:

$2A\,(000),\,(0\tfrac{1}{2}\tfrac{1}{4})$

$2B\,(\tfrac{1}{2}\tfrac{1}{2}0),\,(\tfrac{1}{2}0\tfrac{1}{4})$ (2.70)

$4C\,(u\tfrac{1}{4}\tfrac{1}{8}),\,(\bar{u}\tfrac{3}{4}\tfrac{1}{8})\,(\tfrac{3}{4}u\tfrac{1}{8}),\,(\tfrac{1}{4}\bar{u}\,-\tfrac{1}{8}),$

where $u \simeq 1/4$.

Table 2.5. Representations of the factor group of the chalcopyrite structure which is defined by permutations of the A- or B-atoms and by the C atoms

	E	C_2	$2S_4$	$2C_2'$	$2\sigma_d'$
D^A	$\begin{pmatrix}1&0\\0&1\end{pmatrix}$	$\begin{pmatrix}1&0\\0&1\end{pmatrix}$	$\begin{pmatrix}1&0\\0&1\end{pmatrix}$	$\begin{pmatrix}0&1\\1&0\end{pmatrix}$	$\begin{pmatrix}0&1\\1&0\end{pmatrix}$
χ^A	2	2	2	0	0
D^C	$\begin{pmatrix}1&&&\\&1&&\\&&1&\\&&&1\end{pmatrix}$	$\begin{pmatrix}0&1&0&0\\1&0&0&0\\0&0&0&1\\0&0&1&0\end{pmatrix}$	$\begin{pmatrix}0&0&1&0\\0&0&0&1\\0&1&0&0\\1&0&0&0\end{pmatrix}$	$\begin{pmatrix}1&0&0&0\\0&1&0&0\\0&0&1&0\\0&0&0&1\end{pmatrix}$	$\begin{pmatrix}0&0&0&1\\0&0&1&0\\0&1&0&0\\1&0&0&0\end{pmatrix}$
χ^C	4	0	0	4	0

The point group of the chalcopyrite is D_{2d}. We present in Table 2.5 the 2×2 representation of the factor group defined by permutations of the A (or B) atoms and the representation defined by the C atoms.

The representations of Table 2.5 can be decomposed into the following irreducible representations of the point group D_{2d}:

$$D^A = \Gamma_1 + \Gamma_2 = D^B$$
$$D^C = \Gamma_1 + \Gamma_3 + \Gamma_5. \tag{2.71}$$

Multiplying these representations by the representations of the polar vector (Γ_4 and Γ_5) we obtain the phonon symmetries listed in (2.72) together with their activities:

ir and Raman active: $4\Gamma_4 + 7\Gamma_5$

Raman active, ir inactive: $\Gamma_1 + 3\Gamma_3$ $\tag{2.72}$

silent $2\Gamma_2$;

one pair of Γ_5 modes and a Γ_4 mode are acoustic phonons.

We note that the wurtzite structure has one fully symmetric optical mode Γ_1 and the same holds true for the chalcopyrite structure. These phonons represent a distortion of the crystal which does not change its symmetry. Such distortions correspond to a free parameter of the unit cell, namely, the parameter u of (2.68, 70). This conclusion is quite general; in order to determine unambiguously the positions of all atoms in the PC, a number of parameters equal to the number of optical phonons of Γ_1 symmetry is required [2.57b].

The acoustic phonons of some of the structures just discussed are "Raman forbidden" (e.g., diamond, CaF_2). Nevertheless, light scattering with their participation, the so-called Brillouin scattering, is observed (Sect. 2.1.14). As we shall see, this scattering is related to third rank tensors of the form (2.60).

2.1.11 Fluctuation-Dissipation Analysis

We saw in (2.56) that the scattering efficiency at the frequency shift ω_R is related to the fluctuations of the polarizability operator $\underline{P}$. By expanding this operator in terms of the normal coordinates of elementary excitations, it is possible to relate the scattering cross section to the fluctuations of these normal coordinates. The frequency spectrum of these fluctuations is easily evaluated with the help of the fluctuation-dissipation theorem [2.58].

Let us consider a variable $X(r, t)$ representing the amplitude of some elementary excitation (e. g., the atomic displacement in the case of phonons). To this variable there will, in general, correspond a generalized force $F(t)$ such that the interaction Hamiltonian of this force with the elementary excitation is

$$\delta H = -X(r, t) F(t). \tag{2.73}$$

We consider the Fourier component of $F(t)$, $F(\omega)$. It will produce a change in X of the same frequency given by

$$\delta X(\omega) = T(\omega) F(\omega), \tag{2.74}$$

where $T(\omega)$ is a complex linear response function. The fluctuation dissipation theorem relates the fluctuations in X induced by a temperature T to the imaginary part of the response function [2.58]:

$$\langle X^*X \rangle_\omega = \frac{\hbar}{\pi} (n + \tfrac{1}{2})_\omega \operatorname{Im}\{T(\omega)\} \tag{2.75}$$

with

$$n = [\exp(\hbar\omega/kT) - 1]^{-1}.$$

Equation (2.75) represents the classical version of the fluctuation-dissipation theorem. Its quantum-mechanical version, obtained by replacing X and X^* by the corresponding operators X and $X^\dagger$, is

$$\langle XX^\dagger \rangle_\omega = \frac{\hbar}{\pi} (n + 1) \operatorname{Im}\{T(\omega)\}$$

$$\langle X^\dagger X \rangle_\omega = \frac{\hbar}{\pi} n \operatorname{Im}\{T(\omega)\}. \tag{2.76}$$

As an example, let us consider a phonon normal coordinate ξ and the corresponding generalized force F. The equation of motion for ξ is

$$\xi(-\omega^2 + \omega_0^2 - i\gamma\omega) = F, \tag{2.77}$$

and hence the response function becomes

$$T(\omega) = \frac{1}{\omega_0^2 - \omega^2 - i\gamma\omega} \ . \tag{2.78}$$

The fluctuation-dissipation theorem (2.76) then yields

$$\langle \xi\xi^\dagger \rangle_\omega = \frac{\hbar}{\pi}(n+1)\frac{2\gamma\omega}{(\omega_0^2 - \omega^2)^2 + \gamma^2\omega^2}$$

$$\langle \xi^\dagger\xi \rangle_\omega = \frac{\hbar}{\pi}n\frac{2\gamma\omega}{(\omega_0^2 - \omega^2)^2 + \gamma^2\omega^2} \ . \tag{2.79}$$

An integration of (2.79) for ω from 0 to ∞ reproduces the results of (2.26). Replacement of (2.79) into (2.23) yields the Raman cross section including the Lorentzian line shape of the Raman spectrum [instead of the simplified δ-function of (2.24)].

As another example, we consider a system capable of propagating fluctuations of the electric charge. Associated with such fluctuations there are longitudinal fields related to the charge fluctuations by Poisson's equation:

$$\varepsilon_0 V \cdot E = -\varrho \ . \tag{2.80}$$

These fluctuations can be excited with an external field (generalized force) E_{ext} perpendicular to a sample surface which is perpendicular to the q of the fluctuation. This external field equals the electric displacement vector inside the sample. The "generalized force" corresponding to E can be considered to be the electric displacement D since $\delta H = VE \cdot D$, where V is the volume of the sample. The response function relating a longitudinal field E to D is the inverse longitudinal dielectric constant of the medium $\varepsilon(\omega)^{-1}$. Using (2.76), we find for the fluctuations of the electric field E

$$V\langle EE^\dagger \rangle_{\omega,q} = -\frac{\hbar}{\pi}(n+1)\,\mathrm{Im}\left\{\frac{1}{\varepsilon(\omega, q)}\right\}$$

$$V\langle E^\dagger E \rangle_{\omega,q} = -\frac{\hbar}{\pi}n\,\mathrm{Im}\left\{\frac{1}{\varepsilon(\omega, q)}\right\} \ . \tag{2.81}$$

Equation (2.81) can be used to calculate the line shape and the efficiency of scattering by excitations involving longitudinal electric fields, such as longitudinal ir-active phonons, plasmons and phonon-plasmon coupled modes [Ref. 2.1, p. 147]. A detailed description of the use of the fluctuation-dissipation theorem in light scattering can be found in [2.59].

2.1.12 Scattering by Longitudinal Ir-Active Phonons: Faust-Henry Coefficient

We saw in Sect. 2.1.10 that if a solid does not have inversion symmetry, some of its $q = 0$ phonons can be both ir and Raman-active. Such was the case of the Γ_{15} optical phonons of zincblende, Γ_1 and Γ_5 of wurtzite and Γ_4 and Γ_5 of chalcopyrite. Their infrared activity implies that these phonons contribute to the low frequency polarizability, and hence to the dielectric constant, a term given by [2.60]

$$\Delta\varepsilon_{\alpha,\beta}(\omega) = \frac{1}{V_c} \sum_{v}^{N_{ir}} \frac{\left(\sum_i e_{iv}^* M_i^{-1/2} \hat{e}_{iv\alpha}\right)\left(\sum_i e_{iv}^* M_i^{-1/2} \hat{e}_{iv\beta}\right)}{\omega_v^2 - \omega^2 - i\omega\gamma}, \tag{2.82}$$

where the sums are extended to all atoms in the PC, of volume V_c, and to all ir-active vibrational modes N_{ir}. The effective dynamical charges e_{iv}^* must fulfill the charge neutrality condition

$$\sum_i e_{iv}^* = 0,$$

where the sum is extended to all atoms in the PC. These charges are different for each ir mode.

Let us consider, for the sake of simplicity, a cubic material with ir-active modes (e.g., ZnS). $\Delta\underline{\varepsilon}$ is then isotropic, i.e., equal to $\mathbb{1} \cdot \Delta\varepsilon$. If a transverse electromagnetic field is applied, it polarizes the medium. The speed of propagation of the mixed excitation which consists of an electric field plus a mechanical oscillation becomes

$$v = \frac{\varepsilon_0^{1/2} c}{\sqrt{\varepsilon_\infty + \Delta\varepsilon(\omega)}}, \tag{2.83}$$

where ε_∞ is the frequency-dependent background electronic dielectric constant of the medium (assumed to be nonmetallic). The dispersion relation of the mixed elementary excitations (the so-called polaritons) is found by solving

$$\omega = qv = \frac{\varepsilon_0^{1/2} c}{\sqrt{\varepsilon_\infty + \Delta\varepsilon(\omega)}} q. \tag{2.84}$$

We rewrite $\varepsilon_\infty + \Delta\varepsilon(\omega)$ using (2.82) and taking $\gamma \simeq 0$ in the form

$$\varepsilon_\infty + \Delta\varepsilon(\omega) = \varepsilon_\infty \prod_v^{N_{ir}} \frac{\omega_{v\mathrm{LO}}^2 - \omega^2}{\omega_{v\mathrm{TO}}^2 - \omega^2}. \tag{2.85}$$

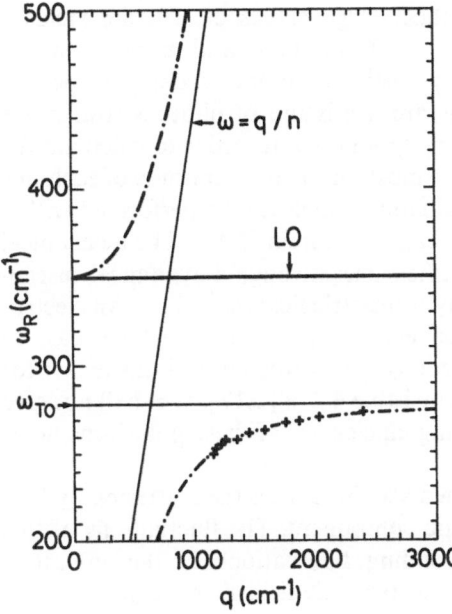

Fig. 2.10. Phonon polaritons in zincblende-type ZnS [2.61]. The points are experimental, the dashed-dotted line calculated

At the frequencies $\omega = \omega_{v\mathrm{LO}}$, the dielectric constant (2.85) vanishes, i.e., $1/\varepsilon$ blows up. According to (2.81), the longitudinal fluctuations in E also blow up. These fluctuations correspond to longitudinal phonons (vibrations along the direction of $\mathbf{q}$) whose frequency has been renormalized from $\omega_{v\mathrm{TO}}$ to $\omega_{v\mathrm{LO}}$ by the long range longitudinal polarization associated with them. The transverse modes are also renormalized by the coupling to the electromagnetic field; their renormalized frequencies, strongly $\mathbf{q}$-dependent, are obtained by solving (2.84). Let us consider the simplest case, that of the zincblende structure with only one set of Γ_{15} optical modes, both ir and Raman-active. The dispersion relation for these so-called "phonon polaritons" is, in the case of only one ir-active mode:

$$q^2 = \frac{\omega^2}{c^2} \frac{\varepsilon_\infty}{\varepsilon_0} \frac{\omega_{\mathrm{LO}}^2 - \omega^2}{\omega_{\mathrm{TO}}^2 - \omega^2}, \tag{2.86}$$

where

$$\omega_{\mathrm{LO}}^2 - \omega_{\mathrm{TO}}^2 = \frac{e^{*2}}{V_c \mu \varepsilon_\infty}. \tag{2.86a}$$

In (2.86a), e^* is the dynamical charge on one of the atoms, V_c the volume of the PC and μ its reduced mass ($\mu^{-1} = M_A^{-1} + M_B^{-1}$). The dispersion relation of (2.86) is plotted in Fig. 2.10 for ZnS with $\omega_{\mathrm{LO}} = 351$ cm^{-1}, $\omega_{\mathrm{TO}} = 279$ cm^{-1} and $\varepsilon_\infty = 5.2$, together with the dispersionless longitudinal modes ($\omega = \omega_{\mathrm{LO}}$) and experimental data for the "lower polariton" branch obtained in the forward Raman scattering configuration [2.61]. In large angle or backscattering experiments $q \simeq 2\omega_L \varepsilon_\infty^{1/2}/c \gg \omega_{TO}$. For these large values of q, (2.86) yields a nearly nondispersive frequency $\omega \simeq \omega_{TO}$ (TO-phonon) and a renormalized photon with the dispersion relation $q = \omega \varepsilon_\infty^{1/2}/c$. In the strongly dispersive region for $q \simeq \varepsilon_\infty^{1/2}\omega_{TO}/c$, the polaritons are a mixture of photons (transverse electric fields) and mechanical vibrations (phonons). In order to calculate the scattering cross section in this region, we must know the amplitude of each one of these components separately, an evaluation which can be performed with a generalized form of the fluctuation-dissipation theorem [2.59]. The mechanical vibration contributes to the scattering efficiencies through a dynamical susceptibility of the type (2.67). The accompanying electric fields contribute an electro-optical term given by the first-order electro-optic tensor which has, in zincblende, only one independent element $\partial \chi_{12}/\partial E_3$ (the form of this tensor for other point group symmetries is shown in Table 4.2 of [2.17]. We shall not give the detailed expressions for the scattering efficiencies of these polaritons here; they can be found in [2.59].

The electro-optic effect just mentioned also influences the scattering by LO-phonons since they have an electrostatic component. The fluctuations of this field are given in (2.81), the corresponding fluctuations of the sublattice displacement $u_A - u_B$ are related to the electrostatic field E through

$$E = -\varepsilon_\infty^{-1} N_c e^*(u_A - u_B); \tag{2.87}$$

N_c = number of PC's per unit volume. The scattering efficiency for Stokes scattering by LO-phonons can thus be written, see (2.51),

$$\frac{\partial S_s}{\partial \Omega \partial \omega_R} = \frac{\omega_s^4 V}{(4\pi^2)c^4} |\hat{e}_s \cdot R_L(q) \cdot \hat{e}_L|^2 \langle EE^\dagger \rangle_{\omega_R}, \tag{2.88}$$

where $\langle EE^\dagger \rangle_{\omega_R}$ is given in (2.81) as a function of $\mathrm{Im}\{\varepsilon(\omega)^{-1}\}$ and the Raman tensor R_L for longitudinal phonons is obtained from the Γ_{15} tensors of Table 2.1 by rotation of the axis so as to obtain the linear combination of phonons which vibrate longitudinally along q:

$$R_L = \hat{q}_x^2 R_x + \hat{q}_y^2 R_y + \hat{q}_z^2 R_z, \tag{2.89}$$

where $\hat{q}$ is the unit vector along q and

$$R_x = \begin{pmatrix} 0 & 0 & 0 \\ 0 & 0 & d \\ 0 & d & 0 \end{pmatrix}; R_y = \begin{pmatrix} 0 & 0 & d \\ 0 & 0 & 0 \\ d & 0 & 0 \end{pmatrix}; R_z = \begin{pmatrix} 0 & d & 0 \\ d & 0 & 0 \\ 0 & 0 & 0 \end{pmatrix}; \tag{2.90}$$

with

$$d = \left(\frac{\partial \chi_{12}}{\partial E_3} - \frac{\varepsilon_\infty}{N_c e^*} \frac{\partial \chi_{12}}{\partial (u_A - u_B)_3} \right).$$

Equation (2.88) can be rewritten in terms of the fluctuations of $u_A - u_B$ using (2.87) to yield

$$\frac{dS_s}{d\Omega d\omega_R} = \frac{\omega_s^4 V_c}{(4\pi^2)c^4} |\hat{e}_s \cdot \underline{R}'_L(q) \cdot \hat{e}_L|^2 \left(\frac{\hbar}{2\mu\omega_{LO}}\right)(n+1)\delta(\omega_R - \omega_{LO}), \tag{2.91}$$

where the tensor $\underline{R}'$ is isomorphic to the $\underline{R}$ of (2.90) with d replaced by

$$d' = \left(-\frac{N_c e^*}{\varepsilon_\infty}\frac{\partial \chi_{12}}{\partial E_3} + \frac{\partial \chi_{12}}{\partial(u_A - u_B)_3}\right)$$

$$= \frac{d\chi_{12}}{d(u_A - u_B)_3}\left(1 - \frac{\omega_{LO}^2 - \omega_{TO}^2}{C\omega_{TO}^2}\right), \tag{2.92}$$

where C is the so-called Faust-Henry coefficient which we rewrite as [Ref. 2.1, Eq. (4.27)]

$$C = \frac{e^*(\partial\chi/\partial u)}{\mu\omega_{TO}^2(\partial\chi/\partial E)}. \tag{2.93}$$

The dimensionless Faust-Henry coefficient is a measure of the relative strength of "mechanical" (deformation potential, as it will be called later) to electro-optic coupling in the electron-phonon interaction. It can be either negative (GaP) or positive (SiC, CdS) and has often absolute values around 0.5 (see Table 2.6). The ratio of the Raman tensor components for LO and TO-scattering is given by

$$\frac{d_{LO}}{d_{TO}} = \left(1 - \frac{\omega_{LO}^2 - \omega_{TO}^2}{C\omega_{TO}^2}\right). \tag{2.94}$$

According to (2.94), the Raman efficiency for LO-scattering is enhanced or quenched with respect to the TO-scattering depending on the sign and magnitude of C ($C < 0$, always enhancement). The values of the enhancement factor $(d_{LO}/d_{TO})^2$ $(\omega_{TO}/\omega_{LO})$ obtained with (2.94) for a few typical semiconductors are listed in Table 2.6 (the additional factor ω_{TO}/ω_{LO} is included so as to take into account differences in the vibrational amplitudes of LO and TO-phonons).

We point out that if local field corrections are included, even TO-phonons can be accompanied by a *longitudinal* (local) field and thus have an electro-optic contribution to the scattering cross section. These contributions, expected to be small, have been calculated by *Ovander* and *Tyn* [2.62].

Besides the possible enhancement of S_{LO} just discussed, a strong enhancement can occur near resonance for polarized scattering. This enhancement, attributed to the so-called *intraband* Fröhlich interaction, is represented by a q-dependent tensor (2.60); it is, therefore, "dipole forbidden". The corresponding resonances, however, can be even stronger than those of dipole allowed effects. These phenomena will be discussed in Sect. 2.2.8.

Table 2.6. Faust-Henry coefficients of several typical semiconductors and corresponding enhancement factor for LO-phonon scattering $(d_{LO}/d_{TO})^2$ $(\omega_{TO}/\omega_{LO})$ obtained with (2.94)

	C (theory)[a]	C (exp)	$\left(\dfrac{d_{LO}}{d_{TO}}\right)^2 \dfrac{\omega_{TO}}{\omega_{LO}} \simeq \left[1 - \left(\dfrac{\omega_{LO}^2 - \omega_{TO}^2}{C\omega_{TO}^2}\right)\right]^2 \dfrac{\omega_{TO}}{\omega_{LO}}$
AlSb	−1.97		
GaP	−0.37	−0.64	2.5
GaAs	−0.83	−0.59	1.5
GaSb	−0.28		
InP	−0.14		
InAs	−0.28		
InSb	−0.66		
ZnS		−0.18[b]	30[b]
ZnSe		−0.7[b]	2.1[b]
		−0.21[c]	8.5[c]
ZnTe		−0.11[b]	7.2[b]
CuCl		−1.05[d]	2.5[d]
		−2.7[e]	1.1[e]
CuBr		−1.2[d]	1.7[d]
CuI		+1.1[d]	0.5[d]

[a] [2.156].
[b] [2.49].
[c] S. Ushioda, A. Pinczuk, E. Burstein, D.L. Mills: [Ref. 2.12, p. 347].
[d] A. Ben-Amar, E. Wiener-Avnear: Appl. Phys. Lett. **27**, 410 (1975).
[e] Like [d], but assuming that the two peaks, at 147 and 159 cm^{-1} belong to the TO phonons. See Z. Vardeny, O. Brafman: Phys. Rev. B**19**, 3276 (1979). Note that this choice improves the systematic variation of C from CuCl to CuI. Note also that a reversal of the sign of a from CuCl to CuI is likely [see S. Ves, M. Cardona: Solid State Commun. **38**, 1109 (1981)].

2.1.13 Second-Order Raman Scattering in Crystals

The expression for the efficiency of second-order Stokes Raman scattering in a crystal is obtained from (2.35)

$$\frac{\partial S_s}{\partial \omega_R \partial \Omega} = \frac{\omega_s^4}{(4\pi)^2 c^4} \sum_{i,j,q} \left| \hat{e}_s \cdot \frac{\partial^2 \chi'}{\partial \xi_i(q) \partial \xi_j(-q)} \hat{e}_L \right|^2$$
$$\cdot \frac{\hbar^2}{4\omega_{i,q}\omega_{j,q}} (n_{iq} + 1)(n_{jq} + 1)\delta(\omega_{iq} + \omega_{jq} - \omega_R), \qquad (2.95)$$

where the second-order Raman susceptibility

$$\frac{\partial^2 \chi'}{\partial \xi_i \partial \xi_j} = V \frac{\partial^2 \chi}{\partial \xi_i \partial \xi_j} \qquad (2.96)$$

is independent of the scattering volume V. The corresponding expressions for antistokes and for difference scattering can be obtained from (2.95) with the help of (2.34).

While for first-order scattering only phonons with $q \simeq 0$ are allowed, the selection rule for q conservation now implies that the sum of the q vectors of the two phonons involved be approximately zero, i.e., $q_i \simeq -q_j$ for Stokes and antistokes scattering and $q_i \simeq q_j$ for difference scattering. Hence, in principle, all phonons can be observed. If the Raman susceptibility $\partial^2 \chi' / \partial \xi_i \partial \xi_j$ is assumed to be independent of q for a given pair of phonon bands i and j, (2.95) reduces to

$$\frac{\partial^2 S_s}{\partial \omega_R \partial \Omega} = \frac{\omega_s^4}{(4\pi)^2 c^4} \sum_{i,j} \left| \hat{e}_s \cdot \frac{\partial^2 \chi'}{\partial \xi_i \partial \xi_j} \cdot \hat{e}_L \right| \cdot$$
$$\frac{\hbar^2}{4\omega_{iq}\omega_{jq}} (n_{iq}+1)(n_{jq}+1) N_{d,ij}(\omega_R), \qquad (2.97)$$

where $N_{d,ij}$ is the combined density of phonon states for the phonon branches i and j. Actually, $\partial^2 \chi' / \partial \xi_i \partial \xi_j$ is not constant for a given pair of bands. A reasonable approximation may be to assume $\partial^2 \chi' / \partial \xi_i \partial \xi_j \propto \omega$ as the susceptibility must vanish for $\omega_R \to 0$ as a result of translational invariance (a uniform translation does not change χ). For overtones, $N_{d,ii}(\omega_R) = N_{d,i}(\omega_R/2)$ and the second-order Raman spectrum contains information about the *density* of one-phonon states $N_{d,i}$ [2.40, 63, 64].

One may actually consider decomposing $N_{d,ij}(\omega_R)$ into components of the various irreducible symmetries corresponding to the Raman tensor (the experimental spectra are easily decomposed in the various irreducible components using, for instance for cubic materials, Table 2.2. The results obtained for Si are given in Fig. 2.11). Such a procedure is, in principle, not very fruitful. The density of states is dominated by general points of the Brillouin zone (BZ) with no special symmetry. Let us consider a *combination* pair of phonons i,j at a given *general* q at which each of the phonons i and j cannot be fundamentally

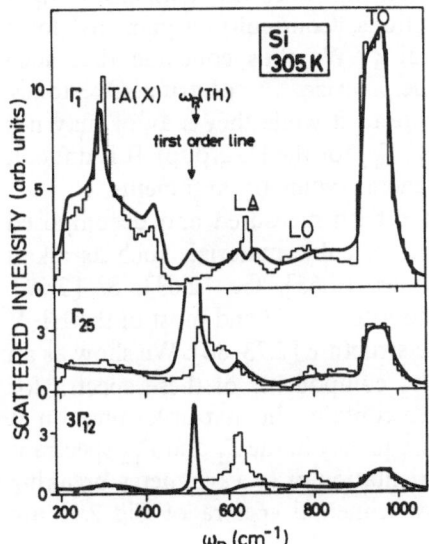

Fig. 2.11. Irreducible components of the second-order Raman spectrum of silicon obtained at 305 K for $\lambda_L = 5145$ Å. The histogram was calculated in [2.40] with the bond polarizability model and the bond charge model lattice dynamics [2.64]

degenerate. When applying all operations of the point group to q, the star of q is generated. This star defines the so-called regular representation of the point group [2.65], a representation with a dimensionality equal to the number of elements g of the point group. All elements of the star of q appear with equal weight in the sum which gives $N_{d,ij}$. By reducing the regular representation we can thus find the irreducible components of $N_{d,ij}$. Actually, it is well known that the regular representation contains *all* irreducible representations $D_i(P)$, each a number of times equal to their dimensionality S_i^M. Hence, one may expect that the second-order Raman spectrum of a given combination band contains all Raman tensor components given in Table 2.1 for each point group, with a weight equal to their dimensionality. This is actually not the case; the tensors belonging to the identity representation Γ_1 produce scattering efficiencies considerably stronger than all others, at least for overtone scattering (for overtone scattering in materials with inversion symmetry, the argument given above yields only *even* representations of the point group; in these materials phonon overtones are *approximately* Raman allowed but ir-forbidden). This means that the susceptibility $\partial \chi' / \partial \xi_i \partial \xi_j$ cannot be independent of q and ω_R. That this must be so can be easily seen by considering high symmetry directions such as [100], [111], and [110] in cubic materials. Along these directions, the argument given above for a general q breaks down and the pairs of phonons do not contain all (even) irreducible representations. The irreducible representations of a pair of phonons and all their equivalents must be obtained by performing the appropriate products of space group representations [2.21]. If the two members of the phonon pair belong to *different* irreducible representations, their product cannot contain the identity representation Γ_1, i.e., cannot contribute to completely polarized scattering.

This will be the case for a *pair* composed of a longitudinal and a transverse phonon in a cubic crystal along the high symmetry directions. Hence in this case, the corresponding $\partial^2 \chi' / \partial \xi_i \partial \xi_j$ must vanish. Since this happens along a large number of high symmetry directions, continuity arguments force $\partial^2 \chi' / \partial \xi_i \partial \xi_j$ to be small also at a general q. We thus conclude that such combinations should not contribute to the *polarized* Γ_1 spectrum. Overtones, on the contrary, should strongly contribute to it while they may or may not participate in the depolarized spectra ($\Gamma_{25'}$, Γ_{12} for the O_h group). It is difficult to obtain more information using only general symmetry arguments.

The second-order Raman spectra have been measured and decomposed into irreducible components for a number of cubic materials such as alkali halides [2.67], alkaline earth chalcogenides [2.68], Ge [2.63], Si [2.69], diamond [2.70], fluorites [2.71] and antifluorites [2.72] and most of the III–V and II–VI semiconductors with zincblende structure [2.73–80]. We show as an example in Fig. 2.11 the Γ_1, $\Gamma_{25'}$, and Γ_{12} components of these spectra for crystalline silicon. The $\Gamma_{25'}$ component also contains the first-order phonon at 520 cm^{-1}. The weak line observed at this frequency in the Γ_1 and Γ_{12} spectra is a residual effect due to either sample misorientation or to a symmetry-breaking mechanism. The sharp features in the experimental spectra of Fig. 2.11 are

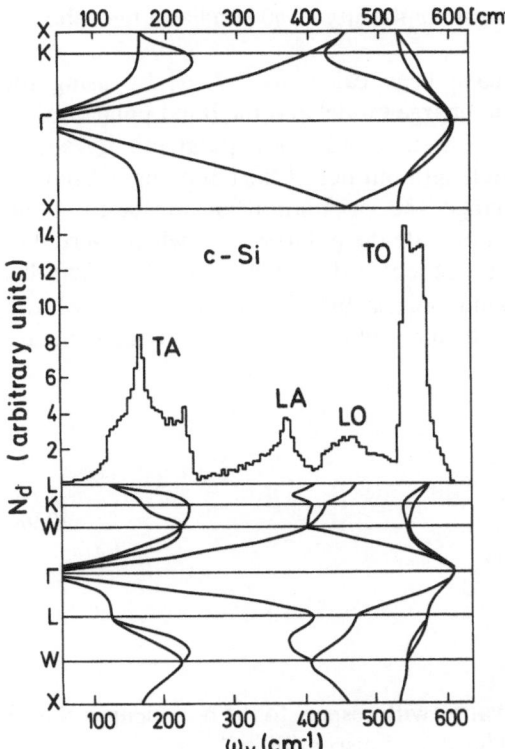

Fig. 2.12. Phonon dispersion and density of states obtained for c-Si with Weber's bond charge model [2.64]

usually interpreted in terms of Van Hove singularities or critical points which occur along high symmetry directions usually near the boundaries of the BZ.

In agreement with the arguments given above, the Γ_1 spectrum of Fig. 2.11 is mainly an overtone spectrum. In order to help its understanding, we show in Fig. 2.12 the phonon dispersion of Si as calculated with Weber's bond charge model [2.65] and the corresponding $N_d(\omega)$. The main features of the Γ_1 spectrum of Fig. 2.11 correspond to 2TA and 2TO overtones, and within these bands sharp structure is found related to the flat dispersion relations along the $X - K$ and $K - W$ lines and the L-critical point. We note that the Γ_{12} spectrum is very weak, probably negligible within the accuracy of the required polarizer and analyser settings and depolarization by surface roughness. The $\Gamma_{25'}$ spectrum, although small, is certainly above this error. It is dominated by the 2TO peak and it does not contain any 2TA structure.

The second-order spectrum of Ge [2.63] is very similar to the one just discussed when measured at small laser frequencies ($\omega_L < 2$ eV) [2.44a]. For $\omega_L \gtrsim 2.2$ eV, a resonant structure appears at the 2TO(Γ) frequency. The second-order spectrum of diamond [2.70] has a similar 2TO(Γ) peak but of nonresonant character. It has been interpreted alternatively as a feature in the phonon density of states due to an anomaly of the dispersion relation near $q = 0$

[2.81], as an anomaly in the Raman susceptibility [2.40] and as a two-phonon bound state [2.82a].

The histograms of Fig. 2.11 have been calculated [2.40] by using the vibrational eigenvectors of the bond charge model and the bond polarizability hypothesis; it is assumed that each Si–Si bond has a polarizability whose magnitude is a function of the bond length but not of the bond angle. Changes in the bond angle, however, change the orientation of the static bond polarizability tensor and thus also modulate the polarizability when referred to fixed coordinate axes. Six parameters are required by such a model: the parallel ($\alpha_{||}$) and the perpendicular ($\alpha_{\perp}$) bond polarizabilities and their first (α') and second (α'') derivatives with respect to the bond length. These parameters are written in the form [2.40]

$$\alpha_v = 4(\alpha_{||} + 2\alpha_{\perp})/3V_c$$
$$\alpha_q = 4(\alpha_{||} - \alpha_{\perp})/3V_c$$
$$\alpha_1 = r_0\alpha'_v \tag{2.98}$$
$$\alpha'_1 = r_0^2\alpha''_v$$
$$\alpha_{25'} = r_0\alpha_q(\ln|\alpha_q/R^2|)'$$
$$r_0\alpha'_{25'},$$

where the prime represents the derivative with respect to the bond length R and r_0 is the equilibrium bond length. The static "susceptibility" α_v is related to the dielectric constant ε_L through

$$\varepsilon_L = \varepsilon_0(1 + 4\pi\alpha_v). \tag{2.99}$$

In (2.98, 99), we have kept the designation "α" for *susceptibilities* in cgs units so as to follow the notation of [2.40]. Hence, of the six parameters in (2.98), one (α_v) is determined by the static dielectric constant. The first-order scattering by $\Gamma_{25'}$ phonons is determined by the susceptibility α_q (this susceptibility has indeed $\Gamma_{25'}$ symmetry, as can be seen by rotating the $\Gamma_{25'}$ tensors of Table 2.1 to bring one axis onto the [111] direction). The Γ_1 second-order scattering is determined by α'_1 (second-order bond elongations) and by α_1 (first-order elongation and first-order rotation).

The susceptibilities α_1 and α'_1 determine the Γ_1 component of the Raman tensor, while α_q, $\alpha_{25'}$, α'_{25} determine $\Gamma_{25'}$ and α_q and $\alpha_{25'}$ determine Γ_{12}. In [2.40], a numerical fit to the spectra observed for Ge, Si, and diamond was made. Only relative Raman efficiencies were available at that time hence α_1/q, α'_1/q, $\alpha_{25'}/q$ and $\alpha'_{25'}/q$ were used as fitting parameters. It is obvious that the spectra of Fig. 2.11 are sufficiently structured to permit a reasonably accurate determination of these parameters. The fitting curves are also shown in Fig. 2.11 and the resulting values of the parameters listed in Table 2.7. The parameter α_v was determined absolutely by fitting the dependence of the

Table 2.7. Parameters obtained by fitting the second-order Raman spectra of diamond, Si, and Ge with the bond polarizability model. Also, values of the elasto-optic constants $p_{11}-p_{12}$ and p_{44} and of the first-order Raman tensor a calculated with this model. Experimental values of the latter and of the Raman polarizability a are given in brackets. From [2.40]

	C	Si	Ge
ε_x	(5.86)	(11.7)	(16.3)
α_1/α_q	4.13	− 46.16	− 57.45
α'_1/α_q	284.65	−180.02	−288.22
$\alpha_{25'}/\alpha_q$	2.13	− 23.08	− 24.39
$\alpha'_{25'}/\alpha_q$	255.6	0.0	−248.78
$p_{11}+2p_{12}$	− 0.16	− 0.058	− 0.28
	(− 0.16)	(− 0.058)	(− 0.28)
$p_{11}-p_{12}$	− 0.283	0.013	0.016
	(− 0.293)	(− 0.167)	(− 0.0095)
			0.011
p_{44}	− 0.172	− 0.0076	0.019
	(− 0.172)	(− 0.082)	(− 0.074)
			(0.012)
α_v	0.387	0.851	1.218
α_q	0.387	− 0.069	− 0.167
$\alpha_\parallel(\text{Å}^3)$	3.293	6.006	9.89
$\alpha_\perp(\text{Å}^3)$	0.0	7.749	15.49
$a(\text{Å}^2)$	+ 3.5	+ 13.95	+ 43.1
	(+ 4.3)	(+ 60)	

dielectric constant on volume measured by application of a hydrostatic pressure. Introducing the elasto-optic coefficients p_{ijkl} defined as [2.64]

$$-(\varepsilon_0/\varepsilon_L)^2 \Delta\varepsilon_{ij} = \varepsilon_0 \sum_{kl} p_{ijkl} e_{kl} \tag{2.100}$$

(e_{kl} are the components of the strain tensor) and the additional definition for the independent components of $\boldsymbol{p}$: $p_{11}=p_{1111}$, $p_{12}=p_{1122}$, $p_{44}=p_{1212}$, we can easily obtain with the bond polarizability model

$$p_{11}+2p_{12} = \left(\frac{\varepsilon_0}{\varepsilon_L}\right)^2 (3\alpha_v - \alpha_1). \tag{2.101}$$

Since α_v can be found from ε_L with (2.99), (2.101) defines α_1 if $p_{11}+2p_{12}$ is known. The value of α_v so obtained enabled the authors of [2.40] to normalize their relative α's. They were then able to compute from these α's the two remaining elasto-optic coefficients

$$p_{11}-p_{12} = -\frac{\varepsilon_0^2}{\varepsilon_L^2} \alpha_q$$

$$p_{44} = -\frac{\varepsilon_0^2}{3\varepsilon_L^2} \{(1-\zeta)\alpha_{25'} + 3\alpha_q\} \tag{2.102}$$

[in (2.102) ζ is the bond rigidity parameter, see Sect. 2.2.6a] and the first-order Raman tensor component a [defined in (2.67)]:

$$a = V_c (r_0 \sqrt{3})^{-1} \alpha_{25'}. \tag{2.103}$$

Several comments about the results of Table 2.7 are in order. For diamond, the Raman measurements and the measurements of the p_{ij}'s have been performed in the visible, well below any electronic absorption ($\omega_L \ll \omega_0$). The data are nondispersive and hence one of the implicit assumptions of the bond polarizability model is fulfilled. Correspondingly, calculated values of $p_{11} - p_{12}$ and p_{44} agree very well with experiment. The calculated magnitude of a also agrees with experimental results although theoretical evidence concerning the sign of a is contradictory for diamond (not for silicon); we are inclined to believe the signs of Table 2.7 (Sect. 2.1.18g). We point out, however, that the *sign* of a in Table 2.7 is obtained from a fit of the Γ_{12} component which is small and inaccurately defined.

We should also point out that recently the linear differential polarizabilities of diamond have been related to those of saturated hydrocarbons [2.82b]. For this purpose, local field corrections are essential.

An interesting feature is the fact that $\alpha_\perp \simeq 0$ for diamond, while for Si $\alpha_\parallel \simeq \alpha_\perp$ and for Ge $\alpha_\perp$ is even larger than $\alpha_\parallel$. This is related to the increasing degree of metallization from diamond to Ge. In diamond, the polarizability is determined mainly by transitions between bonding and antibonding sp^3 hybrid orbitals [2.40]. These transitions are polarized parallel to the bonds. In Si, and even more so in Ge, a number of dehybridized p^3 bonding to p^3 antibonding transitions take place. The lowest of these transitions in the solid are polarized perpendicular to the corresponding orbital.

While the values of a obtained in Table 2.7 for Si and Ge agree reasonably well with microscopically calculated ones and also, in the case of Si, with experimental ones (Sect. 2.1.18), this agreement is lacking for $p_{11} - p_{12}$ and p_{44}. This results from the fact that the gaps or dispersion mechanisms responsible for the Raman susceptibility are not the same as those responsible for the elasto-optic constants [2.40]. We shall come back to this point in Sect. 2.3.4.

The bond polarizability model just described breaks down as one goes from Ge to more ionic tetrahedral semiconductors such as the II–VI compounds. A detailed analysis has been performed [2.73] for the sequence ZnS, ZnSe, ZnTe using the overlap shell model of *Bruce* and *Cowley* [2.83] to describe the lattice dynamics. The nonlinear polarizabilities are represented, within this model, by nonlinear spring constants linking the electronic shell to the atomic cores. Two such constants are required, one for the cation and one for the anion. With such a model, the fits to experiments shown in Fig. 2.13 were obtained. The anharmonicity of the anion contributes strongly to the spectrum of ZnS but its contribution decreases throughout the sequence ZnS→ZnSe→ZnTe. Strong nonlinear polarizabilities have been invoked to explain a number of effects observed for chalcogenides, especially oxides [2.84].

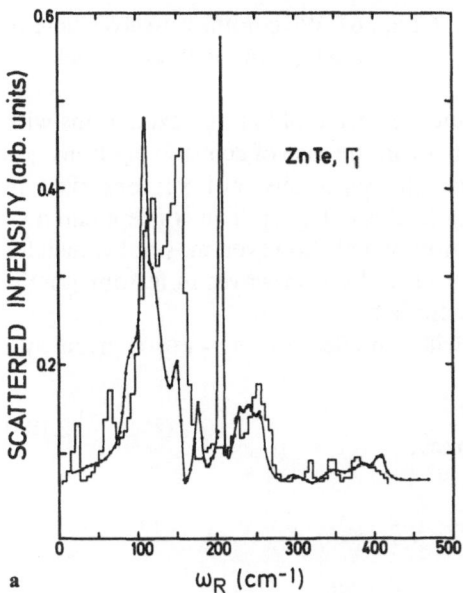

a

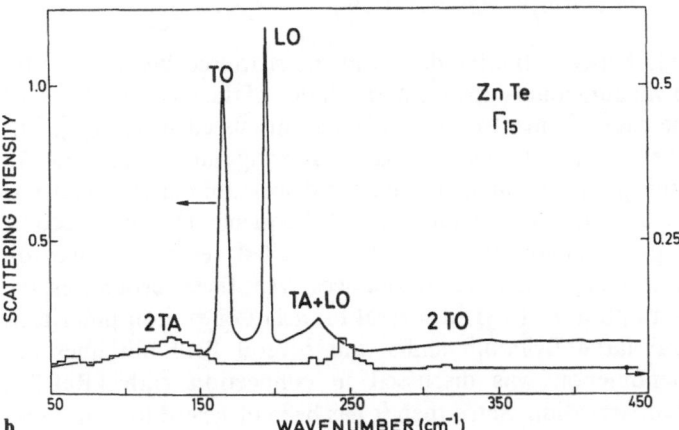

b

Fig. 2.13a, b. Γ_1 and Γ_{15} components of the second-order Raman tensor of ZnTe for $\omega_L = 1.97$ eV at room temperature. The Γ_{12} component is negligible. The histogram represents a fit with an anharmonic shell model [2.73]. The integrated scattering efficiency of the allowed first-order TO-line in the Γ_{15} spectrum at 160 cm^{-1} is 6.5×10^{-5} m^{-1} ster^{-1} (Sect. 2.1.18)

The anharmonic polarizability of the Zn ion contributes considerably to the spectra. In the case of ZnTe (Fig. 2.13), the observed spectra can be accounted for solely in terms of this polarizability.

2.1.14 Brillouin Scattering

There has been a great amount of activity in the past 10 years in the field of Brillouin scattering, especially for opaque materials such as semiconductors, metals and metallic glasses, thanks to the instrumentation developed mainly by *Sandercock*. These advances will be described in [Ref. 1.2, Chap. 6]. Earlier work was reviewed by *Pine* in [Ref. 2.1, Chap. 6]. We confine ourselves here to presenting the underlying general principles in the light of the theory developed above.

By Brillouin effect we understand the scattering of light by excitations with a linear dispersion relation $\omega_B = v \cdot q$ where, in the case of acoustic phonons, v is the appropriate speed of sound. Acoustic plasmons, not yet observed in Brillouin scattering, should also belong to this category. The concept can also be applied to very low frequency excitations which, however, may not vanish for $q = 0$. To this category belongs the scattering by spin waves in ferromagnetic, ferrimagnetic and antiferromagnetic materials.

The kinematics of the first-order Brillouin effect for $\omega_B = v \cdot q$ is given by

$$\begin{aligned} \pm q &= k_L - k_s \\ \pm v \cdot q &= \omega_B = \omega_L - \omega_s \end{aligned} \quad \begin{cases} + \text{ for Stokes} \\ - \text{ for antistokes}. \end{cases} \tag{2.104}$$

From (2.104) we obtain

$$\omega_B = \pm 2k_L v \sin \frac{\Theta}{2} = \pm \frac{4\pi v n_L}{\lambda_L} \sin \frac{\Theta}{2} = \pm \frac{2\omega_L v n_L}{c} \sin \frac{\Theta}{2}, \tag{2.104a}$$

where Θ is the angle between the incident and the scattered beam, λ_L is the laser wavelength in vacuum and n_L the refractive index of the medium, assumed to be isotropic. The case of anisotropic media was discussed in [2.28b]. The maximum Brillouin shift ω_B is obtained for backscattering: $|\omega_B| = 4\pi n_L v/\lambda_L$. We should mention at this point that in an anomalous dispersion region (electronic resonance), n_L can reach very high values and the frequency shift can become rather large. This phenomenon leads to the so-called resonant polariton Brillouin scattering near sharp excitonic resonances. Near these resonances, the excitons combine with photons to yield several branches of exciton-polaritons (Sect. 2.1.12). Several large Brillouin shifts and several Brillouin lines are observed. This phenomenon was discussed in connection with [Ref. 2.1, Fig. 6.8] prior to its observation. Since then it has been observed for a number of materials (e.g. GaAs, CdS, CuBr, etc.). It is discussed in detail in [Ref. 1.2, Chap. 6].

For opaque materials, k_L is "smeared out" as a result of the nonvanishing imaginary part of the refractive index n_i, in other words, k_L possesses within the material an imaginary part $k_{Li} = (\omega/c)n_i$. Because of the linear dependence of ω_B on k_{Li}, the "broadening" k_{Li} of k_L is translated into a broadening of ω_B:

$$\Delta\omega_B \simeq \frac{4\omega_L v n_i}{c} \sin \frac{\Theta}{2}. \tag{2.105}$$

This broadening, which is usually asymmetric, was observed by *Sandercock* in his original measurements for Ge and Si and used to obtain the absorption coefficient at the wavelength λ_L [2.85]. Several attempts have been made to calculate the exact Brillouin line shape associated with this broadening by decomposing the electromagnetic fields into Fourier components with k real. This decomposition must be performed carefully, taking into account the correct boundary conditions for the electromagnetic field and the phonons at the vacuum-medium interface. For this reason, [Ref. 2.1, Eq. (6.20)] seems to be incorrect [2.86]. The correct expression is [2.86], see also [Ref. 1.2, Eq. (5.42)],

$$S(q) \propto \frac{q^2}{[(k_L' + k_s')^2 - (k_L'' + k_s'')^2 - q^2]^2 + 4(k_L' + k_s')^2 (k_L'' + k_s'')^2}, \tag{2.106}$$

where $k_{s,L}' + ik_{s,L}'' = \omega_{s,L} n_{s,L}/c$ are the complex wave vectors of the incident and scattered radiation. The difference between (2.106) and (6.20) of [2.1], however, is small if $k_L'' + k_s'' \ll k_L' + k_s'$.

A similar phenomenon is obtained when observing Brillouin scattering in thin films. In this case and for backscattering perpendicular to the film, the q does not have to be conserved (no translational symmetry in this direction). The size-quantized elementary excitations have q vectors of magnitude

$$q = \frac{\pi}{d} \kappa, \quad \kappa = 1, 2, 3 \dots$$

and frequencies

$$\omega_B = \frac{\pi}{d} \kappa v, \tag{2.107}$$

where d is the film thickness. Hence, a series of lines at the frequencies of (2.107) should be observed. For $\lambda_L \gg d$, the even excitations of (2.107) should only couple weakly to the light. Results obtained with a five-pass Fabry-Perot interferometer for magnons in amorphous $Fe_{80}B_{20}$ are shown in Fig. 2.14

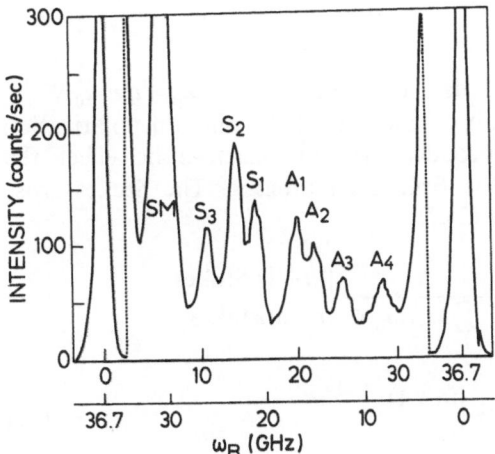

Fig. 2.14. Brillouin spectrum of magnons in an amorphous $Fe_{80}B_{20}$ sample 106 nm thick. S_K and A_K label Stokes and antistokes peaks of order K [see (2.107)] and SM is a Stokes surface magnon. Note the absence of antistokes surfaces magnon [2.87]

[2.87]. These results are rather similar to those found for polymetric chains, the so-called longitudinal acoustic modes (LAM) of, for instance, polyethylene [2.88a].

When attempting to calculate the Brillouin scattering cross section or efficiency with (2.47) and its antistokes counterpart, one realizes that for excitations with $q=0$, $d\chi/d\xi=0$, as acoustic phonons with $q=0$ correspond to uniform translations which do not change χ. In crystals with full cubic symmetry, correspondingly, the $q=0$ acoustic phonons have Γ_{15} symmetry and are thus Raman forbidden (Table 2.1). Hence, the Brillouin efficiency must come from terms of first order in q in the expansion of $d\chi/d\xi$ in the power series of q, see (2.60). Let us consider an acoustic vibration of wave vector q. The displacement of an atom of position vector R is

$$u = \frac{1}{\sqrt{N}} e^{i(q \cdot R - \omega_B t)}. \tag{2.108}$$

The eigenvectors $e_i(q)$, defined in (2.43, 44), give the direction of the vibration u. They have the magnitude

$$|e_i|^2 = \frac{M_i}{\sum_i M_i} \tag{2.109}$$

so that the amplitude of the displacement calculated with (2.44) is the same for all atoms. For each q there are three orthogonal directions of e which can be obtained by solving the corresponding equations for the propagation of elastic waves [2.47]. The fluctuation amplitude of u is given by (for each mode)

$$\langle uu^\dagger \rangle_{\omega_B} = \frac{\hbar}{2\left(\sum_i M_i\right)\omega_B N_c}(n+1)$$

$$\langle u^\dagger u \rangle_{\omega_B} = \frac{\hbar}{2\left(\sum_i M_i\right)\omega_B N_c} n. \tag{2.110}$$

To the wave (2.108) corresponds a strain wave $e_{jk} = q_j e_k u_0 N^{-1/2}$ $\exp[i(qR - \omega_B t)]$, where e_k is the kth component of the phonon polarization vector. It is this strain which produces via the elasto-optic effect the fluctuations in χ responsible for the Brillouin scattering. The Stokes cross section thus becomes in analogy to (2.47):

$$\frac{d\sigma_S}{d\Omega} = \frac{\varepsilon_L^4 \omega_s^4 V q^2}{\varepsilon_0^4 (4\pi)^2 c^4} |\hat{e}_s \cdot (p:\hat{q}e) \cdot \hat{e}_L|^2 \frac{\hbar}{2\left(\sum_i M_i\right)\omega_B N_c} \begin{cases} (n+1) \text{ Stokes} \\ n \text{ antistokes} \end{cases}$$

$$= \frac{\varepsilon_L^4 \omega_s^4 V \hbar \omega_B}{2\varepsilon_0^4 (4\pi)^2 c^4 \varrho v^2} |\hat{e}_s \cdot (p:\hat{q}e)\hat{e}_L|^2 \begin{cases} (n+1) \\ n. \end{cases} \tag{2.111}$$

In (2.111), $\underset{\sim}{p}$ is the fourth rank photoelastic tensor defined in (2.100) which must be contracted with the unit vector along q ($\hat{q}$) and the phonon polarization vector e. ϱ is the density of the crystal. We point out that for Brillouin scattering, except at the lowest temperatures, $\hbar\omega_B \ll kT$ and $n \simeq kT/\hbar\omega_B \simeq n+1$; Stokes and antistokes intensities are equal (but only for phonons, not for magnons, see Fig. 2.14). We also note that although (2.111) arises from a susceptibility of the type (2.60), the *magnitude* of q does not appear explicitly as it is compensated by the factor ω_B^{-1} which enters in the fluctuations of the vibrational amplitude and the factor of ω_B^{-1} which arises from n for $\hbar\omega_B \ll kT$.

The evaluation of (2.111) for an arbitrary direction of $\hat{q}$ requires the solution of the wave propagation problem. The second rank tensors $(\underset{\sim}{p}:\hat{q}e)$ and the corresponding velocities for the three principal directions of propagation in cubic crystals have been listed in [2.17]. Similar results have been tabulated also for a number of lower symmetry crystal classes [2.88b, c]. In anisotropic crystals, not only $\underset{\sim}{p}$ contributes to the scattering but also the anisotropic χ, namely, through rotations in the coordinate axis produced by the non-symmetric components of the strain tensor. Replacement in (2.111) of the dynamic susceptibility $(\varepsilon_L/\varepsilon_0)^2 (\underset{\sim}{p}:qe)_{kj}$ by $\chi_{ke}(\hat{q}_e e_j - \hat{q}_j e_e)/2$ yields the contribution of local rotations to the Brillouin scattering. This effect has been discussed in detail by *Nelson* and *Lax* (2.28a, b].

We treat here as an example of the discussion above the Brillouin scattering efficiencies for propagation along $\hat{q} = [100]$ in a cubic crystal. The eigenvectors in this case are given by symmetry. They are $e = [100]$ for LA modes and $e = [010]$, $[001]$ for TA modes. The corresponding velocities of sound [2.47] and the tensor $(\underset{\sim}{p}:\hat{q}e)$ are, as a function of the elastic constants c_{11}, c_{12}, and c_{44} and the elasto-optic coefficients,

$$\text{LA}: \quad v = \sqrt{\frac{c_{11}}{\varrho}}; \quad (\underset{\sim}{p}:\hat{q}e) = \begin{pmatrix} p_{11} & & \\ & p_{12} & \\ & & p_{12} \end{pmatrix}$$

$$\text{TA(010)}: \quad v = \sqrt{\frac{c_{44}}{\varrho}}; \quad (\underset{\sim}{p}:\hat{q}e) = \begin{pmatrix} 0 & p_{44} & 0 \\ p_{44} & 0 & \\ 0 & 0 & \end{pmatrix} \qquad (2.112)$$

$$\text{TA(001)}: \quad v = \sqrt{\frac{c_{44}}{\varrho}}; \quad (\underset{\sim}{p}:\hat{q}e) = \begin{pmatrix} 0 & 0 & p_{44} \\ 0 & 0 & 0 \\ p_{44} & 0 & 0 \end{pmatrix}.$$

Equations (2.112) indicate that for backscattering on a [100] surface, the LA phonons produce *polarized* scattering while the TA phonons do not couple; they can be observed in 90° scattering with the light incident along [011] and scattered along [01$\bar{1}$].

Materials without inversion symmetry exhibit the so-called piezoelectric effect; a strain produces an electric polarization which is accompanied by an electric field. This field produces, in turn, a change in the susceptibility through the first-order electro-optic effect in a manner similar to that discussed in Sect. 2.1.12, and also a slight renormalization of the speed of sound similar to the TO→LO renormalization. Hence, an electro-optic contribution to the scattering cross section results for piezoelectric acoustic phonons. The explicit expression for this contribution is given in [Ref. 2.17, Eq. (8.45)]. The piezoelectric effect also plays an important role in resonant polariton scattering. It can induce scattering by piezoelectric phonons in *forbidden* configurations. Such scattering, which is usually polarized, has been recently observed in CdS [2.89] and other materials. In highly ionic materials such as CuBr, it is stronger than that related to the elasto-optic tensor [2.90].

Brillouin scattering usually becomes resonant as the frequency of the laser light approaches that of critical points in electronic interband transitions in a manner similar to that discussed in Sects. 2.1.5, 8. In fact, the theoretical considerations given in those sections for the Raman susceptibility can be applied to the resonant behavior of p. Of particular interest is the behavior of the elasto-optic tensor p when approaching the lowest energy gap. Such behavior can be investigated with static piezobirefringence experiments. From such work, the dispersion of the elasto-optic parameters of a large number of tetrahedral semiconductors has been measured [2.91]. With the exception of very small band gap materials (InSb, InAs) and of diamond, $(p_{11} - p_{12})$ and p_{44} have an antiresonant behavior with a zero and a change of sign slightly below the gap [see (2.33a) and Fig. 2.56]. At this point, the corresponding Brillouin cross section goes through a minimum close to zero. These effects can be studied particularly well for Brillouin scattering by phonons generated with the acoustoelectric effect [2.92]. We show in Fig. 2.15 such measurements performed on ZnSe together with a fit of the observed antiresonance with (2.33a) [2.93].

As Brillouin spectrometers have become more sensitive, measurements for strongly absorbing samples have become possible. Under these conditions, one must question the role of the surface proximity in the scattering process. For scattering by acoustic phonons, the surface must be viewed as corrugated as a result of the phonon excitation. For a static corrugation (≡ripple), the surface acts as a diffraction grating and the reflected beam does not obey Snell's law. The direction of the reflected beam is determined by k-conservation *only parallel to the surface* ($k_{\parallel}$; $k_{\perp}$ need not be conserved as there is no translational symmetry perpendicular to the surface).

We thus must have

$$k_{s,\parallel} = k_{L,\parallel} \pm q_{\parallel}, \tag{2.113}$$

where $q_{\parallel}$ is the wave vector of the ripple. In most experiments $k_{s,\parallel}$ and $k_{L,\parallel}$ are coplanar; most calculations and the discussion here are restricted to this case. The formulas for the reflectivity of a static corrugated surface can be also applied to a dynamic ripple. The only difference is that in the latter case a Brillouin frequency shift results. In a cubic material, symmetry requires that incident and scattered fields be both either in the scattering plane or perpendic-

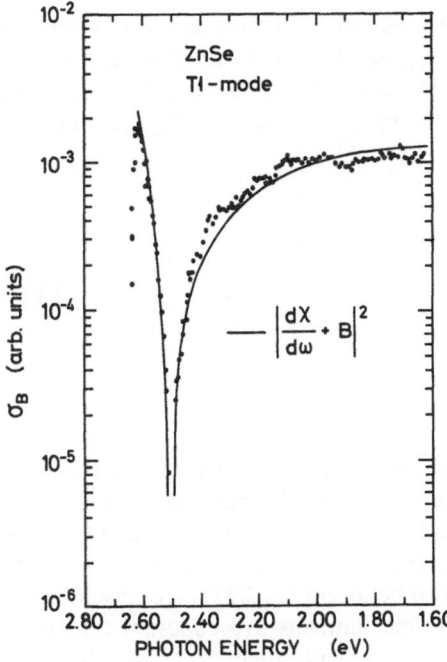

Fig. 2.15. Dispersion of the Brillouin cross section measured in ZnSe for TA electroacoustic phonons propagating along [110] and polarized along [1$\bar{1}$0] at room temperature. The solid line is a fit with (2.33a) [2.93]

ular to it for scattering by surface ripple. This is not the case, in general, for the elasto-optic mechanism discussed above. Also, for incident polarization in the plane of incidence, the scattered light must vanish whenever the scattering geometry is near Brewster-angle conditions ($\Theta_L \simeq -\Theta_s \simeq \arctan n$) [2.94]. Around this Θ_L backscattering ($\Theta_s \simeq \Theta_L$) has a maximum for that polarization [2.94, 95]. With these facts, one can test whether a given type of Brillouin scattering is due to surface ripple or to a direct modulation in the susceptibility (elasto-optic mechanism). Perhaps a more spectacular test is provided by the fact that the elasto-optic coefficients p_{ijkl} are strongly resonant near a direct absorption edge, leading to the resonant behavior of the corresponding Brillouin cross sections shown, for instance, in Fig. 2.15. The ripple mechanism depends on the reflection coefficients which have the same type of singularity as χ, considerably weaker than that of p_{ijkl} [2.94]. Another striking characteristic feature of surface ripple scattering is the fact that it persists when the material is covered by a very thin evaporated metallic film of reflectivity $\simeq 1$ (e.g. Al) [2.95].

The differential cross sections for scattering by surface ripple can be written as [2.96]

$$\frac{\partial^2 \sigma_\perp}{\partial \Omega \partial \omega} = \frac{\omega_L^4 A^2 \cos^2 \Theta_L \cos^2 \Theta_s}{\pi^2 c^4} \left| \frac{k_L^z - \ell_L^z}{k_s^z - \ell_s^z} \right|^2 \langle |u^z(0)|^2 \rangle_{q_x, \omega}$$

$$\frac{\partial \sigma_\parallel}{\partial \Omega \partial \omega} = \frac{\omega_L^4 A^2 \cos^2 \Theta_L \cos^2 \Theta_s}{\pi^2 c^4} \left| \frac{(\varepsilon \ell_s^x \ell_L^x - k_s^z k_L^z (\varepsilon/\varepsilon_0 - 1))}{(\varepsilon \ell_L^z + k_L^z)(\varepsilon \ell_s^z - k_s^z)} \right|^2 \langle |u^z(0)|^2 \rangle_{q_x, \omega},$$

(2.114)

where A is the illuminated area and $u^z(0)$ the normal amplitude of vibration at the surface. The angular brackets $\langle \; \rangle_{q_x, \omega}$ signify the Fourier component of the temporal frequency ω and of spatial frequency along the direction of surface propagation x. The k's and ℓ's are defined by [2.96] (see also Fig. 6.8 of [1.2])

$$
\begin{aligned}
\ell_L^x &= k_L^x = (\omega_L/c) \sin \Theta_L \\
\ell_L^z &= -(\omega_L/c) \cos \Theta_L \\
k_L^z &= -(\omega_L/c)(\varepsilon_L/\varepsilon_0 - \sin^2 \Theta_L)^{1/2} \\
\ell_s^x &= k_s^x = -(\omega_s/c) \sin \Theta_s \\
\ell_s^z &= (\omega_s/c) \cos \Theta_s \\
k_s^z &= (\omega_s/c)(\varepsilon_L/\varepsilon_0 - \sin^2 \Theta_s)^{1/2} .
\end{aligned}
\tag{2.115}
$$

The corresponding expression for surface elasto-optic scattering can be found in [2.97]. The calculation of the cross sections for surface scattering with (2.114) requires the evaluation of the ripple amplitude $\langle |u^z(0)|^2 \rangle_{q_x, \omega}$. This can be done by using the fluctuation-dissipation theorem after evaluation of the appropriate response function. Explicit expressions can be found in [2.97]. We should point out that the ripple and the elasto-optic mechanisms are coherent and their scattering *amplitudes* must be added before squaring them to obtain the cross section. In metals, the surface ripple mechanism turns out to be dominant. In semiconductors, the Rayleigh surface waves produce scattering which contains a mixture of surface ripple and elasto-optic mechanisms [2.97].

2.1.15 Light Scattering in Amorphous and Disordered Materials

The general principles of light scattering in amorphous materials were discussed by *Brodsky* [Ref. 2.1, Chap. 5] in 1975. We summarize here some of the advances which have occurred in this field since then. Considerable material on this subject can also be found in the Proceedings of the International Conferences on Amorphous and Liquid Semiconductors, in particular, those of the years 1977 [2.98], 1979 [2.99], and 1981 [2.100].

Since the demonstration by *Spear* and *Le Comber* in 1975 [2.101] of the possibility of doping n- and p-type amorphous Si (a-Si), this material and its analog a-Ge have received a great amount of attention. Because of its economy of preparation, a-Si has found applications as an optoelectric material for the production of photovoltaic solar cells [2.102], xerographic receptors [2.103], and display devices [2.104]. Raman scattering, together with ir spectroscopy, has played a prominent role in the characterization of the material [2.105].

Amorphous Si is nowadays prepared mainly by cathode sputtering [2.106] or by glow discharge decomposition of SiH_4 [2.101] and, more recently, of SiF_4 [2.107]. Similar methods can be used to prepare a-Ge. The material prepared with SiH_4 contains a certain amount of hydrogen known now to be essential

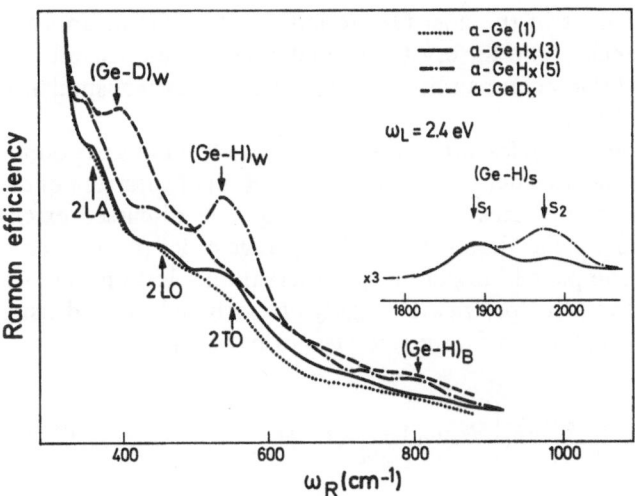

Fig. 2.16. Raman spectra of pure *a*-Ge and two *a*-GeH$_x$ samples; (3): 7at.% H, (5):10 at.% H for $\omega_R > 300$ cm^{-1}. The insert corresponds to the Si-H bond stretching modes of samples (3) and (5) [2.108]

for the electrically active doping process. This hydrogen saturates randomly distributed open bonds (dangling bonds) which would otherwise inhibit the electrical activity of the dopants. For this purpose, hydrogen can also be replaced by fluorine [2.107] which offers the advantage of the higher stability of the Si-F bond.

The Si-H bond exhibits a few interesting vibrational modes, among them the bond stretching mode at 2000–2100 cm^{-1} [2.105] (these bands occur at 1900–2000 cm^{-1} for *a*-Ge:H as shown in Fig. 2.16 [2.108]). The 2000 cm^{-1} component of this mode (1900 cm^{-1} for Ge-H) is usually attributed to single Si-H bonds while that at 2100 cm^{-1} ($\sim$2000 for GeH$_2$) is attributed to Si-H$_2$ groups [2.105].

While these vibrations can be seen in Raman spectra, ir spectroscopy is a more sensitive tool for their investigation. Moreover, the strength of the ir bands due to Si-H bonds can be used to estimate the hydrogen concentration of the sample in a nondestructive way [2.109]. In this manner it has been found that the best material for optoelectronic applications contains 5–10 at.% of hydrogen. Raman spectroscopic studies of the Si-H vibrations are, nevertheless, of importance for samples deposited on opaque (e.g. metallic) substrates.

We show in Fig. 2.16 a portion of the Raman spectrum of four *a*-Ge samples with different amounts of H (or D). The hydrogenated samples exhibit the bond stretching structure $S_1 - S_2$ discussed above for *a*-Si. They also show the "bond wagging" bands at 565 cm^{-1}. These bands shift approximately by the factor $2^{-1/2}$ (the square root of the mass ratio) for the deuterated samples, thus confirming their assignment. The other structure shown in Fig. 2.16 arises from two-phonon overtones of the host Ge-Ge lattice; it occurs at approximately the same frequencies as in the crystalline materials and its strength can be interpreted with the bond polarizability model [2.40] in a manner similar to that discussed in Sect. 2.1.13.

Perhaps the most interesting feature of the Raman spectrum of amorphous materials and glasses is the disorder induced first-order Raman spectrum of the fundamental vibrations of the network. These spectra were discussed extensively in [Ref. 2.1, Chap. 5]. The drastic loss of long-range order (translational symmetry) implies the complete lifting of the q-conservation selection rule. All modes become allowed and the spectra are roughly proportional to the density of vibrational states $N_d(\omega_R)$. For a-Si, we can write [see (2.51)],

$$\frac{\partial S_s}{\partial \Omega \partial \omega_R} = \frac{\omega_s^4}{(4\pi)^2 c^4} \left| \hat{e}_s \cdot \left\langle \frac{d\chi'}{d\xi} \right\rangle \cdot \hat{e}_L \right|^2 N_d(\omega_R) \frac{\hbar}{2\omega_R} (n+1), \tag{2.116}$$

where $\left\langle \dfrac{d\chi'}{d\xi} \right\rangle$ represents an average Raman susceptibility. This Raman susceptibility must vanish for $\omega \to 0$ as a result of translational invariance. A simple proportionality of the Raman susceptibility to ω_R [an extra factor of ω_R^2 in (2.116)] represents the experimental data rather well [2.110].

Two mechanisms responsible for the lifting of q-conservation have been identified [2.111]. One of them is of *mechanical* origin, due to the fact that the eigenvectors of the vibrations in the translationally disordered solid are not plane waves of definite q. The other is *"electrical"* in nature; even if there were no mechanical disorder fluctuations in $d\chi/d\xi$, due, for instance, to fluctuations in density, one should find (a) elastic scattering of the type discussed in Sect. 2.1.3, and (b) a combination of this elastic scattering of wave vector q and scattering by phonons of an *arbitrary* wave vector $-q$. Hence all q vectors become allowed through the good offices of the susceptibility fluctuations. In vitreous silica, the contribution of this *electrical* disorder to the low frequency Raman spectra is one order of magnitude larger than that of the *mechanical disorder* [2.111]. This type of analysis should be extended to other systems. Existing theories for a-Ge and a-Si [Ref. 2.1, Chap. 5] consider only the mechanical disorder explicitly.

We show in Fig. 2.17 the polarized and depolarized spectra of pure a-Si in the region of scattering by one-phonon and two-phonon overtones. The one-phonon spectrum has four bands corresponding to the TA, LA, LO, and TO-bands of Fig. 2.12. The average $D = \sigma_d/\sigma_p$ ratio [see (2.14)] is 0.53, nearly constant throughout the whole spectrum and smaller than previously reported values (~ 0.8, see [2.1], Chap. 5). This ratio fulfills (2.14), as required by symmetry considerations. For an *isotropic* solid, such as a-Si, the symmetries of the Raman tensor can be Γ_1 (a multiple of the unit matrix $= a$) and $J = 2$, see (2.12). The depolarized spectrum represents the β^2 component of the $J = 2$ Raman tensor of (2.12) (actually $3\beta^2/15$). The polarized spectrum represents $(a^2 + 4\beta^2/15)$.

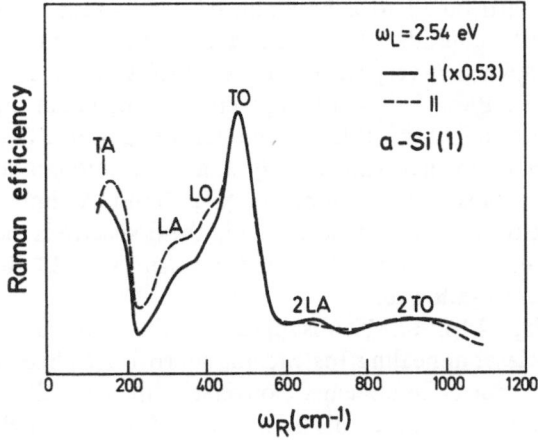

Fig. 2.17. First and second-order Raman spectra of pure a-Si. Both polarized and depolarized spectra are given. Compare with density of states in Fig. 2.12 [2.108]

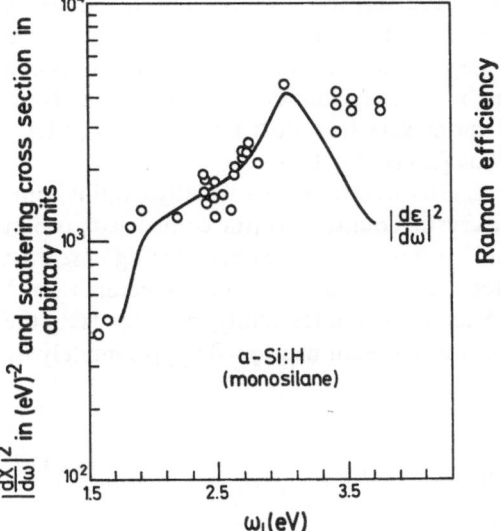

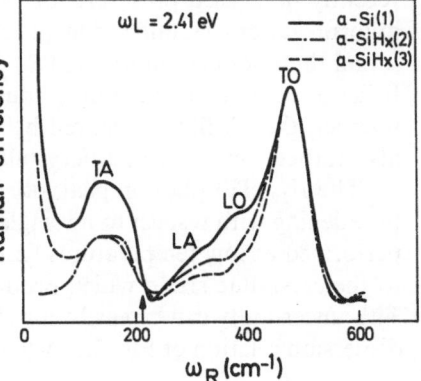

Fig. 2.19. Polarized spectra of pure a-Si(1) and two a-Si:H samples [a-Si(3) 11 % at H]. [2.108]. The arrow at 210 cm^{-1} indicates the position of the a-Si:H quasilocal mode

Fig. 2.18. Resonance in the integrated Raman cross section of a-Si (points, see Fig. 2.17) and fit with $|d\chi/d\omega|^2$ (solid line) obtained from the optical constants of [2.112]. From [2.108]

Amorphous Si has a strong, although broad, dielectric resonance in the visible with a peak in χ_i at $\omega_0 \sim 3$ eV. This behavior leads to a resonance in the Raman scattering efficiency near ω_0 as shown in Fig. 2.18. The broad resonance observed can be fitted rather well with the expression $|d\chi/d\omega|^2$, numerically evaluated from the $\chi(\omega)$ spectra of [2.112]. This expression represents a good approximation to the first term in (2.30).

Figure 2.19 shows the first-order Raman spectra of these a-Si samples, including two hydrogenated ones [2.108]. Hydrogenation changes these spectra relatively little, the main change being possibly a decrease in the LA and

LO-bands. This fact has been attributed to a breaking of the six-fold rings characteristic of the tetrahedral materials [2.113] but a quantitative understanding of this matter is still lacking. A very weak feature appears in Fig. 2.19 at 2.2 eV for the hydrogenated samples. This feature appears very strong in the corresponding ir absorption spectra [2.114]. It has been interpreted as due to a "quasilocal" mode of the Si-H bond involving mainly angular motions and affecting a large number of Si atoms (~ 50) around the Si-H bond. A similar feature appears in a-Ge:H. The reason why it is so strong in the ir spectra is the fact that the basic vibrations of the Si-Si bond are ir-*forbidden* (Sect. 2.1.10) while those of the Si-H bond are ir-allowed.

The Raman spectra of Figs. 2.16, 17, 19 disappear as the material is crystallized through thermal or laser annealing. Instead, the sharp TO(Γ) line at 520 cm^{-1} for c-Si and 300 cm^{-1} for c-Ge appears. Conversely, the crystalline line gives way to the amorphous spectrum as disorder is introduced through ion bombardment in a single crystal [Ref. 2.1, Fig. 5.6]. This feature has recently been used profusely for the characterization of a-Si and for studying amorphization-crystallization processes [2.115]. In this manner, it has been found that n-doped fluorinated Si, prepared at relatively low temperatures and believed hitherto to be amorphous, is actually microcrystalline. In the same manner, thin Si films prepared by the plasma transport method [2.116] have also turned out to be microcrystalline [2.116, 117].

The Γ_{25}, TO-phonon peak of *recrystallized* a-Si does usually exhibit some broadening with respect to its single crystal counterpart (the comparison is best performed at low temperatures [2.117]). This additional broadening is related to the crystallite size d which produces an uncertainty in the q-vector $\simeq 2\pi/d$. This uncertainty can be easily converted into an uncertainty in ω_R by using the dispersion relation of Fig. 2.12 which we represent near $q = 0$ approximately by

$$\Delta\omega_R = - Aq^2$$

with   (2.117)

$$A \simeq 200 \text{ cm}^{-1}\text{Å}^2 .$$

Hence, an additional disorder-induced broadening of 5 cm^{-1} corresponds to an uncertainty in q, $\Delta q \simeq -0.15\,\text{Å}^{-1}$ which corresponds to a grain size $d = 2\pi/\Delta q = 40\,\text{Å}$. This broadening is asymmetric [Ref. 2.118, Fig. 4], with a broader wing towards lower q's.

As already mentioned, a-Si:F is also receiving attention as a potential optoelectronic material. We show in Fig. 2.20 the one-phonon ir-absorption spectrum of a fluorinated and a pure a-Si sample compared with the Raman spectrum of the fluorinated sample. We notice in this figure that the fluorination substantially enhances the ir absorption, a fact which is not surprising as the Si-Si bond is infrared forbidden and becomes infrared allowed when F, strongly electronegative, is attached to one of the Si atoms. Fluorination produces a shift of the TO-peak in the ir-spectrum of $+30$ cm^{-1}. This is also a

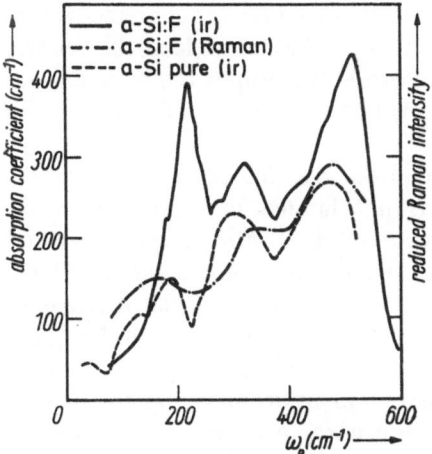

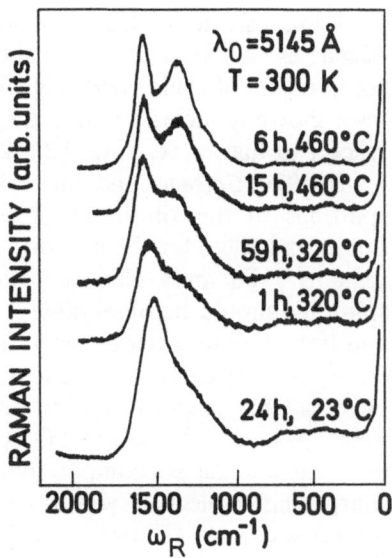

Fig. 2.20. The ir absorption spectrum of a-Si:F (10% F) showing local mode at 210 cm^{-1} and shift and increase in strength of the TO-mode with respect to that of a pure a-Si sample. The Raman spectrum of a-Si:F with an unshifted TO-peak, is also shown [S. C. Shen, C. J. Fang, M. Cardona: Phys. Stat. Sol. (b) **101**, 451 (1980)]

Fig. 2.21. Raman spectra of an evaporated a-C film, subsequently annealed at several temperatures [2.121]

typical induction effect [2.119] produced by the attachment of the fluorine: because of the ir-enhancement, the ir spectrum is mostly produced by Si-Si:F groups. The basic stretching vibration of these groups is up-shifted by the presence of the strongly electronegative fluorine. This explanation is confirmed by the fact that the TO-peak in the Raman spectrum of a-Si:F is indeed unshifted (Fig. 2.20).

Raman scattering has also recently contributed to the clarification of the structure of amorphous carbon (a-C). Depending on the deposition conditions (especially onto substrates at 77 K), this material exhibits a hardness and transparency which reminds of diamond [2.120]. Hence, the presence of four-fold coordinated atoms in these films has been suggested [2.120]. Their Raman spectra (Fig. 2.21) have a shoulder at 1350 cm^{-1} which sharpens up and increases upon annealing. These spectra, however, do not show any traces of the Raman spectrum of diamond. Although the 1350 cm^{-1} peak may be related to the 1332 cm^{-1} peak of diamond, this hypothesis must be discarded as the peak of a-C is too strong to be that of diamond [2.121]. The 1500 peak of Fig. 2.21 corresponds rather well to a peak in the density of states of graphite and so does the peak at 1350 [2.121]. The results for the unannealed films of Fig. 2.21 have therefore been interpreted as due to randomly oriented planar graphitic elements of about 20 Å in size. Annealing produces the appearance of microcrystalline graphitic islands which result in the sharpening up of the Raman structure. The authors of [2.121] thus conclude the lack of fourfold coordination in these films in spite of their hardness and transparency.

Raman scattering has remained an important tool for the characterization of glasses. A recent fit to the spectra of vitreous Se has yielded the relative composition of chains and rings. By this method the proportion of *chains* has been shown to increase above the glass transition temperature ($T_g \simeq 31\,°C$) with respect to that of Se_8 rings [2.122].

In 1976, *Galeener* and *Lucovsky* pointed out that the existence of LO-TO splittings in the vibrational spectra was not limited to crystalline solids. Equation (2.86a) for the splitting does not imply translational symmetry and remains valid for a glass, provided one includes a damping term. LO-TO splittings should be observable if $e^{*2}/\mu V \varepsilon_\infty$ is large (large oscillator strength) and if the damping frequency $\gamma < \omega_{LO} - \omega_{TO}$. The frequencies ω_{TO} and ω_{LO} are quite generally defined as the frequencies at which $\mathrm{Im}\{\varepsilon(\omega)\}$ and $-\mathrm{Im}\{\varepsilon^{-1}(\omega)\}$ have maxima, respectively. For many glasses and amorphous semiconductors, $\gamma \gtrsim \omega_{LO} - \omega_{TO}$ and the LO-TO splittings are hard to observe. They become observable in glasses composed of well-defined molecular units with relatively sharp quasi-molecular vibrations. Such is the case for B_2O_3 [2.123] whose Raman spectrum is shown in Fig. 2.22 together with $\mathrm{Im}\{\varepsilon(\omega)\}$ and $-\mathrm{Im}\{\varepsilon^{-1}(\omega)\}$, the latter obtained from ir measurements. The LO($1550\,cm^{-1}$) − TO($1260\,cm^{-1}$) splitting of the high frequency mode can be easily seen in the Raman spectra. A force constant analysis of these spectra suggests a dominance of B_3O_3 ("bor-oxyl") rings interconnected at the boron sites by oxygen (see insert in Fig. 2.22).

Another interesting structural problem whose solution is greatly aided by Raman spectroscopy concerns glasses of the type $(GeSe_2)_x(As_2Se_3)_{1-x}$ [2.124, 125]. For the compositions $x \gtrless 1/2$, these glassy materials have a structure consisting mainly of edge-sharing or corner-sharing $GeSe_4$ tetrahedra. The symmetric stretching mode of these tetrahedra is seen as a very sharp line at $202\,cm^{-1}$. This line has a sharp companion at $219\,cm^{-1}$ which escapes simple interpretation in terms of vibrations of the tetrahedra [2.124]. *Phillips* et al. [2.125] have suggested that this "companion" line is due to vibrations of "outrigger" elements formed by two chalcogen atoms. Outriggers related to the $AsSe_3$ units also exist.

In order to derive from (2.116) the relationship between the scattering efficiency and the density of vibrational states, we have assumed that $\langle d\chi'/d\xi \rangle$ is a smooth function of ω. This implies that the vibrational modes are strongly localized; if vibrational modes extend over many atoms (long-range correlation), coherence effects produce rapid variations and cancellations of $\langle d\chi'/d\xi \rangle$ similar to those implied by the k-selection rule in crystals. It has recently been suggested that long-range correlations exist, in particular, for the symmetric stretching modes of oxygen in SiO_2-type glass [2.126]. This leads to a narrow line, similar to that of the corresponding crystal, which can be seen especially well in GeO_2 (Fig. 2.24) [2.127]. As in the case of Si, where for crystalline material the Raman mode occurs at a frequency for which the density of states vanishes, Fig. 2.24 indicates that the strongest Raman band occurs at the lower flank of the density of states curve for polarized scattering.

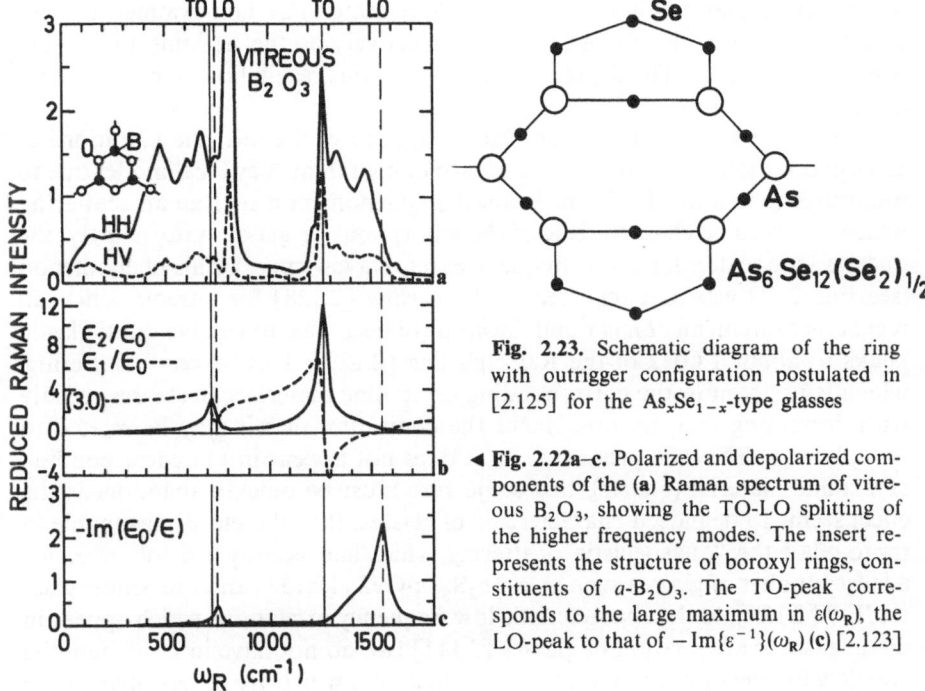

Fig. 2.23. Schematic diagram of the ring with outrigger configuration postulated in [2.125] for the As_xSe_{1-x}-type glasses

◄ Fig. 2.22a–c. Polarized and depolarized components of the (a) Raman spectrum of vitreous B_2O_3, showing the TO-LO splitting of the higher frequency modes. The insert represents the structure of boroxyl rings, constituents of a-B_2O_3. The TO-peak corresponds to the large maximum in $\varepsilon(\omega_R)$, the LO-peak to that of $-\mathrm{Im}\{\varepsilon^{-1}\}(\omega_R)$ (c) [2.123]

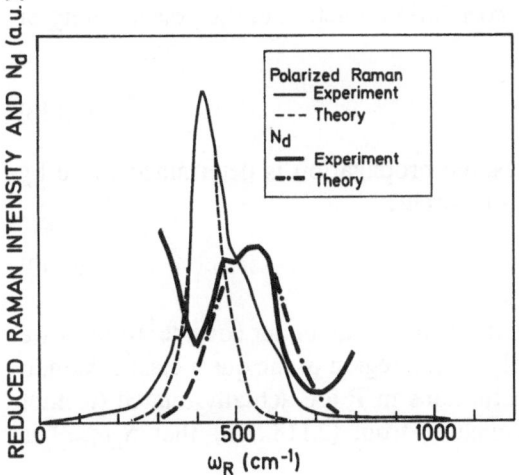

Fig. 2.24. Polarized and depolarized Raman spectra of a-GeO_2 together with corresponding experimental (neutron scattering) and theoretical density of phonon states N_d [2.127]

The depolarized scattering, however, reproduces the density of phonon states well, similar effects are observed in a-GeTe resulting from long-range correlation of the symmetric stretching mode. In [2.127], this result is attributed to long range correlations in the spacing of the SiO_2 units. Under these conditions the *symmetric* components of the polarizability tensors interfere with each other in a manner similar to that in the crystalline state. The non-symmetric components, however, do not interfere constructively as the building blocks are oriented at random. The degree of generality of this result, however, is not clear at present.

We have so far discussed the Raman spectra of the intrinsic vibrations of amorphous materials and glasses and also of high frequency local modes due to impurities (e.g. a-Si:H). There is another phenomenon in Raman scattering which seems to be characteristic of the amorphous or glassy state, namely, the quasi-elastic scattering at low frequencies ($\omega_R \lesssim$ a few cm^{-1}). This phenomenon (see Fig. 2.25) was first reported by *Winterling* [2.128] for vitreous silica. In recent measurements, *Fleury* and *Lyons* have been able to resolve quasi-elastic peaks to about 1 GHz of the Rayleigh line [2.129]. This "excess" scattering, which is not seen in the corresponding crystalline materials, decreases rapidly with decreasing temperature. Hence the conjecture that it may be related to two-phonon difference processes. As it does not appear in the corresponding crystalline material (quartz), this conjecture must be quickly abandoned; the effect seems to be indeed characteristic of glasses. It is, therefore, reasonable to try to relate this "quasi-elastic" scattering, which has recently been observed for a large number of glasses such as a-As$_2$S$_3$, a-GeS$_2$ [2.132] and the Schott glass LaSF-7 [2.129], to the characteristic low frequency excitations which appear in the ultrasonic attenuation of glasses [2.111] (we do not have in mind here the very low frequency excitations or tunneling modes which are responsible for the linear term in the specific heat [2.131a, b]). As shown in (2.56), the Raman scattering efficiency is given by the correlation function of the polarizability or susceptibility:

$$S_R(q, \omega) \propto \langle P_{\alpha\beta}(r, t) P_{\gamma\delta}^\dagger(0, 0) \rangle_{q,\omega}. \tag{2.118}$$

Correspondingly, the loss for ultrasonic propagation is determined to be by virtue of the fluctuation-dissipation theorem:

$$S_u(q, \omega) \propto \langle \varrho(r, t) \varrho^\dagger(0, 0) \rangle_{q,\omega}, \tag{2.119}$$

where ϱ is the strain tensor for the ultrasonic mode under consideration. If we extend the continuous elastic model into the region of the quasi-elastic Raman scattering and assume that the fluctuations in P are actually related (to first order) to fluctuations in ϱ, we conclude from (2.118, 119) that $S_R(q, \omega)$ is proportional to $S_u(q, \omega)$.

In order to test this result, we show in Fig. 2.26 the temperature dependence of the quasi-elastic Raman scattering in vitreous silica as compared with that of the ultrasonic attenuation measured by the width of the Brillouin line. Below ~ 200 K, a reasonable correlation exists but not at higher temperatures. The reason for this discrepancy is not known [2.111].

Another characteristic feature of light scattering by glasses is the presence of a very strong elastic line. The Brillouin line, basically similar to that of the

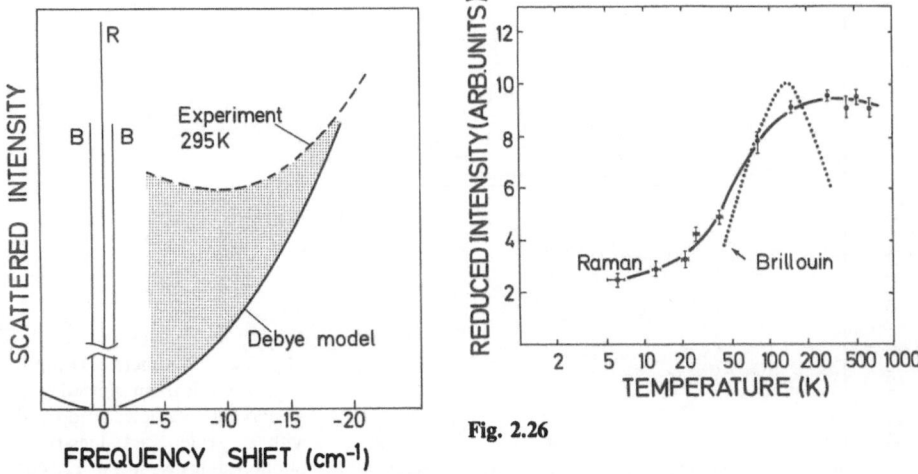

Fig. 2.25. Quasi-elastic Raman spectrum of vitreous SiO_2 at low frequencies showing also the Brillouin (B) and Rayleigh (R) lines [2.128]

Fig. 2.26. Temperature dependence of the quasi-elastic Raman efficiency in vitreous silica (suprasil W1) at $\omega_R = 5\ cm^{-1}$ compared with that of the Brillouin line width (dotted line) [2.111]

crystalline material except for the possible additional width of Fig. 2.26, has a strength typically only a few percent of that of the elastic line [2.132]. It can be either polarized (e.g., in SiO_2 glasses [2.133]), probably related to the density fluctuations of (2.19a), or depolarized (e.g., in $a\text{-}B_2O_3$), thus reflecting anisotropy in the basic building blocks [2.134]. The strength of the elastic scattering is largely responsible for the difficulties in investigating quasi-elastic phenomena. In particular, no observations of scattering by the very low frequency "tunneling modes" responsible for the specific heat anomaly [2.131a, b] have been published (for negative results see [2.131c]).

2.1.16 Defect-Induced Raman Spectra in Crystalline Materials

The effect of defects on the Raman spectra of crystals parallels those discussed above for amorphous solids. Two types of phenomena are observed: local modes due to the defects and defect-induced Raman-forbidden modes, in particular, those due to the violation of the k-conservation selection rule. The latter can be used advantageously to observe the whole density of phonon states in the first-order Raman spectra.

There is extensive literature discussing the Raman spectra due to vibrational local modes of defects in alkali halides (color centers) [2.135]. It will not be reviewed here. In the case of semiconductors, relatively large concentrations of defects are needed ($\sim 10^{18}$) to see vibrational local modes as a result of the opacity of the materials in the visible and of the competition of second-order host spectra. The case of boron in Si has been studied in great detail [2.136]. Of

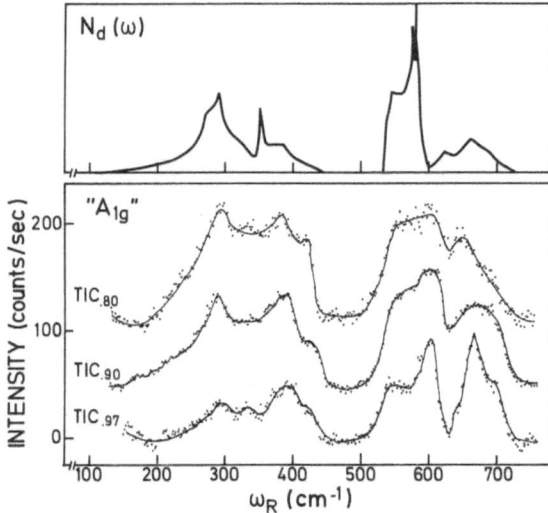

Fig. 2.27. Symmetric component of the Raman efficiency of several TiC$_x$ samples compared with the calculated density of phonon states N_d in TiC [2.140]

particular interest is the fact that the cross section per atom is much larger for the local mode than for the basic Raman $\Gamma_{25'}$ mode of the Si lattice. The cross section increases as the impurity concentration decreases and reaches ~ 200 times the $\Gamma_{25'}$ cross section at a B-concentration of 10^{18} cm^{-3}. This large cross section has been attributed to the electrostatic charge of the B-impurities (acceptors) which couples very strongly with the electronic states and thus produces large values of the corresponding Raman susceptibility [2.136]. The effect appears to bear a strong relationship to the enhancement of Raman scattering in molecules adsorbed on metal surfaces [Ref. 1.3, Chaps. 6 and 7]. As the B concentration N_B increases, the increasing density of free holes screens the Coulomb potential of the vibrating B atoms and the cross section (per impurity atom) decreases like $N^{-1/3}$ [2.136].

Defect-induced forbidden Raman modes have been observed in a large number of crystals. The defects can be either impurities (in particular mass defects) or vacancies. As examples, we mention here the work on alkali halides, in particular, NaCl-KCl [2.137], semiconductor alloys (e.g., In$_{1-x}$Ga$_x$P [2.138]), rare-earth chalcogenides (GdS [Ref. 1.3, Chaps. 3 and 4]), intercalated layer compounds [Ref. 1.2, Chap. 2], and a number of metals such as the A15 compounds (V$_3$Si, Nb$_3$Sn [2.139] and TiC [2.140]). The scattering induced by spin disorder in rare-earth chalcogenides, [Ref. 1.3, Chaps. 3 and 4], can also be considered as being in this category.

As an example of disorder-induced Raman scattering, we show in Fig. 2.27 the symmetric scattering spectra of several TiC$_x$ samples reported recently by *Klein* et al. [2.140] compared with the density of one-phonon states. The agreement between both types of results is rather striking.

On the theoretical side, we should mention a general formulation of the problem of Raman scattering in disordered systems which has appeared in the literature [2.141]. The disorder affects both the electronic states responsible for the polarizability and the phonons. In the formulation of [2.141], both types of effects can be treated separately. The effect of disorder on the electronic states simply blurs resonances [roughly equivalent to increasing γ in (2.30)]. The phonon disorder transforms the sharp Raman phonons, resulting from k-conservation in crystals, into phonon densities of states. An explicit calculation of the latter type of effect has been performed for the case of mass defects (e.g., Ge in Si) [2.142]. Resonance effects can introduce serious distortions of the densities of phonon states in the observed Raman spectra if one assumes that the electronic states are not affected by the disorder [e.g., if $d\chi'/d\xi$ in (2.116) is not a smooth function of ω_L but contains resonances].

The defect-induced first-order Raman phonon spectrum can, sometimes, be related to the second-order spectra [2.143] through a phenomenological argument. This happens whenever the defects produce a random fluctuation in the regular position of the atoms, such as in the case of substitutional isoelectronic impurities of an atomic size much smaller than that of the replaced host atom. In this case, we can compare the scattering process to that of second-order Raman scattering. The latter is produced by fluctuations in the product $\xi'\xi$ as the crystal is thermally agitated (2.34). The former also results from bilinear fluctuations $\xi\xi_d$ where ξ fluctuates due to the thermal agitation and ξ_d is disorder-induced. The corresponding Stokes scattering efficiency for disorder-induced first-order scattering can be easily obtained from (2.95) by means of the replacement

$$\frac{\hbar}{2\omega_{jq}}(n_{jq}+1)\rightarrow\langle\xi_d^2(j,q)\rangle, \tag{2.120}$$

where $\langle\xi_d^2(j,q)\rangle$ represents the disorder-induced fluctuation in the normal coordinate, related to atomic displacements by (2.44). After the substitution (2.120), (2.95) must be summed over j and q. The mechanism just discussed is similar to the "electrical disorder" of Sect. 2.1.15: ξ_d produces an "electrical disorder" $(d\chi/d\xi_d)$. In the electrically-disordered material, any phonon is in principle Raman active.

2.1.17 Stimulated Raman Scattering and Third-Order Susceptibilities

We have so far discussed the phenomenon of *spontaneous* Raman scattering. The scattering process can also be *driven* by an electromagnetic field at the frequency ω_s. The phenomenon, called stimulated Raman scattering, bears the same relationship to the spontaneous effect as the *stimulated* to the *spontaneous* emission of radiation (Einstein relations, [2.144]). The spontaneous effect is equivalent to a stimulated one, stimulated by the zero-point fluctuations in the *electromagnetic* field.

The stimulated Raman effect is readily described in terms of the third-order susceptibility $\chi^{(3)}$ [2.145]:

$$P_\alpha(\omega_s) = \varepsilon_0 \chi^{(3)}_{\alpha\beta\gamma\delta}(\omega_1, \omega_2, \omega_3) E_\beta(\omega_1) E_\gamma(\omega_2) E_\delta(\omega_3), \tag{2.121}$$

with $\omega_s = \omega_1 + \omega_2 + \omega_3$. In (2.121) we have assumed summation with respect to repeated indices. The frequencies ω_1, ω_2, ω_3 are, for stimulated Raman scattering, a permutation of the frequencies ω_L, $-\omega_L$, and ω_s: the frequency ω_L is then "annihilated" and the frequency ω_s is generated, driven (i.e., stimulated) by one of the three fields on the rhs of (2.121).

The third-order susceptibility $\chi^{(3)}_{\alpha\beta\gamma\delta}(\omega_1, \omega_2, \omega_3)$ fulfills some rather straightforward invariance relations with respect to the permutation of the three frequencies. In a cubic material it has only *four* independent elements [2.145]:

$$\chi_{1111}, \chi_{1122}, \chi_{1212}, \chi_{1221}. \tag{2.122}$$

If the material is isotropic (liquids, amorphous materials), the components of (2.122) must also fulfill the isotropy condition:

$$\chi_{1111} = \chi_{1122} + \chi_{1212} + \chi_{1221}. \tag{2.123}$$

It is then customary to rewrite (2.121), the constitutive relation for the case of stimulated Raman scattering, as

$$P(\omega_s) = \varepsilon_0 \chi^{(3)}_{\text{RAM}} |E(\omega_L)|^2 E(\omega_s). \tag{2.124}$$

For the isotropic case we have [2.145]

$$\chi^{(3)}_{\text{RAM}} = 6[\chi^{(3)}_{1122}\hat{e}_s + (\chi^{(3)}_{1212} + \chi^{(3)}_{1221})\hat{e}_L(\hat{e}_L \cdot \hat{e}_s)]. \tag{2.124a}$$

In (2.124) we have not bothered to write the frequencies as arguments of χ. For polarized scattering ($\hat{e}_L \| \hat{e}_s$), (2.124) yields

$$\begin{aligned}
\chi^{(3)}_{\text{RAM}} &= 6(\chi^{(3)}_{1122} + \chi^{(3)}_{1212} + \chi^{(3)}_{1221}) \\
&= 6\chi^{(3)}_{1111}
\end{aligned} \tag{2.125}$$

and for depolarized scattering ($\hat{e}_L \perp \hat{e}_s$)

$$\chi^{(3)}_{\text{RAM}} = 6\chi^{(3)}_{1122}. \tag{2.126}$$

Equation (2.124) can be rewritten as a correction to the linear susceptibility χ_s and the refractive index n_s:

$$\begin{aligned}
\varepsilon_s(E_L) &= \varepsilon_s(0) + \varepsilon_0 \chi^{(3)}_{\text{RAM}} |E(\omega_L)|^2 \\
n_s(E_L) &= n_s(0)[1 + \tfrac{1}{2}\chi^{(3)}_{\text{RAM}} |E(\omega_L)|^2 / n_s^2].
\end{aligned} \tag{2.127}$$

From (2.127) we derive the propagation equation for E_s:

$$E_s(x) = E_s(0) \exp \left\{ i \frac{n_s x}{c} [1 + \tfrac{1}{2} \chi_{RAM}^{(3)} |E(\omega_L)|^2 / n_s^2] \right\},$$ (2.128)

where x is the distance along the direction of propagation. We assume that $n_s(0)$ is real and derive from (2.128) the following law for the intensity I_s as a function of x:

$$I_s(x) = I_s(0) e^{-g_R x},$$ (2.129)

where the Raman gain g_R (power gain for unit length) is given by

$$g_R = (-\text{Im}\{\chi_{RAM}^{(3)}\}) \frac{\omega_s |E(\omega_L)|^2}{c n_s}.$$ (2.130)

The gain g_R for stimulated Raman scattering will be positive whenever $\text{Im} \chi_{RAM}^{(3)}$ is negative. This gain can be related to the cross section for *spontaneous* Raman scattering in a manner similar to that used in the derivation of Einstein's relationship [2.144]. We consider the field E_s as having thermal fluctuations which can be calculated with the fluctuation-dissipation theorem (Sect. 2.1.11). E_s is related to the relevant polarization P_s through the polarizability χ_s:

$$P_s = \varepsilon_0 \chi_s(E_L) E_s$$
$$\delta H_{E,P} = \frac{V}{2} P_s E_s.$$ (2.131)

Equations (2.131) permit the application of the fluctuation-dissipation theorem (2.75) in order to calculate the fluctuations in P_s (for photons, the occupation number $n \simeq 0$ at usual temperatures). Using (2.127), we find

$$\langle P_s P_s^\dagger \rangle_{\omega_s} = \frac{2\varepsilon_0 \hbar}{\pi V} (\text{Im}\{\chi_{RAM}^{(3)}\}) |E_L|^2.$$ (2.131a)

Substituting this equation into (2.1), we find

$$\frac{\partial^2 S}{\partial \Omega \partial \omega_R} = \frac{\omega_s^4 \hbar n_s}{4\pi^3 \varepsilon_0 c^4 n_L} (-\text{Im}\{\chi_{RAM}^{(3)}\})$$ (2.132)

and using (2.130),

$$\frac{\partial^2 S}{\partial \Omega \partial \omega_R} = \frac{\omega_s^3 \hbar n_s^2}{4\pi^3 \varepsilon_0 c^3 n_L} (g_R / |E_L|^2).$$ (2.133)

In handling the above equations one has to keep in mind that the Raman efficiencies, as much as the gain and $\chi_{RAM}^{(3)}$, depend on the polarization chosen for incident and scattered fields. For isotropic materials this dependence is given by (2.124a). Equation (2.132) expresses the fact that the imaginary part of the tensor $\chi_{\alpha\beta\gamma\delta}^{(3)}(\omega_L, -\omega_L, \omega_s)$ is related, to a proportionality constant, to the tensor $I_{\alpha\beta\gamma\delta}$ defined in (2.55).

The tensor $\chi_{\alpha\beta\gamma\delta}^{(3)}(\omega_1, \omega_2, \omega_3)$ can also be used to describe a number of other nonlinear optical phenomena such as frequency tripling, two-photon absorption and coherent antistokes Raman scattering (CARS). The phenomenon of CARS has been reviewed in [Ref. 2.1, Chap. 7], in [2.145], and in Chap. 4 of this volume. It is obtained from (2.121) for $\omega_1 = \omega_2 = \omega_L$, $\omega_3 = -\omega_{st}$, where ω_{st} is a Stokes-Raman frequency. In this case, one finds from (2.121) the polarization P_α at the frequency $\omega_{as} = \omega_L + \omega_L - \omega_{st} = \omega_L + \omega_R$, where ω_R is the Raman shift of the Stokes frequency ω_{st}. Hence, the outgoing beam will be at the *antistokes* frequency.

The CARS polarization is given by a constitutive equation similar to (2.124) with $\chi_{RAM}^{(3)}$ replaced by the CARS susceptibility $\chi_{CARS}^{(3)}$ [the result of performing the required frequency permutations in $\chi_{\alpha\beta\gamma\delta}^{(3)}$]. For plane waves, the intensity I_{as} of the CARS wave as a function of the intensity of the two pump beams, the laser beam I_L and the Stokes beam I_{st} [2.145], is found to be

$$I_{as} = \frac{\varepsilon_0^{-2}\omega_{as}^2}{n_L^2 n_{as} n_{st} c^4} |\chi_{CARS}^{(3)}|^2 I_L^2 I_{st} l^2, \tag{2.133a}$$

where l is the path length and the k-conservation condition $k_{as} = 2k - k_{st}$ has been assumed to be fulfilled exactly. We notice that while the Raman cross section depends on the imaginary part of $\chi^{(3)}$, the CARS intensity depends on the square of its magnitude. In the Raman process, energy is dissipated in the form of phonons, while in the CARS process, this is not the case (see Fig. 2.28).

We show the diagrams of the photon transitions responsible for $\chi_{RAM}^{(3)}$ and $\chi_{CARS}^{(3)}$ in Fig. 2.28. We note that both processes are rather similar, the only difference being the fact that in CARS, there are two different intermediate

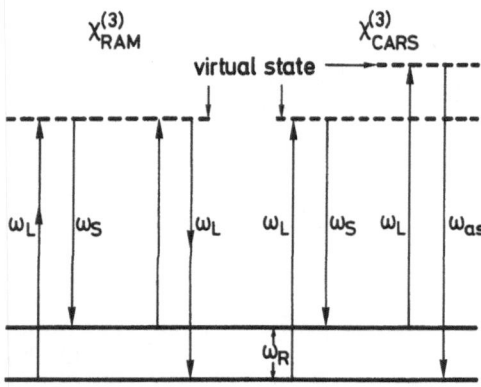

Fig. 2.28. Diagrams of the transitions responsible for $\chi_{RAM}^{(3)}$ and $\chi_{CARS}^{(3)}$

states while in Raman, both intermediate states are the same. Hence, if this intermediate state is far away from electronic resonances, $\chi^{(3)}_{RAM}$ and $\chi^{(3)}_{CARS}$ should be basically the same (except for differences in the permutations of the frequencies arising from the fact that for CARS, two frequencies are equal).

The Raman contribution to the $2\omega_L - \omega_s$ three-wave mixing, while dominant for $\omega_{st} \simeq \omega_L - \omega_R$, is not the only possible mechanism for the mixing. There will be, for instance, a number of excitations involving instead of the Raman excited states of Fig. 2.28, other nonresonant excited states. The total nonresonant contribution to the mixing is usually lumped together into a term $\chi^{(3)e}$, mostly of electronic origin. This term will interfere with the Raman term for $\omega_L - \omega_{st} = \omega_R$. For the case of diamond, the total nonlinear CARS susceptibility becomes [2.146]

$$\chi^{(3)}_{ijkl}(\omega_1, \omega_1, -\omega_2) = \chi^{(3)e}_{ijkl} + \frac{\pi N_c}{6\hbar} \frac{\delta_{ij,\sigma}\delta_{kl,\sigma} + \delta_{il,\sigma}\delta_{kj,\sigma}}{\omega_R - (\omega_1 - \omega_2) + i\gamma} \left(\frac{\hbar}{M\omega_R}\right)|a|^2 (4\pi\varepsilon_0),$$

(2.133b)

where M is the atomic mass, a the "Raman tensor component" or polarizability defined in (2.67) and $\delta_{ij,\sigma} = 0$ if two or more indices are equal, $= 1$ otherwise. The direction σ is the direction of polarization of one of the three degenerate $\Gamma_{25'}$ phonons ($\sigma = 1, 2, 3$) and γ is their line-width parameter.

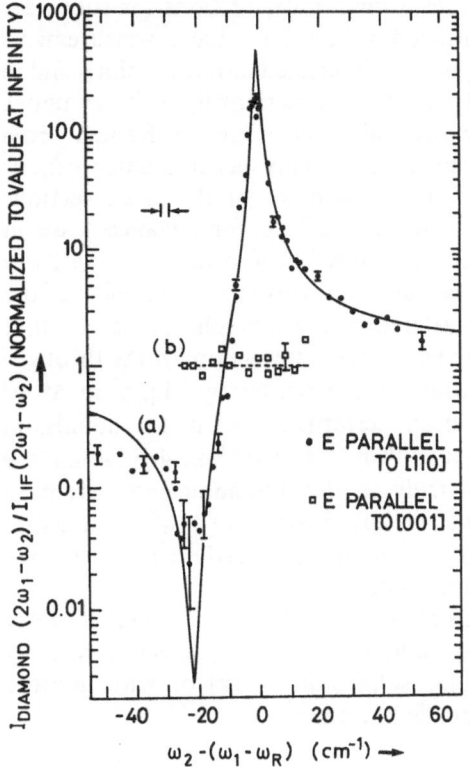

Fig. 2.29. CARS spectrum of the $\Gamma_{25'}$ phonon of diamond. (a) for the three polarizations parallel to [100]. (b) for $\hat{e}_L \| [100]$, $\hat{e}_{st}$ and $\hat{e}_{as} \| [010]$. The fit was made with (2.133) [2.146]

If $\gamma \ll \omega_R$ and if $\chi_{ijkl}^{(3)e}$ is approximately real, interference effects between the Raman term and the nonresonant electronic term in (2.133b) appear. The resulting resonance and antiresonance in the CARS intensity have been observed, among other materials, for diamond [2.146]. We show typical results in Fig. 2.29. Since a is known experimentally (see next section), a fit to the CARS spectra results in the determination of all independent components of $\chi^{(3)e}$ [2.146]. As seen in Fig. 2.29, $\chi^{(3)e}$ (the nonresonant background) is rather isotropic. By these methods, the following values were determined for diamond:

$$\chi_{1111}^{(3)e} = 6.4 \times 10^{-22}\,\text{SI}, \quad \chi_{1221}^{(3)e} = 2.4 \times 10^{-22}\,\text{SI}, \quad \chi_{1122}^{3(e)} = 2.56 \times 10^{-22}\,\text{SI}.$$

These susceptibilities, in SI units, have been obtained by multiplying the cgs data of [2.146] by the conversion factor 1.4×10^{-8}. The isotropy condition

$$2\chi_{1122}^{(3)e} + \chi_{1221}^{(3)e} = \chi_{1111}^{(3)e} \tag{2.134}$$

is rather well satisfied, as expected from the results of Fig. 2.29.

2.1.18 Absolute Scattering Cross Sections

a) General Principles

Experimental measurements of absolute Raman cross sections for solids are rare in the literature. As already mentioned in Sect. 2.1.7, these measurements are delicate and most authors seem to be only interested in Raman shifts and in resonant behavior (i.e., in ω_L-dependence of cross sections in arbitrary units). Recently, however, a number of theoretical calculations of Raman cross sections for solids have appeared (Sects. 2.2.6, 7). This has stimulated experimental work on absolute cross sections. A number of the cross sections determined earlier have turned out to be wrong [see, for instance, GaP in [2.26a]: the cross section is $30\times$ what it is now believed to be (Table 2.8)]. Absolute determinations of scattering efficiencies involve a measurement of the incident intensity and the intensity scattered in a given solid angle $d\Omega$ (that subtended by the entrance slit of the spectrometer), correction for the throughput of the spectrometer and evaluation of the scattering length so as to convert S^* into S (2.52). This has been performed mostly for liquids, in particular, benzene ($dS/d\Omega = 7.2 \pm 0.6 \times 10^{-6}\,\text{m}^{-1}\,\text{sterad}^{-1}$ at 4880 Å for the 992 cm^{-1} phonon [2.147, 148]). A few truly *absolute* measurements have been published for solids (diamond, $CaCO_3$, $NaNO_3$, quartz) in [2.149]. But, as we have already mentioned, these data for diamond are nearly a factor of two smaller than those generally accepted nowadays.

We shall next discuss the *relative* techniques in which the Raman efficiency is measured by comparison with a standard scatterer (benzene; recently diamond which is more convenient for solid state work) or with another scattering phenomenon whose efficiency is known.

Table 2.8. Stokes scattering efficiencies and tensor components a (2.67) measured for a number of crystals and for benzene

Material	ω_L [eV]	$dS/d\Omega$ $[10^{-5}\,Sr^{-1}\,m^{-1}]$	Scattering configuration $\hat{e}_L, \hat{e}_s$	a [Å^2]	Comments	Raman frequency [cm^{-1}]
Benzene[a]	2.54	0.72 ± 0.06	Polarized	–	Right angle scattering	992
Nitro-benzene[b]	1.79	0.8 ± 0.2	Polarizied	–	Inverse Raman effect	1345
Diamond[c]	2.41	6.5 ± 0.8	[100], [010]	+4.3 ± 0.6	Recommended value	1332
Silicon[d]	1.9	1.68 ± 50	[100], [010]	+60 ± 20	Recommended value	520
Germanium[e]	2.18	28,000	[100], [010]	640	Relative to Si 2.2.18.2	300
GaAs[f]	1	13 ± 3	[110], [110] (TO)	63 ± 10	2.2.18.6	269
GaAs[g]	1.2	18	[110], [110] [TO]	50	2.2.18.2	269
GaP[f]	1	4	[110], [110] (TO)	35 ± 6	2.2.18.6	367
GaP[h]	1.92	30 ± 5	[110], [110] (TO)	23 ± 5	2.2.18.3	367
GaP[h]	1.92	39 ± 4	[100], [010] (LO)	26 ± 2	2.2.18.3	403
InP[f]	1	6.4	[110], [110] (TO)	41	Calculated	304
AlSb[f]	1	20	[110], [110] (TO)	77	Calculated	318
ZnS[h]	1.92	0.38 ± 0.1	[100], [010] (LO)	2.46 ± 0.4	2.2.18.3	353
ZnSe[h]	1.83	2.2 ± 0.2	[100], [010] (LO)	7.1 ± 0.6	2.2.18.3	254
ZnTe[h]	1.83	22 ± 4	[100], [010] (LO)	24 ± 5	2.2.18.3	206
ZnO[i]	2.54	0.32	[001], [001] (TO)		Relative benzene	380
BeO[i]	2.54	0.51	[001], [001] (TO)	24 ± 5	Relative to benzene	678
CdS[i]	2.54	0.18	[001], [001] (TO)	24 ± 5	Relative to benzene	234
Se[j]	1.17	200	[001], [001]	Trigonal		237
Mg$_2$Si[e]	2.60	115,000	[110], [110]	680	Relative to Si	258
Mg$_2$Ge[e]	2.60	143,000	[110], [110]	760	Relative to Si	258

Table 2.8. (continued)

Material	ω_L [eV]	$dS/d\Omega$ [$10^{-5}Sr^{-1}m^{-1}$]	Scattering configuration $\hat{e}_L, \hat{e}_s$	a [$Å^2$]	Comments	Raman frequency [cm^{-1}]
Mg_2Sn^e	2.18	7000	[110], [110]	240	Relative to Si	221
$CaCO_3^k$	2.41	1.2 ± 0.3	[100], [100] + [010]	–	Absolute measurement	1086
$NaNO_3^k$	2.41	4.9 ± 1	[100], [100] + [010]	–	Absolute measurement	1051
$Quartz^k$	2.41	0.2 ± 0.05	[100], [100] + [010]	–	Absolute measurement	466
CaF_2^h	2.41	0.084 ± 0.04	[1$\bar{1}$0], [1$\bar{1}$0]	0.47 ± 0.09	2.2.18.3	322
SrF_2^h	2.41	0.14 ± 0.03	[1$\bar{1}$0], [1$\bar{1}$0]	0.61 ± 0.05	2.218.3	285
BaF_2^h	2.41	0.28 ± 0.05	[1$\bar{1}$0], [1$\bar{1}$0]	0.84 ± 0.08	2.2.18.3	240
MgF_2 MnF_2 FeF_2 CoF_2	See ^{e}				Relative to benzene	
$SrTiO_3^m$	2.41	172 ± 9	[100], [100] + [010]		2.2.18.3	Second order
KI^m	2.41	0.21 ± 0.03	[100], [100] + [010]		2.2.18.3	Second order
KBr^m	2.41	0.88 ± 0.09	[100], [100] + [010]		2.2.18.3	Second order

[a] J. G. Skinner, W.G. Nielson: J. Opt. Soc. Am. **58**, 113 (1968).
[b] L.J. Hughes, L.K. Steenhoek, E.S. Young: Chem. Phys. Lett. **58**, 413 (1978).
[c] See Fig. 2.30.
[d] M.H. Grimsditch, M. Cardona: Phys. Stat. Sol. (b) **102**, 155 (1980)
[e] S. Onari, R. Trommer, M. Cardona: Solid State Commun. **19**, 1145 (1976).
[f] C. Flytzanis: Phys. Rev. **B6**, 1278 (1972).
[g] M. Cardona, M.H. Grimsditch, D. Olego: In *Light Scattering in Solids*, ed by J.L. Birman, H.Z. Cummins, and K.K. Rebane (Plenum Press, New York 1979) p. 249.
[h] J.M. Calleja, H. Vogt, M. Cardona: Phil. Mag. (to be published).
[i] C.A. Arguello, D.L. Rousseau, S.P.S. Porto: Phys. Rev. **181**, 1351 (1969).
For the efficiencies of other ZnO Raman active modes and their dispersion see J.M. Calleja, M. Cardona: Phys. Rev. **B16**, 3753 (1977).
[j] A. Mooradian: In *Laser Handbook*, ed. by T. Arecchi, E.O. Schulz-Dubois (North-Holland, Amsterdam 1972) p. 1409.
[k] V.S. Gorelik, M.M. Sushchinskii: Sov. Phys., Solid State **12**, 1157 (1970).
[l] J.L. Sanvajol, A.M. Bon, C. Benoit, R. Almairac: J. Phys. C **11**, 1685 (1978).
[m] M.H. Grimsditch: Solid State Commun. **25**, 389 (1978).

b) Relative Methods for Determining Scattering Efficiencies: Sample Substitution

This method has two versions. In one of them, the sample is simply replaced by the standard by means of a lateral displacement [2.54]. As a standard one may use diamond, calcite, or fluorite. The Stokes efficiency of diamond is $dS_s/d\Omega = (6.5\pm0.8)\times10^{-5}$ m^{-1} sterad^{-1} at 5145 Å for polarisations $\hat{e}_L\|[100]$, $\hat{e}_s\|[010]$ in backscattering (see Table 2.8 for other scattering efficiencies). We show in Fig. 2.30 the polarizability measured by many authors for diamond and the results of a theoretical calculation (Sect. 2.2.6d). From the values of a, the scattering efficiency can be obtained with

$$\frac{\partial S_s}{\partial\Omega} = \left(\frac{\omega_s}{c}\right)^4 \frac{N\hbar}{2\mu\omega_v} \sum_j |\hat{e}_s\cdot\mathbf{R}_j\cdot\hat{e}_L|^2(1+n), \tag{2.134a}$$

where $\mathbf{R}_j$ are the $\Gamma_{25'}$ tensors of Table 2.1 for $d=a$. As seen in Fig. 2.30, diamond should constitute an excellent scattering standard. The data must, as usual, be corrected for collection angle and scattering length (2.52–54). In the second method, a sandwich is made with the sample and the transparent standard in front [2.146, 150]. Both the vibration line of interest and the standard are observed simultaneously without changes in the geometric configuration, thus eliminating sources of error. As a standard one can use, in principle, an evaporated CaF$_2$ film for which the scattering efficiency has been recently determined (see Table 2.8). In all cases, one has to be careful to avoid errors due to vignetting and to the depth of focus of the spectrometer + collecting lens.

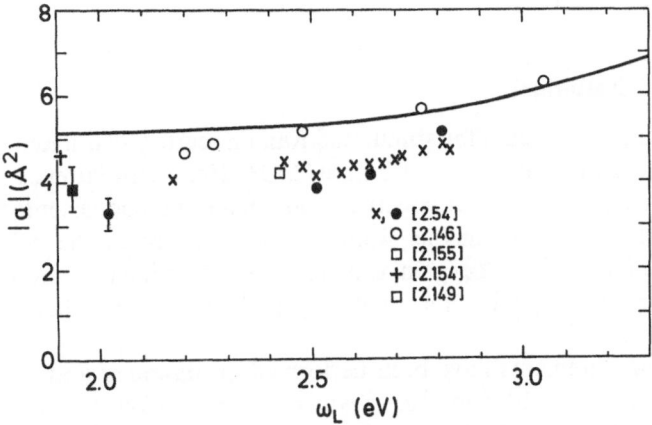

Fig. 2.30. Compilation of measurements of the Raman tensor and its dispersion for diamond. The solid line represents the equation $a(\text{Å})^2 = -6.5$ g $(\omega_L/5.6\,\text{eV})$ discussed in [2.54] and in Sect. 2.2.6 d. According to this equation, a should be *positive* although according to [2.157], a should be negative

c) Brillouin-Raman Efficiency Comparison

We mentioned in Sect. 2.1.7 that the Raman scattering efficiency for liquids and gases can be determined by comparison with that of either the Rayleigh or the rotational lines. The latter is found from parameters (i.e., the *static* polarizability tensor) which are available from other measurements. A similar technique has proven to be very powerful for solids, namely, the comparison of the Raman with the Brillouin efficiencies [2.151]. The latter can be calculated with (2.111) provided the appropriate elasto-optic constants are known. These constants have been measured in separate experiments (piezobirefringence, refractive index versus hydrostatic pressure, scattering by ultrasonic standing waves). The method yields additional dividends: there are (*in cubic materials*) 3 independent elasto-optic constants and only one (or one linear combination) is needed to determine the Raman efficiency which, for a given phonon, depends only on one Raman tensor component. Once this is done, the other elasto-optic constants can be found by looking at other Brillouin configurations. If they were previously known this can be used as a consistency check. It actually happens often that $p_{11} - p_{12}$ and p_{44} are known rather accurately through piezobirefringence experiments. Through the manipulations described above, one can then obtain the "hydrostatic" parameter $p_{11} + 2p_{12}$ as has been done recently for Si [2.152] in the region from 1.9 to 2.7 eV in which, because of the opacity of the material, it was previously unavailable. The Raman-Brillouin technique has been recently used for CaF_2, SrF_2, BaF_2, GaP, ZnS, ZnSe, ZnTe, Si, $SrTiO_2$ and the second-order spectra of the alkali halides (Table 2.8).

We should mention, in closing this section, a novel technique used for determining elasto-optic constants based on the uniaxial stress-induced Raman effect for a forbidden configuration. The main contribution to this effect arises, away from resonances, from the polarization scrambling induced by the stress via the elasto-optic effect [2.153].

d) Stimulated Raman Scattering

As discussed in Sect. 2.1.17, the gain for stimulated Raman scattering is a function of the spontaneous scattering cross section, see (2.133). Hence, the latter can be obtained if the former is measured. Intensity measurements may be easier for the stimulated than for the spontaneous Raman effect because of the collimation of the scattered beam. The gain can also be determined, without changing the geometry, by changing the scattering path in a wedge-shaped sample.

In [2.154], such measurements have been performed for diamond and the value of the efficiency $S = 2.7 \times 10^{-5} \, \mathrm{m}^{-1} \, \mathrm{sr}^{-1}$ was obtained at 6940 Å, somewhat larger than that obtained with other determinations at comparable ω_L's. Also, in such stimulated experiments for low laser power I_L, only the spontaneous effect is seen. As I_L increases, and provided there is some *feedback* mechanism such as the reflection of polished faces perpendicular to the I_L

beam, the material breaks up into optical oscillations (Raman laser, see [Ref. 2.1, Chap. 7]). This occurs whenever [2.154]

$$F\,e^{(g-\alpha)l} = 1\,,$$ (2.134b)

where α is the absorption coefficient, l the sample thickness and F the feedback coefficient $\simeq$ to the reflectivity (for diamond $F \simeq 0.17$). Hence, (2.134b) enables us to determine g from the threshold for oscillations observed for propagation along [111] in diamond at $I_L \simeq 1100\ \mathrm{MW/cm^2}$. From (2.134b) and the value of $\alpha \simeq 0$, the authors of [2.154] found $g = 8.2\ \mathrm{cm^{-1}}$. From this value and $I_L = 1100\ \mathrm{MW/cm^2}$, one evaluates with (2.130) the scattering efficiency $S = 2.5 \times 10^{-5}\ \mathrm{m^{-1}\,sr^{-1}}$ at $\lambda = 6940\,\text{Å}$, in good agreement with other determinations. Nevertheless [2.149] yields for the same configuration $2.8 \times 10^{-5}\ \mathrm{m^{-1}\,sr^{-1}}$ at $5145\,\text{Å}$, which corresponds after correcting for ω^4, to $8.5 \times 10^{-6}\ \mathrm{m^{-1}\,sr^{-1}}$ at $6940\,\text{Å}$! We believe this value to be incorrect. *McQuillan* et al. [2.154] obtained $|a| = 4.6\,\text{Å}^2$ from their value of S in agreement with other measurements (Fig. 2.30) and with the average value recommended in Table 2.8 $(a = 4.3 \pm 0.6\,\text{Å}^2)$.

e) Electric-Field-Induced Infrared Absorption

A rather ingenious method was used by *Anastassakis* and *Burstein* [2.155] to determine a for diamond by measuring the *electric-field-induced* forbidden infrared absorption produced by $\Gamma_{25'}$ phonons. If we use (2.85, 86) to describe this absorption, we must set $e^* = 0$ for diamond when the applied electric field is zero. For a finite field along [100] we have

$$\frac{\partial e_y^*}{\partial E_x} = \frac{\partial M_y}{\partial E_x \partial u_z} = \varepsilon_0 V_c \frac{\partial \chi_{yz}}{\partial u_z} = a\,,$$ (2.134c)

where M is the electric dipole moment per primitive cell. Hence, by measuring the field-induced absorption, for the configuration under consideration proportional to $|\partial e_y^*/\partial E_x|^2 E_x^2$ (i.e., to $a^2 E_x^2$), we can determine a. In this manner, values of a between 3.4 and 4.4 Å^2 were found in [2.155].

f) LO/TO Intensity Ratio in Zincblende

In polar materials, ir active Raman phonons split into an LO and a TO component as discussed in Sect. 2.1.12. Let us consider the simplest possible case, namely, that of a zincblende-type crystal. The LO and the TO Raman intensities are not determined exclusively by the polarizability a. An additional parameter, namely, the Faust-Henry coefficient C of (2.93) is needed. The ratio of the LO to the TO intensity is given by the square of (2.94) multiplied by a factor which takes care of the scattering geometry. For a (100) face (see

Table 2.2) only LO-phonons are allowed, while for a (110) face only TO-phonons are allowed. Hence, these faces are not good to determine, without changing the scattering geometry, the ratio d_{LO}/d_{TO} (we assume back scattering). For the (111) surface, however, we see simultaneously both LO and TO-phonons and therefore C can be easily determined from the ratio of their intensities. Once this is done, $d\chi/du$, and hence a (2.67) can be found with (2.93), provided the effective charge e^* and the electro-optic coefficient $\partial\chi/\partial E$ is known. The effective charge is related to the LO-TO splitting through (2.86a). In this manner, *Flytzanis* obtained for the polarizability a in the limit of $\omega_L \ll$ the E_0 energy gap, $63 \pm 10 \, \text{Å}^2$ for GaAs and 30 ± 7 for GaP [2.156].

g) Sign of the Raman Polarizability a

The measurement of the Raman efficiency yields, according to (2.51, 134a), only the magnitude of the Raman susceptibility or equivalently, in the case of diamond, zincblende, or fluorite, the magnitude of the polarizability a. Its sign, however, also has a clear physical meaning. The sign of a represents whether the polarizability along (111) increases ($a > 0$) or decreases ($a < 0$) when the separation between the two atoms at (000) and $a/4$(111) is increased along (111). A first principles calculation, such as that carried out in [2.157] (also described in Sect. 2.2.6) yields, if reliable, the sign of a. The sign of a acquires experimental relevance whenever the associated Raman line interferes co-herently with a Raman continuum. In this case, the line shape becomes asymmetric and the direction of the asymmetry is determined, in part, by the sign of a. These types of phenomena are discussed in Chap. 2 of [1.3]. By these experimental means it has been established that for Si, $a > 0$, in agreement with calculations [2.157, 158]. The theoretical expression for diamond

$$a(\text{Å}^2) = -6.5 \, G(\omega_L/\omega_g), \tag{2.135}$$

with $\omega_g = 5.6 \, \text{eV}$ and G the function given below in (2.174), also yields $a > 0$ as G is negative for $\omega_L < \omega_g$ [2.54] (note, however, that according to the calculation in [2.157], the sign of a should be negative for diamond). The *magnitude* of (2.135) has been plotted in Fig. 2.30 together with all experimental determinations of $|a|$ available for diamond. Theory, as much as the analysis of the ratio of LO to TO intensities in terms of (2.94), also yield for most III–V and II–VI compounds a positive a. Similarly a recent simple-minded theoretical calculation for CaF_2, SrF_2, and BaF_2 yields for these materials $a > 0$ [2.49] (Sect. 2.2.7). We note that the differential polarizability of the stretching mode of most linear molecules (H_2, N_2, CS_2) has also been calculated to be positive [2.158a]; in view of our results we suspect this to be a rather general fact. The problem of the sign of the "Brillouin" tensor components is much easier. While the Brillouin intensities yield only the magnitudes of elasto-optic constants, direct piezobirefringence (i.e. *interferometric*) experiments also yield their sign. For a recent compilation of elasto-optic constants, their signs, and their dispersion, see [2.91].

2.2 Quantum Theory

In Sect. 2.1 we treated the phenomenon of light scattering by matter from a classical point of view, introducing here and there some quantum mechanical concepts which were needed for rounding off the discussion. Various aspects of the quantum theory of light scattering have been treated in many standard works [2.21, 159] including Chaps. 1, 3, 4 in [2.1]. We discuss here the main quantum theoretical concepts needed for an understanding of light scattering in simple solids, in particular, semiconductors. This includes a number of facts which are only found scattered through the recent literature. Emphasis is placed on unifying the various approaches and in showing their mutual relationship and, when appropriate, their equivalence. The equivalence to the dielectric formulation of Sect. 2.1 is also emphasized. A more formal, Green's functions formulation of the theory of light scattering will be found in [Ref. 1.2, Chap. 5].

2.2.1 Hamiltonian for Molecules and Solids in Interaction with an Electromagnetic Field

This Hamiltonian can be written as

$$\mathcal{H} = \mathcal{H}_e + \mathcal{H}_i + \mathcal{H}_{ei} + \mathcal{H}_p + \mathcal{H}_{ep} + \mathcal{H}_{ip}, \tag{2.136}$$

where $\mathcal{H}_e$ is the Hamiltonian of the electrons, $\mathcal{H}_i$ that of the ions, $\mathcal{H}_p$ that of the photons, $\mathcal{H}_{ei}$ representing the electron-ion interaction, $\mathcal{H}_{ep}$ that of electrons and photons and $\mathcal{H}_{ip}$ the ion-photon interaction. It is customary to treat part of (2.136) exactly and the rest by perturbation theory. Depending on what part is treated exactly, the various possible approximations arise.

In one type of treatment $\mathcal{H}_e$ is renormalized into $\mathcal{H}'_e$ by using the adiabatic approximation for $\mathcal{H}_e + \mathcal{H}_i + \mathcal{H}_{ei}$. $\mathcal{H}'_e + \mathcal{H}_p + \mathcal{H}_{ep}$ is then solved exactly to obtain the so-called exciton-polariton states, analogous to the phonon-polariton states described in Sect. 2.1.12. The phonons and the *electron-phonon interactions* arising from the adiabatic approximation are then treated by perturbation theory. This treatment becomes necessary whenever the optical absorption spectrum contains sharp and strong lines (excitons, *strong coupling* to photons). The entities which scatter within the solid are no longer the photons but the corresponding polaritons (mixed photon-excitons). The scattering phenomenon depends then in a very explicit way on the boundary conditions at the solid-vacuum interface which affect the transformation of the observable incident and scattered photons into the polaritons. These boundary conditions are still the object of considerable controversy [2.160].

The polariton picture, as a strong coupling description, becomes necessary to treat light scattering for ω_L or ω_s near strong excitonic resonances (resonant polariton scattering). Its theory has been reviewed by *Bendow* [2.161]. Because of the difficulties involved in the boundary conditions, it is not easy to identify resonant polariton phenomena in Raman scattering. In Brillouin scattering, however, resonant polariton effects are seen as strong deviations of (2.105) which involve only the kinematic conservation laws (independent of boundary conditions). This type of phenomena are described in Chap. 7 of [1.2] and will not be discussed here any further.

Another type of approximation, particularly useful for simple molecules, is to treat $\mathscr{H}_e + \mathscr{H}_i + \mathscr{H}_{ei}$ exactly within the adiabatic (or the harmonic) approximation [2.162]. The photon hamiltonian $\mathscr{H}_p$ is treated exactly and the electron-photon interaction between mixed electronic vibrational states by perturbation theory. The ion-photon interaction, or its renormalized version the phonon-photon interaction, which produces the so-called ionic Raman effect [2.163] is usually negligible except at very low laser frequencies. At these frequencies, however, the phenomenon becomes inobservable on account of the ω_s^4 law and of problems associated with sources and detectors. At visible frequencies ($\omega \gg \omega_v$), the ionic polarizabilities (2.82) have relaxed away and thus their contribution to the Raman tensors are negligible except for the effect on LO-phonon intensities discussed in Sect. 2.1.12.

The third type of treatment [2.164] consists of solving $\mathscr{H}_e + \mathscr{H}_i + \mathscr{H}_{ei}$ and calculating electron and phonon eigenstates and eigenvectors with the adiabatic approximation. The ground and the excited electronic states are then taken to be those of the ground state equilibrium positions of the ions R_0 and the electron-phonon interaction (i.e., the changes of these states and their energies with R) is treated by perturbation theory. This approximation is generally appropriate for solids and large molecules. In these cases, the interaction of *each* of the large number of delocalized phonons with the partly delocalized electronic states is small and the use of perturbation theory is justified.

In order to simultaneously describe spontaneous and stimulated light scattering phenomena, it is convenient to treat the photons in second-quantized notation. The photons are bosons with creation and annihilation operators $a_k^\dagger$ and a_k, respectively. Their Hamiltonian $\mathscr{H}_p$ becomes

$$\mathscr{H}_p = \sum_k \tfrac{1}{2}\omega(k)(a_k^\dagger a_k + a_k a_k^\dagger). \tag{2.137}$$

Equation (2.137), like all subsequent equations, has been written in *atomic units*, i.e., for $\hbar = |e| = m = 1$. In these units the speed of light is the reciprocal of the fine structure constant $\hbar c/e^2 = 137$. The unit of energy is the double-Rydberg or Hartree (27.2 eV) and the unit of length the Bohr (0.53 Å). The permittivity of vacuum equals, in this case, $(4\pi)^{-1}$ ("rationalized" atomic units). For simplicity we have omitted from (2.137) the photon polarization index (for each k there are two orthogonal polarization modes).

By taking the thermodynamic expectation value of (2.137), we find

$$\langle \mathscr{H}_p \rangle_k = \omega(k)(n_k + \tfrac{1}{2}) = V\varepsilon_0 \langle E^2 \rangle_k$$

$$= V\frac{\omega^2}{c^2}\langle A^2 \rangle_k, \tag{2.138}$$

where n_k is, in thermal equilibrium, the statistical factor of (2.25) with ω_v replaced by $\omega(k)$ and V the volume under consideration. Equations (2.138) enable us to calculate the average value (fluctuations) of the electric field and the vector potential $\langle E^2 \rangle_k$ and $\langle A^2 \rangle_k$ either in thermal equilibrium or for a given laser excitation. For $\omega(k)$ in the visible or near infrared and without high laser excitation, $n_k \simeq 0$ except at unusually high temperatures. For spontaneous scattering experiments the scattered photons fulfill these conditions. If a strong laser beam of wave vector k is present, n_k deviates from the thermal equilibrium value. Its value is then the average number of photons pumped by the laser in the mode k [given as a function of the average field by (2.138)]. If this laser has a frequency equal to that of a *scattered beam*, stimulated scattering results. The electric field operator at the point r, $\hat{E}(r)$ and correspondingly, $\hat{A}(r)$, can be obtained by inspection of (2.137, 138) and using Maxwell's equations:

$$\hat{A}(r) = \sum_k \left(\frac{1}{2\varepsilon_0 V \omega(k)} \right)^{1/2} \hat{e}_k (a_k e^{ik \cdot r} + a_k^{\dagger} e^{-ik \cdot r})$$

$$\hat{E}(r) = i \sum_k \left(\frac{\omega(k)}{2\varepsilon_0 V} \right)^{1/2} \hat{e}_k (a_k e^{ik \cdot r} - a_k^{\dagger} e^{-ik \cdot r}),$$

(2.139)

where $\hat{e}_k$ represents the polarization vectors of the electric fields and the equations apply to photons propagating in free space. For a medium with a nondispersive and real dielectric constant (or with a relatively small imaginary part), this dielectric constant, actually resulting from the term $\mathscr{H}_{ep}$ in (2.136), can be taken into account through renormalization of (2.139) simply by replacing ε_0 by ε. Any strongly dispersive contribution to ε must be treated, in principle, in the polariton picture mentioned above.

The term $\mathscr{H}_{ep}$ in (2.136) is easily obtained from the kinetic energy contribution to $\mathscr{H}_e$:

$$\tfrac{1}{2} \sum_j [(p_j + \hat{A}(r_j)]^2 = \tfrac{1}{2} \sum_j (p_j)^2 + \tfrac{1}{2} \left(\sum_j p_j \cdot \hat{A}(r_j) + \hat{A}(r_j) \cdot p_j \right) + \tfrac{1}{2} \sum_j [\hat{A}(r_j)]^2 .$$

Hence,

$$\mathscr{H}_{ep} = \mathscr{H}_{ep}' + \mathscr{H}_{ep}''$$

$$\mathscr{H}_{ep}' = \tfrac{1}{2} \sum_j (p_j \cdot \hat{A}(r_j) + \hat{A}(r_j) \cdot p_j)$$

(2.140)

$$\mathscr{H}_{ep}'' = \tfrac{1}{2} \sum_j [\hat{A}(r_j)]^2 .$$

In (2.140) the summation index j runs over all electrons. The electron-photon Hamiltonian $\mathscr{H}_{ep}$ has two terms: $\mathscr{H}_{ep}'$, linear in the electromagnetic field, and a quadratic term $\mathscr{H}_{ep}''$. The expressions of $\mathscr{H}_{ep}$ in terms of a_k and $a_k^{\dagger}$ are easily obtained by replacing (2.139) into (2.140). We find for Stokes scattering

$$
\mathcal{H}'_{ep} = \sum_j \left(\frac{1}{2V\varepsilon_0}\right)^{1/2} [\omega_L^{-1/2} a_{k_L} e^{i k_L \cdot r_j} \hat{e}_L \cdot (p_j + \tfrac{1}{2} k_L)
$$
$$
+ \omega_s^{-1/2} a_{k_s}^\dagger e^{-i k_s \cdot r_j} \hat{e}_s \cdot (p_j - \tfrac{1}{2} k_s)]
$$
$$
\mathcal{H}''_{ep} = \sum_j \frac{1}{2V\varepsilon_0 \omega_s^{1/2} \omega_L^{1/2}} \hat{e}_s \cdot \hat{e}_L e^{i(k_L - k_s) \cdot r_j}. \tag{2.141}
$$

2.2.2 Electronic Raman Scattering

Let us first consider the case of a free-electron-like metal and neglect the presence of phonons. The electronic part of the Hamiltonian and its wave functions are:

$$
\mathcal{H}_e = \sum_k \tfrac{1}{2} \omega_c(k)(c_k^\dagger c_k + c_k c_k^\dagger)
$$
$$
\phi(k) = \frac{1}{\sqrt{V}} e^{i k \cdot r}; \ \omega_c(k) = \frac{k^2}{2}, \tag{2.142}
$$

where $c_k^\dagger$, c_k represent creation and annihilation operators for electrons. The state of the system is determined by the temperature and the Fermi "energy" ω_F. We shall perform a calculation of the scattering efficiency due to electronic excitations in the case $T \simeq 0$. A generalization to finite temperatures requires the multiplication of the result by $[1 + n(\omega_R)]$, where $n(\omega_R)$ is the Bose-Einstein factor which corresponds to the scattering energy ω_R [Ref. 2.1, Eq. 4.10].

The term $\mathcal{H}'_{ep}$ of (2.141) produces electronic scattering of light in second-order perturbation theory while $\mathcal{H}''_{ep}$ does it in first-order. A straightforward calculation of the corresponding scattering probabilities yields a ratio of the first-order to the second-order term:

$$
\frac{I''_s}{I'_s} \simeq \left|\frac{\omega_L}{\omega_F}\right|^2. \tag{2.143}
$$

For standard heavily doped semiconductors, $\omega_F \simeq 0.1$ eV, while $\omega_L \simeq 2$ eV in ordinary experiments. Hence the $\mathcal{H}''_{ep}$ first-order perturbation term dominates the scattering efficiency *if interband processes are neglected* (we shall consider them below).

Let us treat the simplest case of one single electron at the bottom of the conduction band (2.142). The spontaneous scattering probability per unit time is obtained by applying the "golden rule" to the Hamiltonian $\mathcal{H}''_{ep}$ (2π multiplied by the square of the matrix element between initial and final state) and multiplying by the density of final states, i.e., the density of photons of frequency ω_s which equals $(8\pi^3)^{-1} k_s^2 dk_s d\Omega$. In this manner, we again find the result of (2.9). This result, obtained for the case of *one electron*, only remains valid for a finite free electron density if this density is small. If the density becomes large, the *longitudinal* perturbation represented by $\mathscr{H}''_{ep}$ in (2.141) is strongly altered by the collective polarization of the free electrons in a manner similar to that described in Sect. 2.2.12 for LO-*phonons*. The result of (2.9) must be modified as shown in [Ref. 2.1, p. 150]. The scattering efficiency becomes [Ref. 2.1, Eq. (4.10b)]

$$\frac{\partial^2 S}{\partial \omega_R \partial \Omega} = -\left(\frac{\omega_s}{\omega_L}\right)^2 (\hat{e}_L \cdot \hat{e}_s)^2 r_0^2 [1+n(\omega_R)] \frac{q_R^2 \varepsilon_0}{4\pi} \operatorname{Im} \frac{1}{\varepsilon(\omega_R \cdot q_R)} . \tag{2.144}$$

where $\varepsilon(\omega_R \cdot q_R)$ represents the dielectric constant of the free electron gas plus any possible background. The *longitudinal* nature of the free electron excitations produced is clearly revealed by the appearance of $\operatorname{Im}\{\varepsilon^{-1}\}$ in (2.144) (Sect. 2.1.12). This function has a maximum at the plasma frequency of the free electron system. The scattering is *polarized* ($\hat{e}_L \| \hat{e}_s$).

As we have shown in (2.143), the term $\mathscr{H}'_{ep}$ is usually negligible for scattering by *free electrons*. However, in real solids, in particular semiconductors, there are strong interband matrix elements of p. In this case, the *second-order* perturbation terms must be included in the scattering *amplitude* (before squaring it to get the efficiency). The result is, in the dipole approximation $k_{L,s} \simeq 0$, similar to (2.9) with $\hat{e}_L \cdot \hat{e}_s$ replaced by

$$\hat{e}_L \cdot \hat{e}_s + \sum_{i \neq c} \langle c|\hat{e}_s \cdot p|i\rangle \langle i|\hat{e}_L \cdot p|c\rangle \left\{ \frac{1}{\omega_i - \omega_c - \omega_L} + \frac{1}{\omega_i - \omega_c + \omega_s} \right\}. \tag{2.145}$$

Equation (2.145) exhibits resonances for $\omega_L \simeq \omega_i - \omega_c$. If ω_L is much smaller than all gaps $\omega_i - \omega_c$ contributing to the sum of (2.145), this expression becomes, by setting $\omega_L \simeq \omega_s \simeq 0$,

$$\hat{e}_L \cdot \underset{\sim}{m}_e^{-1} \cdot \hat{e}_s,$$

where

$$\underset{\sim}{m}_e^{-1} = \mathbb{1} + \sum_{i \neq c} \frac{2\langle c|p|i\rangle \langle i|p|c\rangle}{\omega_i - \omega_c} . \tag{2.146}$$

The above relationship between the electron *effective mass tensor* $\underset{\sim}{m}_e$ and the matrix elements of p is the well-known $k \cdot p$ relation for the effective mass [2.48]. In cubic tetrahedral semiconductors with a conduction band minimum at $k=0$, the effective mass tensor becomes isotropic and the scattering remains polarized (for an anisotropic m_e^{-1}, unpolarized scattering is possible). Usually one single term dominates in the sum over i. For GaAs, InSb and other tetrahedral semiconductors with minima at $k=0$, one has

$$m_e \simeq 1 + \frac{2P^2}{\omega_0}, \tag{2.147}$$

where ω_0 is the $k=0$ gap in eV and $2P^2 \simeq 25$ eV.

We would like to discuss, before closing this section, the extremely resonant case $\omega_i - \omega_c \simeq \omega_L$ in (2.145) and use as an example for $\omega_i - \omega_c$ the so-called $E_0 + \Delta_0$ gap of GaAs, i.e., for ω_c, the spin-orbit-split Γ_7 valence band state of this material. The wave functions of this state have $J = 1/2$, $J_z = \pm 1/2$ symmetry [2.48]:

$$(\tfrac{1}{2}, \tfrac{1}{2}) = \frac{1}{\sqrt{3}}(X + iY)\downarrow + \frac{1}{\sqrt{3}}Z\uparrow$$

$$(\tfrac{1}{2}, -\tfrac{1}{2}) = \frac{1}{\sqrt{3}}(X - iY)\uparrow - \frac{1}{\sqrt{3}}Z\downarrow. \tag{2.148}$$

Since the wave functions of (2.148) are a mixture of spin-up and spin-down they can, when used as intermediate state i in (2.145), connect states of spin-up with states of spin-down in the conduction band, i.e., produce spin-flip scattering. In order to do this we must have $\hat{e}_L \perp \hat{e}_s$, that is, for instance, $\hat{e}_s$ along z and $\hat{e}_L$ along x [the conduction band is s-like and hence $\hat{e}_s$ or $\hat{e}_L$ must be along x to couple to the X part of (2.148)]. Near resonance, (2.145) is proportional to

$$\frac{2P^2/3}{\omega_{E_0 + \Delta_0} - \omega_L}(\hat{e}_L \times \hat{e}_s), \tag{2.149}$$

where, as mentioned earlier, $2P^2 \simeq 25$ eV. The scattering efficiency is proportional to the square of (2.149). Note that the Raman tensor which corresponds to (2.149) is *completely antisymmetric*, a vector product being equivalent to an antisymmetric tensor.

Resonant spin-flip scattering near $E_0 + \Delta_0$ for n-type GaAs is discussed in [2.165] and [Ref. 1.3, Chap. 2]. In Fig. 2.31, we illustrate the discussion above with the resonant behavior observed for a sample with $n = 7 \times 10^{17}$ electrons $\times$ cm^{-3}. Note that the depolarized scattering of (2.149) is only possible in this case of isotropic effective mass because the resonance condition singles out only one term in the sum of (2.145).

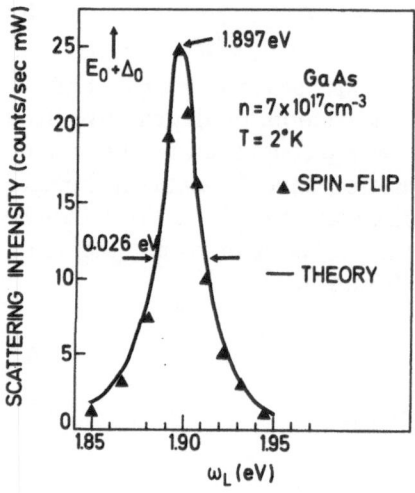

Fig. 2.31. Resonant spin-flip scattering observed for ω_L near the $E_0 + \Delta_0$ gap of GaAs [2.165]

2.2.3 Scattering by Phonons: Frank-Condon Formulation

Let us consider light scattering bringing a solid or a molecule from a ground state Ψ_0 to an excited state Ψ_f and treat these states in the adiabatic approximation, i.e., let us write Ψ as a product of an electronic wave function $\psi(r, R)$ times a vibrational function $\phi_v(R)$:

$$\Psi(r, R) = \psi(r, R)\phi_v(R), \tag{2.150}$$

where v denotes the vibrational state under consideration. We shall treat here scattering by phonons from a vibrational state n_0 to a state n_f. In this case, the electronic wave functions of the initial and final state are the same and the Hamiltonian $\mathscr{H}_{ep}''$ does not contribute to the scattering. The corresponding matrix element is zero because of the orthogonality of ϕ_0 and ϕ_v. Hence, the sole contribution to the scattering arises from $\mathscr{H}_{ep}'$. Using the golden rule in the same manner as in Sect. 2.2.2 (second-order perturbation theory), the Stokes cross section becomes

$$\frac{\partial^2 \sigma_s}{\partial \omega_R \partial \Omega} = \frac{\omega_s^2}{\omega_L^2} r_e^2 \left\langle \left| \sum_{i, n_i} \langle 0, n_f | e^{-i k_s \cdot r} \hat{e}_s \cdot (p - \tfrac{1}{2} k_s) | i, n_i \rangle \right. \right.$$

$$\left. \cdot \langle i, n_i | e^{i k_L \cdot r} \hat{e}_L \cdot (p + \tfrac{1}{2} k_L) | 0, n_0 \rangle \frac{1}{\omega_{i, n_i} - \omega_{0, n_0} - \omega_L} \right.$$

$$\left. + \text{nonresonant term (NRT)} \right|^2 \delta(\omega_{0, n_f} - \omega_{0, n_0} - \omega_R) \right\rangle_{n_0, n_f}, \tag{2.151}$$

where the angular brackets represent the thermodynamic average over n_0 and n_f. In (2.151) we have written only the so-called resonant terms explicitly. The nonresonant terms (NRT) are obtained from the resonant terms by permuting $\hat{e}_L$, k_L with $\hat{e}_s$, k_s and changing $-\omega_L$ into $+\omega_s$.

Equation (2.151) can be simplified somewhat by making the dipole approximation, i.e., by assuming $k_s \simeq k_L \simeq 0$. It is then possible to transform the scattering efficiency expression into one containing matrix elements of the electronic coordinate r instead of the operator p. This involves a simple gauge transformation which consists of adding to the vector potential $A\hat{e}$ the term $-\text{grad}\,\Theta$ with $\Theta = (\hat{e} \cdot r)A \exp[i(k \cdot r - \omega t)]$. The transformed Hamiltonian contains no vector potential and, in its stead, the scalar potential $(\partial/\partial t)(\hat{e} \cdot r)A \exp[i(k \cdot r - \omega t)]$. The scattering cross section thus becomes

$$\frac{\partial^2 \sigma_s}{\partial \omega_R \partial \Omega} = r_e^2 \omega_s^4 \left\langle \left(\sum_{i, n_i} \langle 0, n_f | \hat{e}_s \cdot r | i, n_i \rangle \times \langle i, n_i | \hat{e}_L \cdot r | 0, n_0 \rangle \right. \right.$$

$$\left. \left. \cdot \frac{1}{\omega_{i, n_i} - \omega_{0, n_0} - \omega_L} + \text{NRT} \right)^2 \delta(\omega_{0, n_f} - \omega_{0, n_0} - \omega_R) \right\rangle_{n_0, n_f}. \tag{2.152}$$

By comparing (2.152) with (2.151), we note that while (2.152) contains the ω_s^4 factor required by the polarizability theory for small ω_L (2.24), (2.151) does not. This apparently paradoxical results leads us to conclude that if one expands (2.151) in a power series of ω_L, all terms up to that of third order in ω_L must vanish. This point, which leads to the so-called *f*-sum rule for the terms of zero order in ω_L and to another similar sum rule for the terms of second order, has been discussed in [2.50]. One must be particularly careful when truncating the sums in (2.151), i.e., when using a finite number of intermediate states as then these sum rules need not be fulfilled and the ω_s^4 dependence may not be found.

Equations (2.152) is particularly useful for the calculation of $\partial^2 S/\partial\omega_R\partial\Omega$ in a localized system such as a molecule: in nonlocalized systems one runs into difficulties as r diverges. Hence, for *periodic* solids it is more convenient to express the scattering efficiency in terms of the matrix elements of p as done in (2.151). A comparison of (2.151, 152) with (2.24, 51) enables us to write explicit expressions for the Stokes "transition polarizability" $\underset{\sim}{\alpha}_s$ and the "transition susceptibility" $\underset{\sim}{\chi}_s$ defined as:

$$\underset{\sim}{\alpha}_s = \frac{\partial\underset{\sim}{\alpha}}{\partial\xi}\langle\xi\xi^\dagger\rangle^{1/2}$$

$$\underset{\sim}{\chi}_s = \frac{\partial\underset{\sim}{\chi}}{\partial\xi}\langle\xi\xi^\dagger\rangle^{1/2}.$$

(2.153)

We find

$$\underset{\sim}{\alpha}_s = \left\langle \sum_{i,n_i} \langle 0, n_0+1|r|i, n_i\rangle \times \langle i, n_i|r|0, n_0\rangle \right.$$
$$\left. \cdot\frac{1}{\omega_{i,n_i} - \omega_{0,n_0} - \omega_L} + \mathrm{NRT} \right\rangle$$

$$\underset{\sim}{\chi}_s = \frac{4\pi V^{-1}}{\omega_s\omega_L}\left\langle \sum_{i,n_i} \langle 0, n_0+1|p|i, n_i\rangle \times \langle i, n_i|p|0, n_0\rangle \right.$$
$$\left. \cdot\frac{1}{\omega_{i,n_i} - \omega_{0,n_0} - \omega_L} + \mathrm{NRT} \right\rangle.$$

(2.154)

In (2.154) we have expressed the transition polarizability $\underset{\sim}{\alpha}_s$ (usually utilized for molecules) as a function of matrix elements of r, and $\underset{\sim}{\chi}_s$ (used for solids) as a function of matrix elements of p. A sum over all electrons is implicit in both equations. Also, an imaginary frequency should be added to the denominators of (2.154) in order to take into account the finite lifetime of the intermediate state (we have omitted this "damping term" throughout most of the equations. It is implicitly understood).

Let us consider the expression for $\underset{\sim}{\alpha}_s$ in (2.154) and use the wavefunction of the Born-Oppenheimer approximation (2.150) for the eigenstates. We define the

R-dependent dipole matrix elements $M_{i0}(R)$ as

$$M_{i0}(R) = \int \psi_i^*(r, R) r \psi_0(r, R) dr. \tag{2.155}$$

By expanding $M_{i0}(R)$ in a power series of R around the equilibrium position R_0, we find

$$M_{i0}(R) = M_{i0}(R_0) + \left(\frac{\partial M_{i0}}{\partial R_\xi}\right)_{R_0} R_\xi + \ldots, \tag{2.156}$$

where R_ξ can be chosen to be a *normal coordinate*, see (2.20). By replacing (2.156) into (2.154), we find [2.162, 166]

$$\underset{\sim}{\alpha} = \langle \underset{\sim}{A} \rangle + \langle \underset{\sim}{B} \rangle + \text{NRT}$$

$$\underset{\sim}{A} = \sum_i M_{i0}(R_0) \times M_{i0}(R_0) \sum_{n_i} \langle n_f | n_i \rangle \langle n_i | n_0 \rangle \frac{1}{\omega_{i,n_i} - \omega_{0,n_0} - \omega_L}, \tag{2.157a}$$

$$\underset{\sim}{B} = \sum_{i,n_i} \left[M_{i0}(R_0) \times \left(\frac{\partial M_{i0}}{\partial R_\xi}\right)_{R_0} \frac{\langle n_f | n_i \rangle \langle n_i | R_\xi | n_0 \rangle}{\omega_{i,n_i} - \omega_{0,n_0} - \omega_L} \right.$$
$$\left. + \left(\frac{\partial M_{i0}}{\partial R_\xi}\right)_{R_0} \times M_{i0}(R_0) \frac{\langle n_f | R_\xi | n_i \rangle \langle n_i | n_0 \rangle}{\omega_{i,n_i} - \omega_{0,n_0} - \omega_L} \right]. \tag{2.157b}$$

The term $\underset{\sim}{A}$ in (2.157a) contains scalar products of vibrational functions in the initial and the intermediate electronic states. For $n_0 = n_f$, (2.157) represents the standard polarizability of the system including mixed vibronic-electronic excitations. For $n_i \neq n_f$, it represents the so-called Frank-Condon Raman polarizability. Away from resonance the energy denominators in (2.157a) can be taken out of the sum over n_i. This sum then becomes, using the completeness relation for vibrational eigenstates,

$$\sum_i \langle n_f | n_i \rangle \langle n_i | n_0 \rangle = \langle n_f | n_0 \rangle = \delta_{0,f} \tag{2.158}$$

and the $\underset{\sim}{A}$-term only contributes to elastic scattering. This situation is no longer true very close to resonance. For *small molecules*, the centers R_0 of the vibrational functions are shifted considerably from the initial state to an excited state and the overlap does not vanish [even without shift, a nonvanishing overlap can result in (2.158) if the vibrational frequencies of initial and intermediate states are different]. Multiphonon processes can thus result from the $\underset{\sim}{A}$-terms near resonance. For solids and large molecules, however, R_0 is nearly the same in ground and excited states and so is the corresponding vibrational frequency. In this case, the vibrational functions of initial and intermediate states are orthonormal and no Raman scattering usually results from the $\underset{\sim}{A}$-terms. In any case, since the wave function of the ground vibrational state is completely symmetric, that of the intermediate vibrational state $\langle n_i |$ and also that of $\langle n_f|$ must be completely symmetric if $\underset{\sim}{A}$ is not to vanish. Hence, only *symmetric* vibrations can be excited with the $\underset{\sim}{A}$-term. It rarely contributes to Raman scattering in solids, except may be, for strongly *localized vibrations* (see [2.166a] and Sect. 2.3.5).

In order to evaluate the $\underset{\sim}{B}$-term (the so-called Herzberg-Teller term), we must examine $(\partial M_{i0}/\partial R_\xi)_{R_0}$. This derivative arises through a mixture of the intermediate state i with other states j produced by the potential change induced by the phonon displacement, i.e., by the electron-phonon interaction Hamiltonian H_{ev}. Using first-order perturbation theory, we can write

$$\left(\frac{\partial M_{i0}}{\partial R_\xi}\right)_{R_0} = \sum_j M_{j0}(R_0)h_{ij}, \tag{2.159}$$

where

$$h_{ij} = \frac{\langle i|\mathscr{H}_{ev}|j\rangle}{\omega_i - \omega_j}.$$

Replacing (2.159) into the expression for $\underset{\sim}{B}$ in (2.157) and symmetrizing with respect to i and j, we find

$$\begin{aligned}
B = \sum_{i,j,n_i} \Bigg\{ &M_{i0} \times M_{j0} \frac{\langle n_f|n_i\rangle \langle n_i|R_\xi|n_0\rangle \langle i|\mathscr{H}_{ev}|j\rangle}{(\omega_i - \omega_j)(\omega_{i,n_i} - \omega_{0,n_0} - \omega_L)} \\
&+ M_{j0} \times M_{i0} \frac{\langle n_f|R_\xi|n_i\rangle \langle n_i|n_0\rangle \langle j|\mathscr{H}_{ev}|i\rangle}{(\omega_i - \omega_j)(\omega_{i,n_i} - \omega_{0,n_0} - \omega_L)} \Bigg\}.
\end{aligned} \tag{2.160}$$

Away from strict resonance, the sums over n_i can be performed by using the completeness relation for vibrational eigenstates. The matrix element $\langle n_f|R_\xi|n_0\rangle = \delta_{n_0, n_f \pm 1}$ then appears in (2.160) and the B-term yields the standard one-phonon Stokes $(n_f = n_0 + 1)$ and antistokes $(n_f = n_0 - 1)$ scattering. No strong deviations from this selection rule are expected near resonance, especially if the lifetime broadening of the intermediate state is $\gtrsim \omega_v$.

Equations (2.160) can be written in a more compact form if we assume that the eigenstates $|n_i\rangle$ and $|n_f\rangle$ are the same as $|n_0\rangle$, i.e., that the potential energy curves for the vibrational eigenstates are the same in the initial as in the excited electronic states (i.e., no shift and no vibrational frequency change. For a clarification of this point, the so-called static approximation, see [2.166b]). We must then have in (2.160) $n_i = n_0$ and $n_f = n_0 + 1$ for Stokes scattering. Replacing $i \rightleftarrows j$ in the second term of the rhs of (2.160), we find for Stokes scattering

$$\underset{\sim}{B} = \sum_{i,j} M_{i0} \times M_{j0} \frac{\langle n_0 + 1|R_\xi|n_0\rangle \langle i|\mathscr{H}_{ev}|j\rangle}{(\omega_{j,n_0} - \omega_{0,n_0} - \omega_L)(\omega_{i,n_0} - \omega_{0,n_0} - \omega_s)}. \tag{2.161}$$

If we make the approximation $\omega_L = \omega_s$, (2.161) can be easily related to the derivative of the linear polarizability with respect to the phonon amplitude, obtained from the A-term of (2.157a) for $n_f = n_0$. In this manner, we finally recover the "classical" expression (2.23). However, near a resonance ($\omega_L \simeq \omega_j$ or $\omega_s \simeq \omega_j$) we cannot assume $\omega_L = \omega_s$; the tensor $\underset{\sim}{B}$ thus possesses an *antisymmetric* component. Hence, we encounter for the first time the antisymmetric Raman tensors for scattering by phonons mentioned in Sect. 2.1.9.

The treatment of the Frank-Condon terms $\underset{\sim}{A}$ and the Herzberg-Teller terms $\underset{\sim}{B}$ given above is basically that of *Albrecht* [2.162]. It does not include nonadiabatic terms such as those in which R_{ς} connects two vibrational excited states. These terms, which also include coupling by *phonons* between the ground state and an excited state, have been discussed by a number of authors [2.167]. They introduce corrections of the order of $\omega_v/(\omega_i - \omega_j)$ which, in some special cases, can be appreciable: their effect is to shift scattering efficiency from the ω_L (*in-going*) to the ω_s (*out-going*) resonance of (2.161) or vice versa, depending on the sign of the relevant energy difference $\omega_j - \omega_i$ [2.166a]. Such effects have been identified in the Raman spectra of metalloporphyrins [2.166a].

For $\omega_j \simeq \omega_i$, (2.161) yields a result equivalent to the first term in the rhs of (2.30) provided one assumes $\omega_L \simeq \omega_s$, i.e., the requirement for the validity of the "classical" treatment. Very near resonance the condition $\omega_L \simeq \omega_s$ will not be valid. In this case, according to (2.161) two separate resonances appear, one for $\omega_L = \omega_j - \omega_0 = \omega_i - \omega_0$ (*in-going* resonance) and another of the same strength for $\omega_s = \omega_j - \omega_0 = \omega_i - \omega_0$ (*out-going* resonance). Both of these resonances occur together at the average frequency $\omega_L \simeq \omega_i - \omega_0 + \omega_v/2$ in most experiments with solids. Sometimes, however, both resonances can be seen separately. An example, observed for GaSe at 80 K [2.168], is shown in Fig. 2.32 together with a fit based on (2.161) with $\omega_j = \omega_i$ and with a broadening parameter for the intermediate state $\Gamma = 7\,meV$. Please note the contrast between this behavior and that encountered for pure electronic scattering (Fig. 2.31) in which only an in-going resonance is observed.

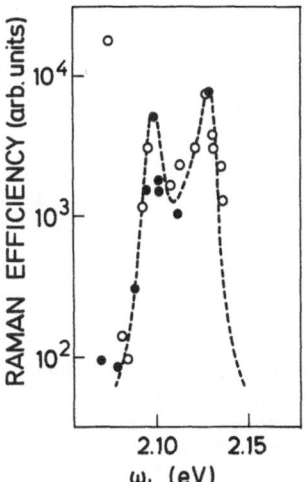

Fig. 2.32. In-going and out-going resonances observed for the 31.4 meV LO-phonon of GaSe at 80 K. The dashed curve is a fit with (2.161) including an imaginary part of ω_i [2.168]

2.2.4 Perturbation Theory for Electron-Phonon Interaction: First-Order Raman Effect

The treatment of Sect. 2.2.3 yields, within the adiabatic approximation, multiphonon scattering near resonance. For the A-processes, for instance, the scattering efficiency is determined by the overlap in the vibrational wave functions (Laguerre polynomials) of the ground and the excited electronic states. It is expected that these effects will be severely modified by $\underset{\sim}{B}$-terms and by nonadiabatic terms, especially in the multiphonon case. Under the assumption of equal potential energy curves for the ground and the excited state (static approximation [2.166b]), the A-terms yield only elastic scattering and the $\underset{\sim}{B}$-terms only one-phonon scattering. This case can be easily handled by treating the electron-phonon interaction completely by perturbation theory. This amounts to the evaluation of the probability for the processes of Fig. 2.33. The resulting six terms are given in (2.36) of [2.1]. These *six terms* amount to all possible third-order perturbation processes in which a photon is destroyed creating an electron-hole pair, a phonon is created (Stokes) while scattering the electron-hole pair, and a photon is created while annihilating the electron-hole pair.

Equation (2.36) of [2.1] is expressed in terms of the matrix elements of p and hence, as already mentioned, does not immediately lead to the ω_s^4 dependence of the scattering efficiency [2.50]. This dependence is easily recovered if each matrix element of p is replaced by [250]:

$$\langle j|p|i\rangle \to \frac{\omega_s^2}{(\omega_i-\omega_j)^2}\langle j|p|i\rangle. \tag{2.161a}$$

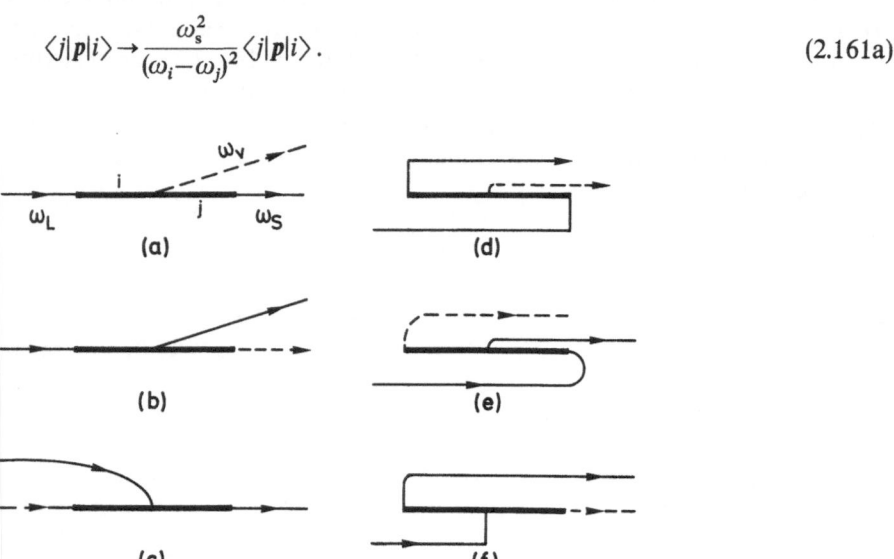

Fig. 2.33a–f. Diagrams which contribute to the Stokes scattering by one phonon in third-order perturbation theory. The corresponding expressions for the Raman susceptibility are given in (2.36) of [2.1]. The thick line represents the electronic excitation (exciton), the thin solid lines the photons and the dashed line the phonon

The first order Hamiltonian for the electron-phonon interaction ($\mathscr{H}_{EL}$ in [Ref. 2.1, Eq. (2.36)]) can be written as, in second-quantized notation,

$$\mathscr{H}_{ev} = \bar{\mathscr{E}}_{lk,v}(b_v^\dagger + b_v)a_l^\dagger a_k, \tag{2.162}$$

where $a_k^\dagger$, a_k represent creation and annihilation operators for electrons, respectively, and $b_v^\dagger$, b_v the corresponding operators for the phonons v. The deformation potential $\bar{\mathscr{E}}_{lk,v}$ can also be written as, see (2.26, 27),

$$\bar{\mathscr{E}}_{lk,v} = \sum_i^{\text{all atoms}} \langle l|\partial H/\partial u_i|k\rangle M_i^{-1/2}\hat{e}_i \sqrt{\frac{1}{2\omega_v}}. \tag{2.163}$$

The electron-phonon deformation potentials are found in the literature either in units of energy (like $\bar{\mathscr{E}}_{lk,v}$) or in energy per unit displacement (like $\langle l|\partial H/\partial u_i|k\rangle$). For materials with the zincblende structure, $\bar{\mathscr{E}}_{lk,v}$ is often renormalized and given as $\mathscr{E}_{lk,v}$ such that

$$\mathscr{H}_{ev} = a_0^{-1} \sqrt{\frac{1}{2N\mu\omega}} \mathscr{E}_{lk,v}(b_v^+ + b_v)a_l^+ a_k, \tag{2.164}$$

where μ is the reduced mass of the unit cell and a_0 the lattice constant. Sometimes the deformation potentials are defined with numerical factors of the order of one, other than those given in (2.164). The worker should identify them before making use of them to obtain *absolute values* of scattering efficiencies.

Figures 2.33a–c represent resonant processes, the most strongly resonant one being that of (a) for $i=j$. Figures 2.33d–f are nonresonant processes: we see that the characteristic of the latter is that the scattered photon is emitted before ω_L is absorbed. The most resonant form (a) with $i=j$ can be easily rewritten near resonance as (compare with [Ref. 2.1, Eq. (2.94)])

$$\underset{\sim}{\chi}_s = \frac{\chi(\omega_L)-\chi(\omega_s)}{\omega_v} \frac{d\omega_v}{d\xi}(n+1)^{1/2}\sqrt{\frac{1}{2\omega_v}} + \text{constant}, \tag{2.165}$$

where $\underset{\sim}{\chi}$ represents the standard linear susceptibility. The constant in (2.165) is the sum of less resonant and nonresonant contributions. Far away from the resonance we can make $\omega_v \to 0$ in (2.165). In this case, the finite difference ratio of (2.165) becomes the derivative of $\chi(\omega)$ with respect to ω for $\omega = \omega_L - \omega_v/2$. These are so-called two-band terms of Sect. 1.5. The formulation then becomes equivalent to the "classical" one of Sect. 1.2 in which the Raman susceptibility is the derivative of the linear susceptibility with respect to the normal coordinates (2.24, 47). The latter formulation is advantageous as it automatically yields all of the six contributions of Fig. 2.33 without having to do any tedious bookkeeping.

a) Parabolic Bands

Let us consider the case of intermediate states which correspond to transitions between a set of parabolic valence and conduction bands in a solid:

$$\omega_c = \tfrac{1}{2}k \cdot \left(\frac{1}{m_c}\right) \cdot k$$

$$\omega_{cv} = \omega_c - \omega_v = \omega_0 + \tfrac{1}{2}k \cdot \left(\frac{1}{m*}\right) \cdot k \qquad (2.166)$$

$$\omega_v = \tfrac{1}{2}k \cdot \left(\frac{1}{m_v}\right) \cdot k \qquad \left(\frac{1}{m*}\right) = \left(\frac{1}{m_c}\right) + \left(\frac{1}{m_v}\right),$$

where $(1/m*)$ is the reduced mass tensor for interband transitions and ω_0 the energy gap. The corresponding linear susceptibility tensor is obtained from (2.154) by neglecting the vibrational structure, taking for the intermediate states v→c excitations and replacing in the prefactor $(\omega_s\omega_L)^{-1}$ by ω_{cv}^{-2} [2.50]. By transforming the sum over i in (2.154) into an integral over ω_{cv}, we find

$$(4\pi)^{-1}\chi(\omega) = \int_0^\infty \frac{P \times P}{\omega_{cv}^2} N_d(\omega_{cv}) \left(\frac{1}{\omega_{cv} + i\Gamma - \omega} + \frac{1}{\omega_{cv} + i\Gamma + \omega}\right) d\omega_{cv}. \qquad (2.167)$$

We have included in (2.167) the factor $(4\pi)^{-1}$ so as to conform to the standard expressions given in the literature *in cgs units* for the susceptibility. We first treat the case of an M_0 *critical point* in which m is positive definite. The combined density of states $N_d(\omega_{cv})$ is then given by (including spin degeneracy)

$$N_d(\omega_{cv}) = \frac{2^{1/2}}{\pi^2} m_d^{3/2}(\omega_{cv} - \omega_0)^{1/2} \qquad \text{for} \quad \omega > \omega_0$$

$$= 0 \qquad \qquad \text{for} \quad \omega < \omega_0. \qquad (2.168)$$

In (2.168), m_d is the reduced density of states effective mass $m_d = (m_1^* m_2^* m_3^*)^{1/3}$, where m_i^* are the three principal components of the reduced mass tensor. The vector P in (2.167) represents an average matrix element of p over a surface $\omega_{cv} = $ constant and the $\times$ represents the tensor product. If assumed independent of ω_{cv}, $P \times P$ can be taken out of the integral in (2.167). Equation (2.167) is then usually evaluated for $\Gamma \to 0$ using the "golden rule"

$$\int_0^\infty \frac{g(x)}{x - a + i\Gamma} dx = \int_0^\infty \frac{g(x)}{x - a} dx + \pi i g(a), \qquad (2.169)$$
$$\Gamma \to 0$$

where $\int$ represents the Cauchy principal part of the integral. We obtain for the imaginary part of $\chi(\omega)$

$$(4\pi)^{-1}\chi_i(\omega) = \frac{2^{1/2}}{\pi}m_d^{3/2}P \times P\omega^{-2}(\omega-\omega_0)^{-1/2} \quad \text{for} \quad \omega > \omega_0$$

$$= 0 \qquad \qquad \qquad \text{for} \quad \omega < \omega_0. \qquad (2.170)$$

The corresponding real part of (2.170) is easily found by noting that χ must be an analytic function of ω which tends to zero for $\omega \to \infty$ and is well behaved for $\omega \to 0$. Its imaginary part must be (2.170). We can easily construct this function by inspection [2.48, 169]:

$$(4\pi)^{-1}\chi(\omega) = \frac{2^{1/2}}{\pi}m_d^{3/2}P \times P\omega^{-2}[2\omega_0^{1/2} - (\omega_0-\omega)^{1/2} - (\omega_0+\omega)^{1/2}]. \quad (2.171)$$

The signs of the square roots of (2.171) are determined by introducing a cut in the complex ω-plane from $\omega = 0$ to $-\infty$ and adding to ω_0 a small imaginary part $i\Gamma$ with $\Gamma > 0$ (always implicit).

Equation (2.171) can be rewritten in reduced frequency units $x = \omega/\omega_0$ in terms of the complex function $F(x)$:

$$(4\pi)^{-1}\chi(\omega) = C F(x) \qquad (2.172)$$

with

$$F(x) = x^{-2}[2 - (1-x)^{1/2} - (1+x)^{1/2}]$$

$$C = \frac{2^{1/2}}{\pi}m_d^{3/2}P \times P\omega_0^{-3/2}.$$

The real part of χ can thus be written

$$(4\pi)^{-1}\chi(\omega) = C f(x) \quad x < 1$$
$$= C x^{-2}[x^{1/2} - (1+x)^{1/2}] \quad x > 1 \qquad (2.173)$$
$$f(x) = x^{-2}[2 - (1-x)^{1/2} - (1+x)^{1/2}].$$

The function $f(x)$, and its analytic continuation for $x > 1$, is shown in Fig. 2.34 together with the corresponding imaginary part of $F(x)$. On account of the assumptions made (constancy of P and parabolicity extending to infinity), (2.172, 173) is only expected to be accurate for ω in the neighborhood of ω_0.

The two-band term in the Raman susceptibility can be easily evaluated from (2.171) with (2.165). We find, in the limit of $\omega_L \simeq \omega_s$,

$$(4\pi)^{-1}\chi_s(\omega) = \frac{C}{2\omega_0}G(x)\frac{d\omega_0}{d\xi}(n+1)^{1/2}(2\omega_v)^{-1/2}$$

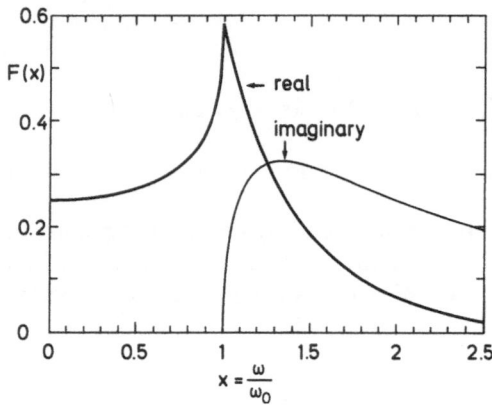

Fig. 2.34. Real and imaginary part of $F(x)$ representing the dependence of the susceptibility on reduced frequency near a 3-dimensional allowed critical point, see (2.172, 173)

with   (2.174)

$$G(x) = x^{-2}[2 - (1-x)^{-1/2} - (1+x)^{-1/2}].$$

The real and the imaginary parts of $G(x)$ are shown in Fig. 2.35.

We note that, strictly speaking, (2.174) is only meaningful for the most dispersive term $(1-x)^{-1/2}$. The other terms are of the same order as three-band terms and therefore they are not to be taken too seriously unless three band terms are included. As seen in Sect. 2.2.3, three-band terms represent the change in P produced by the phonon distortion ξ. Even if we assume P to be independent of ξ, m_d in (2.172) can be a function of ω_0, see (2.146). If the mass m_d is determined by the gap ω_0 according to a relationship of the type (2.146), $\mathcal{C}$ becomes independent of ω_0 and (2.174) must then be modified to be [2.169].

$$(4\pi)^{-1}\chi_s(\omega) = \frac{\mathcal{C}}{2\omega_0}[G(x) + 3F(x)]\frac{d\omega_0}{d\xi}(n+1)^{1/2}(2\omega_v)^{-1/2}.$$ (2.175)

The main difference between (2.175) and (2.174) is that (2.175) gives $\chi_s(0) = 0$, while a finite value for $\chi_s(0)$ is obtained from (2.174) as $G(0) = -3/4$ and $F(0) = 1/4$. However, as already mentioned, these constant "backgrounds" are not always meaningful. We show in Fig. 2.36 the function $|G(x)|^2$ which represents the two band resonance in the Raman efficiency according to (2.174) and the corresponding effect obtained without assuming $\omega_s = \omega_L$, i.e., the function [see (2.165)]

$$\Delta^{-2}\left|(1+\Delta)^{-3/2}F\left(\frac{x}{1+\Delta}\right) - (1-\Delta)^{-3/2}F\left(\frac{x}{1-\Delta}\right)\right|^2 \quad \text{for} \quad \Delta = \frac{\omega_v}{2\omega_0} = 0.04.$$
                                                                                          (2.176)

Figure 2.36 clearly indicates the types of errors which may be committed by using the "classical" theory. These errors disappear whenever a broadening of

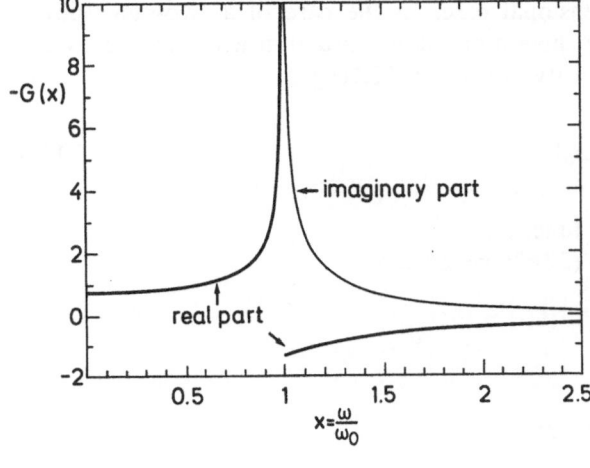

◀ **Fig. 2.35.** Real and imaginary parts of the function $G(x)$ of (2.174)

Fig. 2.36. Function $|G(x)|^2$ which represents, according to (2.174), the two-band resonance in the Raman scattering for a three-dimensional allowed direct gap in the case $\omega_v \ll |\omega_L - \omega_0|$. Also, the result of the more accurate expression (2.176) without this simplification for $\Delta = \omega_v/\omega_L = 0.04$

▼

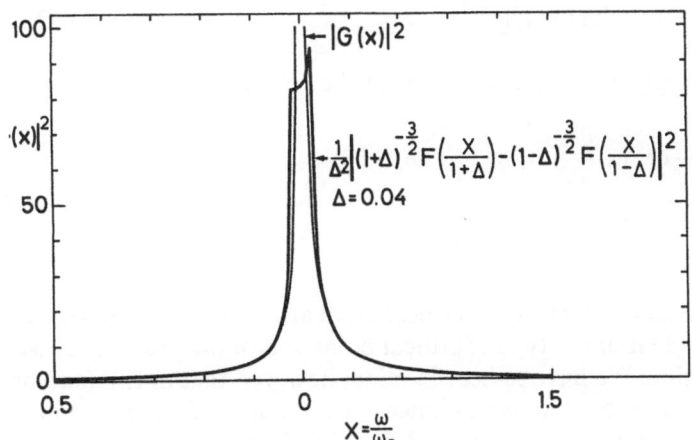

Fig. 2.36.

the electronic states $\Gamma > \omega_v$ is included. We note that very near resonance $(x \simeq 1)$ $|G(x)|^2 \sim |1 - x|^{-1}$.

The above treatment was performed for an M_0 critical point, i.e., around a minimum in ω_{cv}. For the other types of critical points (M_m) the treatment is basically the same except that the resulting F and G must be multiplied by $(i)^m$ [2.48]. Hence, the shape of the resulting Raman resonance, proportional to $|G|^2$, is the same regardless of the index m of the critical point. Also, weak excitonic interaction can be included by replacing m by $m + \alpha$ where $0 < \alpha < 1$ [2.48]. While this interaction strongly alters the absorption and reflection spectra, it does not change the two-band Raman resonance very near the critical point.

We have treated above the case of a parabolic *three-dimensional* interband critical point. Nonparabolicity can also be easily included in the treatment by means of the $\mathbf{k} \cdot \mathbf{p}$ method [2.169]. We shall also find occasion to use the results for the parabolic two-dimensional case. In the case of a two-dimensional minimum, equivalent to the three dimensional situation with $m_{\parallel}^* = \infty$ over a length of k-space Δk, the density of states is [2.169]

$$N_d(\omega_{cv}) = \frac{\Delta k}{\pi} m_d H(\omega_{cv} - \omega_0), \qquad (2.177)$$

where H represents the step function.

By replacing (2.177) into (2.167), we obtain

$$(4\pi)^{-1}\chi(\omega) = \underset{\sim}{D} \times F^{(2)}(x),$$

where

$$\underset{\sim}{D} = \frac{\Delta k}{\pi} m_d \omega_0^{-2} P \times P$$

$$F^{(2)}(x) = -x^{-2} \ln(1 - x^2) \qquad (2.178)$$

and, correspondingly, for two-band terms in the "classical" approximation

$$(4\pi)^{-1}\underset{\sim}{\chi_s} = \frac{\underset{\sim}{D}}{\omega_0} G^{(2)}(x) \frac{d\omega_0}{d\xi} (n+1)^{1/2} (2\omega_v)^{-1/2}$$

$$G^{(2)}(x) = \frac{-2}{1 - x^2}. \qquad (2.178')$$

In this case, the Raman efficiency very near resonance $x \simeq 1$ is proportional to $|1 - x|^{-2}$, independent of the type of critical point and of the presence of weak excitonic interaction. We have to bear in mind, however, that in two (and one) dimensions, contrary to the three-dimensional case, excitonic interaction always produces bound states [2.48] and that the simple replacement of m by $m + \alpha$ mentioned above does not take into account the presence of this bound state properly. Broadening of the intermediate state may, nevertheless, wash out this bound state and restore the approximate validity of (2.178').

We should briefly treat, for completeness, the one-dimensional parabolic case applicable to organic one-dimensional solids and also to the Penn model of tetrahedral semiconductors [2.170]. In this case,

$$N_d = \frac{1}{\pi} \left(\frac{m^*}{2}\right)^{1/2} (\omega - \omega_0)^{-1/2}$$

$$(4\pi)^{-1}\underset{\sim}{\chi} = \left(\frac{m^*}{2}\right)^{1/2} P \times P \omega_0^{-2} G(x)$$

$$(4\pi)^{-1}\underset{\sim}{\chi_s} = \frac{1}{2} \left(\frac{m^*}{2}\right)^{1/2} P \times P \omega_0^{-3} J(x) \frac{d\omega_0}{d\xi} (n+1)^{1/2} (2\omega_v)^{-1/2}$$

$$J(x) = x^{-2}[2 - (1-x)^{-3/2} - (1+x)^{-3/2}]. \qquad (2.179)$$

2.2.5 Review of the Optical Properties of Tetrahedral Semiconductors

The lowest absorption edge of tetrahedral semiconductors can be either direct (GaAs, InSb, InP, CuCl...) or indirect (Ge, Si, GaP, AlSb, diamond). Indirect edges result in weak, phonon-aided (strongly temperature-dependent) absorption. This absorption edge is only relevant to Raman scattering in which the phonons aiding it also participate as emitted or absorbed phonons. It therefore contributes to scattering by two phonons near X in GaP and Si [2.171]. However, it does not seem to contribute to one-phonon scattering. The first direct gap is at the center of the Brillouin zone and, with the exception of Si, diamond and SiC (and possibly some large gap III–V's), it corresponds to transitions from a p-like valence state ($\Gamma_{25'}$ in Ge, Γ_{15} in GaAs) to an s-like conduction state ($\Gamma_{2'}$ in Ge, Γ_1 in GaAs) (Fig. 2.37). This is the so-called E_0 gap which can be handled in detail analytically [2.169]. Within pseudopotential theory, the two states forming the E_0 gap of germanium are obtained by diagonalizing 2×2 matrices [2.169]. The threefold degenerate $\Gamma_{25'}$ valence states have xy, yz, zx symmetry. They are connected to the conduction states by the following matrix of p [2.169]:

$$P = \langle xy|p_z|c\rangle = \langle yz|p_x|c\rangle = \langle zx|p_y|c\rangle \simeq \frac{2\pi}{a_0}. \tag{2.180}$$

The $\Gamma_{25'}$ state splits into two as a result of spin-orbit interaction: Γ_8^+ and Γ_7^+. We thus obtain two absorption edges separated by the spin-orbit splitting Δ_0. The Γ_8^+ state is fourfold degenerate. Its symmetry properties under the operations of the cubic group are the same as those of $J = 3/2$ states ($J_z = \pm 3/2$, $\pm 1/2$). Γ_7^+ is twofold degenerate and symmetrywise equivalent to $J = 1/2(J_z = \pm 1/2)$, see (2.148). The average matrix elements of p connecting each of the $J = 3/2$ and $J = 1/2$ states with the corresponding $\Gamma_{2'}$ state are equal to $P/\sqrt{3}$. By using the results of Sect. 2.35, we obtain for the E_0 edge [2.169]

$$(4\pi)^{-1}\chi(E_0) \simeq \frac{C_0''}{4\pi} F(x); \quad x = \omega/\omega_0$$

$$(4\pi)^{-1}\chi(E_0 + \Delta_0) \simeq \frac{C_0''}{8\pi} F(x'); \quad x' = \omega/\omega_0 + \Delta_0, \tag{2.181}$$

where $C_0'' \simeq P^{-1} \simeq \left(\dfrac{a_0}{2\pi}\right)$. For most materials of the family, $P^{-1} \simeq 1.8$ in atomic units. An attempt to fit the *real* part of χ with (2.181) plus a constant (to account for other polarization mechanisms) yields values of C_0'' between 2 and 5. The enhancement has been attributed to excitonic interaction [2.172].

The E_0, $E_0 + \Delta_0$ absorption edges are relatively weak. They reach within 0.1 eV of the edge absorption coefficients of the order of 10^4 cm^{-1}. The next higher gaps are the so-called $E_1 - E_1 + \Delta_1$ gaps (absorption coefficients $\simeq 10^5$ cm^{-1}) which occur between the two highest, spin-orbit-split bands along the $\Lambda = \{111\}$ directions and the lowest conduction band. These valence band

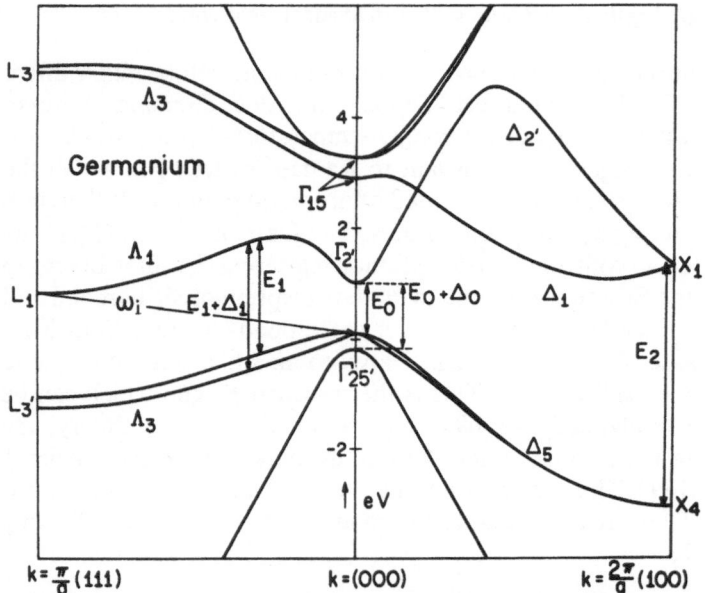

Fig. 2.37. Band structure of germanium showing the indirect transitions ω_i, the lowest direct gaps $E_0 - E_0 + \Delta_0$, the $E_1 - E_1 + \Delta_1$ gaps and some contribution to the E_2 gap [2.48]

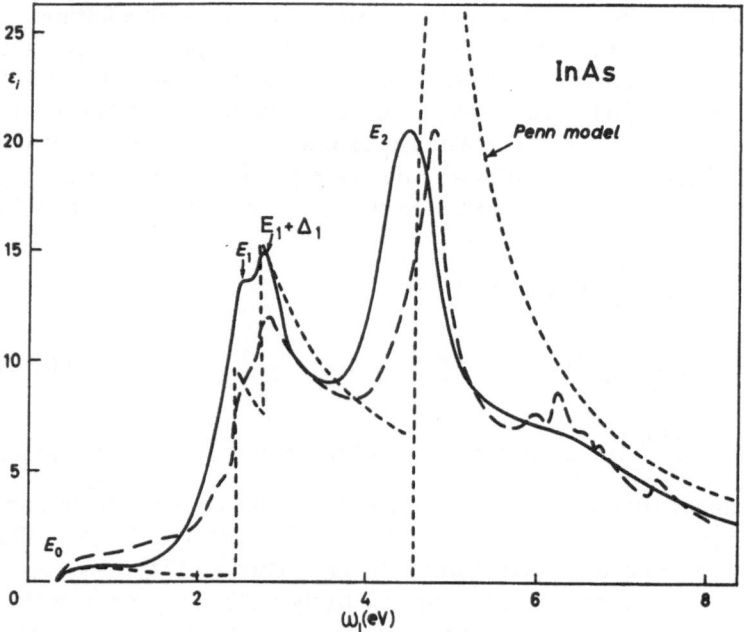

Fig. 2.38. Imaginary part of $\chi(\omega)$ for InAs showing E_0, E_1, and E_2 critical points. The long-dashed curve is theoretical and was obtained from a complete band structure calculation and *BZ* integration [2.173]. The short-dashed curve results from simplified parabolic density of states models [Ref. 2.169, Fig. 8]. The solid line is experimental

states have Λ_3 single group and $\Lambda_{4,5} - \Lambda_6$ double group symmetries. The conduction band is $\Lambda_1(\Lambda_6)$. They are nearly parallel throughout most of the BZ and hence produce quasi-two-dimensional critical points. An example is shown in Fig. 2.38 for InAs including the experimental χ_i spectrum, the fit with simple parabolic bands such as (2.181) for $E_0 - E_0 + \Delta_0$, the relation (2.185) below for $E_1 - E_1 + \Delta_1$, and the results of a more elaborate full band structure calculation [2.173].

The treatment of the parameters of the $E_1 - E_1 + \Delta_1$ gap [note that $\Delta_1 \simeq (2/3)\Delta_0$] can be found in [2.169]. We shall reproduce the principal facts and results here. For a given [111] direction the valence bands have, for $\Delta_1 = 0$, symmetries $\bar{X}$ and $\bar{Y}$, where $\bar{X}$, $\bar{Y}$ are perpendicular to [111], and the conduction bands are completely symmetric under the C_{3v} operations. We thus have the nonzero matrix elements of P:

$$|\langle \bar{X}|p_x|c\rangle| = |\langle \bar{Y}|p_y|c\rangle| = \bar{P} \simeq P. \tag{2.182}$$

The spin-orbit interaction splits Λ_3 and yields the wave functions

$$\Lambda_{4,5}: \frac{1}{\sqrt{2}}(\bar{X} + i\bar{Y})\uparrow; \frac{1}{\sqrt{2}}(\bar{X} - i\bar{Y})\downarrow$$

$$\Lambda_6: \frac{1}{\sqrt{2}}(\bar{X} + i\bar{Y})\downarrow; \frac{1}{\sqrt{2}}(\bar{X} - i\bar{Y})\uparrow. \tag{2.183}$$

Hence, the matrix elements of $p_{x,y}$ between each of the $\Lambda_{4,5} - \Lambda_6$ bands and the allowed conduction bands now become $P/\sqrt{2}$. The contribution of the $E_1 - E_1 + \Delta_1$ gaps to $\chi(\omega)$ can be calculated with (2.178), taking into account that there are eight equivalent {111} directions. Choosing the polarization of the light parallel to [1$\bar{1}$0], we find that the total contribution to χ is 8/3 times the contribution of the [111] direction.

The reduced effective masses for the $E_1 - E_1 + \Delta_1$ edge can also be easily calculated with the $\boldsymbol{k} \cdot \boldsymbol{p}$ method (2.146). We find [2.169]

$$\frac{1}{m_d(E_1)} \simeq \frac{3P}{\omega_1 + \Delta_1/3}; \quad \frac{1}{m_d(E_1 + \Delta_1)} = \frac{3P}{\omega_1 + 2\Delta_1/3}, \tag{2.184}$$

where ω_1 is the energy of the E_1 critical point. Using these results and $\Delta k = \pi\sqrt{3}/a_0$ (i.e., the bands are parallel along [111] almost throughout the whole BZ), (2.178) yields

$$(4\pi)^{-1}\chi(E_1) = -\frac{4\sqrt{3}(\omega_1 + \Delta_1/3)}{9\pi a_0 \omega_0^2} F^{(2)}(1 - x^2)$$

$$(4\pi)^{-1}\chi(E_1 + \Delta_1) = -\frac{4\sqrt{3}(\omega_1 + 2\Delta_1/3)}{9\pi a_0 \omega^2} F^{(2)}(1 - x'^2) \tag{2.185}$$

$$x = \frac{\omega}{\omega_1}, x' = \frac{\omega}{\omega_1 + \Delta_1}.$$

The results obtained with (2.185) for the imaginary part of χ in InAs are shown in Fig. 2.38. We note that the strength of the transitions, which is explained well by (2.185), has not been used as an adjustable parameter.

In order to interpret the gross features of Fig. 2.38, the Penn gap E_2 should now be discussed [2.170]. This gap corresponds to the strongest absorption peak in Fig. 2.38 at $\simeq 4.2$ eV. An analysis of the full band structure calculation indicates that these transitions spread over a wide area of the Brillouin zone, in particular over most of the square faces ($W-K-X$ points) and their surroundings [2.169]. They are thus difficult to describe with simple models but the fact that the bands are nearly parallel over the surface of the BZ (actually, more correctly, over the surface of the *Jones zone* JZ [2.174]) suggests an interpretation based on a one-dimensional dependence in k perpendicular to the boundary of the JZ. This is the so-called Penn model which is discussed in detail in [2.169, 174]. The corresponding $\chi(\omega)$ is described by a function of the form (2.179). These transitions determine the (real) value of $\chi(0)$ and are closely related to most *average* thermodynamic or chemical properties [2.175]. The corresponding $\chi(0)$ is given by

$$\chi(0) \simeq \omega_p^2/\omega_g^2 , \qquad (2.186)$$

where $\omega_p = (4\pi N_v)^{1/2}$ is the plasma frequency of the valence electrons and ω_g the "Penn" gap. The contribution of the Penn model to $\chi(\omega)$ in InAs is also shown in Fig. 2.38.

We should mention before closing this section that the lowest direct gap of diamond (E_0') is $\Gamma_{25'} \rightarrow \Gamma_{15}$. The small spin-orbit splitting can be neglected in this case. The selection rules for transitions between $\Gamma_{25'}(\overline{xy}, \overline{yz}, \overline{zx})$ and $\Gamma_{15}(x, y, z)$ are

$$\langle \overline{xy}|p_x|y\rangle = \langle \overline{yz}|p_y|z\rangle = \langle \overline{zx}|p_z|x\rangle$$
$$= \langle \overline{xy}|p_y|x\rangle = \langle \overline{yz}|p_z|y\rangle = \langle \overline{zx}|p_x|z\rangle = P' \simeq \frac{2\pi}{a_0}.$$

A detailed treatment of $\chi(\omega)$ for this edge can be found in [2.176].

2.2.6 Contributions to the Raman Tensor for First-Order Raman Scattering by Phonons

The calculations below are performed in the classical or "quasistatic" approximation ($\omega_v \ll |\omega_0 - \omega_L + i\Gamma|$). The symmetry of the Raman phonon of germanium is $\Gamma_{25'}$ (equivalently, Γ_{15} in zincblende) and is the same as that of a [111] strain. Hence, the treatment of the Raman tensor is the same as that of the elasto-optic constant p_{44} which contributes to the Brillouin scattering efficiency of all phonons whose velocities contain the elastic constant c_{44}. The other elasto-optic constants $p' = p_{11} - p_{12}$ and $p_H = p_{11} + 2p_{12}$ have nonequivalent treatments [2.169] (see Sect. 2.2.12).

a) $E_0, E_0 + \Delta_0$-Edge

We first discuss the case $\Delta_0 \simeq 0$ or, more accurately, $|\omega_L - \omega_0| \gg \Delta_0$. This case can be treated by using purely orbital wave functions $X = yz$, $Y = xz$, $Z = yx$ (for $\Gamma_{25'}$). The $Z = xy$ phonon couples the wave functions X with Y. In order to diagonalize the corresponding $\mathscr{H}_{ev}$, we use as wave functions $\dfrac{1}{\sqrt{2}}(X \pm Y)$ and Z.
The matrix elements of $\mathscr{H}_{ev}$ thus become

$$
\begin{aligned}
\left\langle \frac{1}{\sqrt{2}}(X+Y)\left|\mathscr{H}_{ev}^{xy}\right|\frac{1}{\sqrt{2}}(X+Y)\right\rangle &= + \frac{\partial \omega_0}{\partial \xi}(n+1)^{1/2}\sqrt{\frac{1}{2\omega_v}} \\
\left\langle \frac{1}{\sqrt{2}}(X-Y)\left|\mathscr{H}_{ev}^{xy}\right|\frac{1}{\sqrt{2}}(X-Y)\right\rangle &= - \frac{\partial \omega_0}{\partial \xi}(n+1)^{1/2}\sqrt{\frac{1}{2\omega_v}} .
\end{aligned}
\tag{2.187}
$$

For historical reasons the electron-phonon interaction constant $\partial \omega_0/\partial \xi$ is usually written in terms of the "deformation potential" d_0 [eV]:

$$
\frac{\partial \omega_0}{\partial \xi} = \frac{d_0}{2a_0}\sqrt{\frac{3}{\mu}},
\tag{2.188}
$$

where μ is the *reduced mass* of the PC.

In order to calculate χ_s we must now determine the effective masses and the corresponding matrix elements of p taking into account the degeneracy of $\Gamma_{25'}$. Let us consider for the purpose of calculating the only independent matrix element of χ_s (see Table 2.1) the case $\hat{e}_L = [100]$, $\hat{e}_s = [010]$ for which we couple to the XY phonon of (2.187) and examine the valence bands along the cubic axes. The reduced masses are, see (2.147),

$$
\begin{aligned}
m_x^*(X) &\simeq \frac{m_e^*}{2} \simeq \frac{P^2}{\omega_0}; \quad m_y^*(X) = m_z^*(X) \simeq m_e^* \simeq \frac{2P^2}{\omega_0} \\
m_y^*(Y) &\simeq \frac{m_e^*}{2} \simeq \frac{P^2}{\omega_0}; \quad m_x^*(Y) = m_z^*(Y) \simeq m_e^* \simeq \frac{2P^2}{\omega_0} \\
m_z^*(Z) &\simeq \frac{m_e^*}{2}\frac{P^2}{\omega_0}; \quad m_x^*(Z) = m_y^*(Z) \simeq m_e^* \simeq \frac{2P^2}{\omega_0},
\end{aligned}
\tag{2.189}
$$

where the subscript indicates the direction of k-space and the letter in brackets labels the band. m_e^* is the mass of the $\Gamma_{2'}$ electrons. The Hamiltonian $\mathscr{H}_{ev}$ is not diagonal with respect to XYZ, hence the bands along the directions of (2.189) should yield three-band terms [the phonon mixes X with Y according to (2.187)]. For $k_{\bar{x}} \| (110)$ and $k_{\bar{y}} \| (1\bar{1}0)$, $\mathscr{H}_{ev}$ is diagonal. In the latter case the masses are the same as those of (2.189), replacing x, y, z by $\bar{x}$, $\bar{y}$, z. Hence, the density of states mass for each band is $m_d = m_e^*/2^{1/3}$. The corresponding matrix elements of p are:

$$\langle \bar{X}|p_x|\Gamma_{2'}\rangle = \frac{P}{\sqrt{2}} = \langle \bar{Y}|p_x|\Gamma_{2'}\rangle$$

$$\langle \bar{X}|p_y|\Gamma_{2'}\rangle = \frac{P}{\sqrt{2}} = -\langle \bar{Y}|p_y|\Gamma_{2'}\rangle .$$

(2.190)

Hence, if we only take into account in our calculation the $k_{\bar{z}}$, $k_{\bar{x}}$, and $k_{\bar{y}}$, directions of k-space, we exclusively find two-band terms. We then encounter again (2.174) with the nonzero element of χ_s:

$$(4\pi)^{-1}\chi_{s,xy}(E_0) = -\frac{C}{4\omega_0}G(x)\frac{d_0}{a_0}\sqrt{\frac{3}{2N\mu\omega_v}}(n+1)$$

(2.191)

with $C = \frac{1}{\pi}m_e^{*3/2}P^2\omega_0^{-3/2}$. Equation (2.191) yields for the Raman polarizability a of (2.67) and Table 2.8

$$a = -\frac{m_e^{*3/2}P^2 d_0 a_0^2}{16\pi\omega_0^{5/2}}\sqrt{\frac{3}{2}}G(x) ,$$

(2.192)

where $m_e^* \simeq \omega_0/2P^2$.

We note that (2.192) suffers from the fact that we have only included in it the directions of k-space for which two-band terms appear. This does not happen, for instance, for k along x or along y. In this case, the function $\omega_0^{-5/2}G(x)$, obtained as a derivative of $\omega_0^{-3/2}F(x)$ with respect to ω_0, must be replaced by a finite difference involving the average difference in valence band energies

$$\omega_0^{-3/2}\frac{F\left(\frac{\omega+\delta/2}{\omega_0}\right) - F\left(\frac{\omega-\delta/2}{\omega_0}\right)}{\delta} ,$$

(2.193)

where δ is approximately the difference between the light and the heavy hole bands at the k-vector for which $\omega_{cv} - \omega_0 \simeq |\omega_L - \omega_0|$. The function (2.193) approximately equals $\omega_0^{-5/2}G(x)$ near resonance and, even if this approximation is not too good, since it must be averaged with the function $G(x)$ corresponding to the $[1, \pm 1, 0]$ directions, it should not significantly alter the result of (2.192). Possible corrections can be lumped into numerical factors of the order of unity. These factors should be the same for a as for the elasto-optic constant p_{44} (Sect. 2.2.12).

From the reasoning above we conclude that the Raman susceptibility is obtained, to a good accuracy, by assuming that the $x \pm y$ valence bands are *rigidly shifted* by the phonon perturbation, the amount $\pm \partial\omega_0/\partial\xi$ given in (2.188), while the band Z does not change.

The result in (2.191) is valid provided $|\omega_L - \omega_0| \gg \Delta_0$. We shall now treat the case of a finite Δ_0 in which two resonances occur, one at E_0 and another at $E_0 + \Delta_0$. The Γ_8^+ valence state is split by the $\Gamma_{25'}$ phonon while Γ_7^+ is not. Hence, E_0 contributes strongly dispersive two-band terms to χ_s while $E_0 + \Delta_0$ only contributes less dispersive three-band terms. The latter are due mainly to the coupling by the phonon of Γ_7^+ with Γ_8^+. The calculation of the two-band and three-band contributions to χ_s near E_0 and $E_0 + \Delta_0$ can be found in [2.169] and will not be repeated here. This calculation is also based on the approximation of rigid splittings and shifts for the Γ_8^+ valence bands under the phonon perturbation, an approximation which is justified by a treatment similar to that given above. In terms of a the result is [2.169]

$$a(E_0, E_0 + \Delta_c) = \frac{C_0'' a_0^2 \sqrt{3}}{128\pi\omega_0} d_0 \left\{ -G(x) + \frac{4\omega_0}{\Delta_0} \left[F(x) - \left(\frac{\omega_0}{\omega_0 + \Delta_0} \right)^{3/2} F(x') \right] \right\},$$

(2.194)

where $x = \dfrac{\omega}{\omega_0}$, $x' = \dfrac{\omega}{\omega_0 + \Delta_0}$, $C_0'' \simeq \dfrac{1}{P}$.

For $\Delta_0 \to 0$, the term in curly brackets in (2.194) tends to $-3G(x)$ and, except for a small difference in the numerical coefficients due to the different ways of averaging the effective masses, (2.194) becomes equivalent to (2.192). The G-term in (2.194) represents the two-band terms associated with $E_0(\Gamma_8^+)$ while the F-terms represent the three-band terms associated with the $\Gamma_8^+ - \Gamma_7^+$ coupling by the phonon.

Using (2.181), (2.194) can be rewritten in terms of the separate contributions of E_0 and $E_0 + \Delta_0$ to the susceptibility $\chi(\omega)$ [under some assumptions these contributions can be extracted from the experimental $\chi(\omega)$]. We find

$$a(E_0, E_0 + \Delta_0) \simeq -\frac{a_0^2 \sqrt{3} d_0}{16} \left\{ \frac{\partial \chi^{(E_0)}(\omega)}{\partial \omega_0} \right. $$
$$\left. - \frac{2}{\Delta_0} \left[\chi^{E_0}(\omega) - 2 \left(\frac{\omega_0}{\omega_0 + \Delta_0} \right)^{3/2} \chi^{E_0 + \Delta_0}(\omega) \right] \right\} (4\pi)^{-1}, \quad (2.195)$$

where $\dfrac{\partial \chi^E(\omega)}{\partial \omega_0} = -\dfrac{\partial \chi^E(\omega)}{\partial \omega} + \dfrac{2}{\omega} \chi^E(\omega)$.

Equation (2.195) expresses a in terms of the experimental spectrum of $\chi(\omega)$ and the deformation potential d_0 and thus it may liberate us to some extent, of the limitations in the approximations which went into deriving (2.194). A fit with (2.194) to the $E_0 - E_0 + \Delta_0$ resonances of GaAs is shown in Fig. 2.8.

The deformation potential d_0 can be easily calculated by using pseudopotential theory. In the case of germanium and silicon (also diamond [2.176]), it can be written in closed form as [2.169]

$$d_0 = -\frac{4\pi}{\sqrt{3}} [\beta^2(v_4 - v_{12}) + \gamma\beta \sqrt{2}(v_3 - v_{11})], \quad (2.196)$$

where v_i are the pseudopotential form factors and β and γ represent the admixture of [111] and [200] plane waves which appears in the $\Gamma_{25'}$ wavefunction ($\beta \simeq 0.835$, $\gamma \simeq 0.55$ for Ge and Si). Equation (2.196) yields $d_0 \simeq +33$ eV for Ge and Si. For the III–V and II–VI zincblende materials, the theory is slightly more complicated as it involves antisymmetric form factors but the resulting values of d_0 are nearly the same [2.73]. Recent calculations by *Vogl* and *Pötz* [2.177] using both the pseudopotential and the tight binding method also yield similar results for a large number of materials of this family. It is interesting to note that the corresponding deformation potential d for a strain along [111] can be written as

$$d = d' - \tfrac{1}{4}\zeta d_0, \tag{2.197}$$

where d' is the strain deformation potential which one would have in the absence of *internal strains* ($\zeta = 0$). For $\zeta = 1$ there are internal strains produced by the fact that the bonds are completely incompressible. For Ge and Si, $\zeta \simeq 0.7$, while for diamond, $\zeta \simeq 0.2$ [2.178].

b) $E_1 - E_1 + \Delta_1$-Edge

This treatment is also similar to that for the p_{44} elasto-optic coefficient given in [2.169]. We must include all equivalent [111] directions and two contributions arise, one due to the "intervalley" splitting by the phonon of the equivalent {111} directions (two-band terms) determined by the deformation potential $d_{1,0}^5$ [2.179], and another due to the "intravalley" coupling by the phonon of the spin-orbit-split $\Lambda_{4,5}$ and Λ_6 valence bands (three-band terms) determined by $d_{3,0}^5$. For a phonon XY, the splitting of the {111} directions due to $d_{1,0}^5$ is

$$[111], [\bar{1}\bar{1}1]; \quad \Delta\omega_1 = \frac{d_{1,0}^5}{6}\sqrt{\frac{1}{2\mu\omega_v}}$$

$$[1\bar{1}1], [\bar{1}11]; \quad \Delta\omega_1 = -\frac{d_{1,0}^5}{6}\sqrt{\frac{1}{2\mu\omega_v}}. \tag{2.198}$$

$d_{1,0}^5$ represents the difference between conduction and valence deformation potentials. As mentioned above, the $(X \pm iY)/2$ (i.e., $\Lambda_{4,5} - \Lambda_6$) valence states are coupled by the deformation potential $d_{3,0}^5$. This coupling can be obtained from the matrix elements

$$\langle \bar{X}_{[111]}|\mathcal{H}_{ev}(XY)|\bar{X}_{[111]}\rangle = -\frac{\sqrt{2}}{3a_0}d_{3,0}^5\sqrt{\frac{1}{2\mu\omega_v}}$$

$$\langle \bar{X}_{[111]}|\mathcal{H}_{ev}(XY)|\bar{Y}_{[111]}\rangle = 0 \tag{2.199}$$

$$\langle \bar{Y}_{[111]}|\mathcal{H}_{ev}(XY)|\bar{Y}_{[111]}\rangle = \frac{\sqrt{2}}{3a_0}d_{3,0}^5\sqrt{\frac{1}{2\mu\omega_v}}$$

and the corresponding matrix elements for the [$\bar{1}$11] direction.

We represent the contribution of E_1 and $E_1 + \Delta_1$ to the linear susceptibility $\chi(\omega)$ by (2.185). We note that the deformation potentials $d_{1,0}^5$ and $d_{3,0}^5$ are to be regarded as averages over the whole [111] direction of the Brillouin zone. Typical dependences of $d_{1,0}^5$ and $d_{3,0}^5$ along the [111] direction can be found in [2.44a]. We point out that for $k \to 0$ the following compatibility relations exist:

$$d_{3,0}^5 = \sqrt{2} d_0, \tag{2.200a}$$

$$(d_{1,0}^5)_{\text{valence}} = -d_0. \tag{2.200b}$$

We must keep in mind, however, that (2.200a, b) are only valid for $k \simeq 0$. Equation (2.200a) remains approximately valid throughout Λ while (2.200b) decreases by a factor of two between Γ and L.

A straightforward calculation yields for the case $\Delta_1 \to 0$ (which applies to Si as in this material, $\Delta_1 = 0.03 \text{ eV} \ll \Gamma \simeq 0.1 \text{ eV}$):

$$a(E_1, E_1 + \Delta_1) = -\frac{a_0^2}{4\sqrt{6}} \left(d_{3,0}^5 + \frac{1}{2\sqrt{2}} d_{1,0}^5 \right) \left[\frac{\omega d\chi(\omega)}{\omega_g d\omega} + \frac{\alpha}{\omega} \chi(\omega) \right] (4\pi)^{-1}. \tag{2.201}$$

In (2.201), we have used the relationship

$$\frac{d\chi}{d\omega_g} = -\frac{\omega}{\omega_g} \frac{d\chi}{d\omega} - \frac{\alpha}{\omega} \chi, \tag{2.202}$$

where α depends on the model used to calculate the variation of the transverse reduced mass $\mu_\perp^*$ with phonon deformation. The case in which the phonon does not affect $\mu_\perp^*$, corresponds to $\alpha = 2$. If the phonon changes $\mu_\perp^*$ in a way proportional to ω_1, we obtain $\alpha = 1$. Actually, the most reasonable model yields $\alpha = 5/3$ as $\mu_\perp^{-1} = m_{e\perp}^{-1} + m_{eh}^{-1} \simeq \frac{3}{2} m_{e\perp}^{-1}$ [2.169]. Within $k \cdot p$ theory, $m_{e\perp}$ is not affected by the phonon perturbation while m_{eh} is, in the same manner as ω_1. This yields $\alpha = 5/3$.

A fit to the $E_1 - E_1 + \Delta_1$ resonance in the first-order efficiency of GaAs is shown in Fig. 2.8 [2.53]. Equation (2.201) is written in terms of the experimental $\chi(\omega)$. It can be written equally well in terms of the analytical function $F^{(2)} = -x^{-2} \ln(1 - x^2)$ by using (2.185). In (2.201), $d_{3,0}^5$ is usually much larger than $\frac{1}{2\sqrt{2}} d_{1,0}^5$ [2.44a].

If the spin-orbit splitting Δ_1 cannot be neglected, we can derive the expression

$$a = \frac{a_0^2}{4\sqrt{6}} \left[\frac{-1}{2\sqrt{2}} d_{1,0}^5 \frac{d\chi(\omega)}{d\omega_1} + 2d_{3,0}^5 \left(\frac{\chi^{(E_1)} - \chi^{(E_1 + \Delta_1)}}{\Delta_1} \right) \right] (4\pi)^{-1}, \tag{2.203}$$

where $\chi^{(E_1)}$ and $\chi^{(E_1 + \Delta_1)}$ represent the separate contributions of E_1 and $E_1 + \Delta_1$ to χ. Equation (2.203) can be transformed into one containing $d\chi/d\omega$, which is determined from experiment, instead of $d\chi/d\omega_1$ by using (2.202). In the limit $\Delta_1 \to 0$, (2.203) tends to (2.201).

c) E_2-Edge

As already mentioned in Sect. 2.2.6, the E_2 edge or Penn gap, while in principle simple (one-dimensional), is difficult to handle in many respects, for instance, for the calculation of its contribution to a. A method has been suggested by *Aslaksen* [2.180] based on (2.186). Let us assume a phonon with $\hat{e}$ along [111] with displacement u. Using (2.186) for $\omega = 0$, we find [see (2.67)]

$$a(0) = \frac{a_0^3}{8} \frac{\sqrt{3}}{2} \frac{d\chi_\parallel}{du} (4\pi)^{-1}$$

$$= -\frac{a_0^2}{2} \chi(0) \frac{d\ln\omega_g}{d\ln r} (4\pi)^{-1} \tag{2.204}$$

where r represents the bond length and $\chi_\parallel$ the susceptibility parallel to the bond. Equation (7) of [2.180] incorrectly includes the change in plasma frequency with r (the phonon does not alter the volume!). Using $d\ln\omega_g/d\ln r = -2.5$ [2.169], (2.204) yields for silicon $a = +32\text{Å}^2$ in excellent agreement with the values extrapolated to $\omega \to 0$ from $a = +60\text{Å}^2$ listed in Table 2.8 for 1.9 eV. Although this result may be somewhat fortuitous, it is remarkable that one can at least predict with this argument the elusive sign of a (Sect. 2.1.18).

d) E_0'-Edge of Diamond

The lowest direct edge of diamond also takes place at Γ but the corresponding conduction state is Γ_{15} (p-like) instead of the $\Gamma_{2'}$ (s-like) state of germanium. This edge is labeled E_0'. In Si, a similar situation arises but E_0' is nearly degenerate with $E_1 - E_1 + \Delta_1$. Because of the larger volume of k-space associated with the latter, E_0' is negligible in Si. The contribution of the E_0' edge of diamond to a has been discussed in [2.176]. The following expression was calculated:

$$a(E_0') = \frac{\sqrt{3}a_0^2}{16} (d_0^{15} - d_0) \frac{d\chi}{d\omega_0} (4\pi)^{-1}. \tag{2.205}$$

The deformation potentials d_0^{15} (for the Γ_{15} conduction state) and d_0 can be calculated by pseudopotential theory (although the use of pseudopotential theory for diamond is questionable); d_0 is given by (2.196) while the corresponding expression for d_0^{15} is

$$d_0^{15} = \frac{4\pi}{\sqrt{3}} (v_4 - v_{12}). \tag{2.205a}$$

Using these expressions and values of v_i found in the literature, one obtains $d_0 = 90$ and $d_0^{15} = -43$ eV. With these values and the measured $\chi(\omega)$ [2.181], one finds for diamond

$$a(E_0') = -6.5g\left(\frac{\omega}{\omega_g}\right) = +4.9\text{Å}^2 \quad \text{for} \quad \omega \to 0. \tag{2.206}$$

This contribution suffices to explain the experimental value and the sign of a for diamond (Table 2.8) [2.54].

2.2.7 Fluorite-Type Materials: CaF_2, SrF_2, and BaF_2

The phonons at the Γ point of these materials have been discussed in Sect. 2.2.10 and their eigenvectors shown in Fig. 2.9. There are two sets of optical phonons: Γ_{15} ir-allowed (split into LO and TO) and $\Gamma_{25'}$ Raman allowed (three-fold degenerate). The latter are rather important in Raman scattering as they are often used as standards in resonance Raman measurements: because of their large E_0 gap (~ 10 eV) the Raman polarizabilities a are nondispersive in the visible. These polarizabilities, and the corresponding efficiencies, have been determined rather accurately [2.49]. They are listed in Table 2.8.

The Raman polarizabilities mentioned above seem due mainly to the effect of the electron-phonon coupling on the edge exciton at the frequency E_0. We represent the contribution to χ of this exciton, related to transitions between the Γ_{15} valence band and the Γ_1 conduction band [2.182], by the Lorentzian

$$(4\pi)^{-1}\chi = \frac{N_{\text{eff}}}{E_0^2 - \omega_L^2}, \tag{2.207}$$

where the "effective valence electron density N_{eff}" is obtained from the absorption spectrum [or rather from the spectrum of $\text{Im}\{\chi(\omega)\}$] with

$$N_{\text{eff}} = \frac{1}{2\pi^2}\int_0^\infty \omega[\text{Im}\{\chi(\omega)\}]\,d\omega. \tag{2.208}$$

In order to evaluate a it is also convenient to consider a longitudinal phonon propagating along [111]. In this case, because of the symmetry, the tensor $\partial\chi/\partial\xi$ is diagonal. We label u_3 as the relative displacement of the two fluorine atoms in the $\Gamma_{25'}$ phonon (see Fig. 2.9). We can easily write for two-band terms

$$a = \frac{a_0^3}{8}\frac{\partial\chi_\parallel}{\partial E_0}\left(\frac{\partial E_0}{\partial u_3}\right)(4\pi)^{-1}$$

$$= \frac{\sqrt{3}a_0^2 d_0}{4}\frac{N_{\text{eff}}E_0}{(E_0^2 - \omega_L^2)^2} = \frac{a_0^2 d_0 N_{\text{eff}}}{4E_0^3}\sqrt{3} \quad (\text{for } \omega_L \ll E_0). \tag{2.209}$$

Table 2.9. Deformation potentials d_0 for optical phonons interacting with the Γ_{15} valence state of CaF_2, SrF_2, and BaF_2. Also, corresponding values of the Raman polarizability a

	d_0 [eV] pseudopot.	d_0 [eV] LCAO	a [Å^2] pseudopot.	a [Å^2] LCAO
CaF_2	45.1	38.8	0.46	0.40
SrF_2	37.8	37.8	0.46	0.46
BaF_2	30.0	34.2	0.38	0.44

The deformation potential of the Γ_{15} valence state $d_0 = -a_0(dE_0/du_3)_{\parallel}$ can be evaluated either with pseudopotential theory or with the LCAO method. From pseudopotential theory we find [2.72]

$$d_0 = \frac{16}{\sqrt{3}} \alpha\beta \frac{V_F}{V_{cell}} (v_3^F - v_{11}^F), \qquad (2.210)$$

where α and β represent the admixture of {111} and {200} waves in Γ_{15}, respectively, V_F and V_{cell} are the volumes of the F ion and that of the CaF_2 primitive cell, and v_3^F, v_{11}^F represent pseudopotential form factors for the fluorine ion. It is also possible to evaluate d_0 using atomic functions for the $2p$ states of fluorine [2.49]. The values of d_0 obtained with these methods are given in Table 2.9 together with the corresponding values of a obtained with (2.209), using for Im$\{\chi(\omega)\}$ the spectra found in the literature [2.183]. The agreement with the experimental data of Table 2.8 is rather good, a fact which lends credibility to the signs of a and d_0 derived from the theory. It is to be hoped that the dispersion of $a(\omega_L)$ given in (2.209) will be tested in the future as measurements with vacuum uv lasers or with synchrotron radiation become possible.

2.2.8 Forbidden LO-Scattering

In Sect. 2.1.12, we treated some peculiarities of scattering by Raman active LO-phonons in materials without a center of inversion. These effects consist of a modification of the scattering efficiency due to electro-optic coupling: the LO-phonons are accompanied by a longitudinal field which modulates χ through the first-order electro-optic effect. Similar effects are, in principle, also encountered in Brillouin scattering by *piezoelectric* acoustic phonons [2.17].

We shall indicate here the quantum-mechanical formulation of this effect and, at the same time, introduce another type of effect which cannot be treated within the macroscopic context of Sect. 2.1.12, i.e., the Fröhlich-interaction-induced forbidden scattering by LO-phonons (and also, correspondingly, Brillouin scattering by piezoelectric acoustic phonons). Let us consider the ir-active phonons of a zincblende-type material (extension to ir-active phonons of other materials such as fluorite is straightforward). The *longitudinal* polarization can be easily written in terms of the normal coordinate ξ of (2.44):

$$P = \frac{\varepsilon_0 e^*}{\varepsilon_\infty V_c^{1/2} \sqrt{\mu}} \left[\xi e^{i(q \cdot R - \omega_{LO} t)} + \xi^* e^{-(i q \cdot R - \omega_{LO} t)} \right]. \tag{2.211}$$

This polarization produces a longitudinal field of magnitude $\mathscr{E}$ which, in turn, produces a potential $-i\mathscr{E}/q$. This potential is actually the electron-phonon interaction Hamiltonian $\mathscr{H}_{ev}$. The exponential factors in (2.111) take care of momentum and energy conservation in the scattering process. The Hamiltonian $\mathscr{H}_{ev}$ can be easily written in second-quantized notation by making the substitutions, see (2.26),

$$\xi = \sqrt{\frac{1}{2\omega_{LO}}}\, b \;,\; \xi^* = \sqrt{\frac{1}{2\omega_{LO}}}\, b^\dagger$$

and introducing the electron creation and annihilation operators $c^\dagger$ and c:

$$\mathscr{H}_{ev} = \frac{C_F}{|q|} (b_q^\dagger + b_{-q}) c_{k-q}^\dagger c_k V^{-1/2}, \tag{2.212}$$

We shall treat only Stokes scattering here, i.e., the $b^\dagger$ term in (2.212). The Fröhlich constant C_F is given by

$$C_F = -i \left(\frac{\omega_{LO}}{2Vc} \right)^{1/2} \left(\frac{1}{\varepsilon_\infty} - \frac{1}{\varepsilon_{rf}} \right)^{1/2}, \tag{2.213}$$

where ε_{rf} is the dielectric constant for $\omega = 0$. In order to obtain (2.213) from (2.111), one has to perform straightforward algebraic manipulations with (2.86a), including the use of the Lyddane-Sachs-Teller relation $\omega_{LO}^2/\omega_{TO}^2 = \varepsilon_{rf}/\varepsilon_\infty$.

The divergent nature of (2.212) for $q \to 0$ is the source of the anomalies we want to discuss. The matrix elements of (2.212) between the electronic states l and j such that $k_j = k_l - q$ (so as to conserve q) are:

$$\langle (n+1), j | \mathscr{H}_{ev} | n, l \rangle = C_F \left[\frac{\delta_{jl}}{|q|} + \langle j | \frac{q \cdot p}{|q|} | l \rangle (1 - \delta_{jl}) \right]. \tag{2.214}$$

In (2.214), δ_{jl} is zero if j and l belong to *different energy bands* (interband scattering), one otherwise. The singular first term in the rhs of (2.214) stems from the orthonormality of Bloch periodic functions within a given band (intraband scattering). The second term, zero for *intraband* scattering, is obtained through a $k \cdot p$ expansion of the wave function. It is not singular and it can be shown that it leads to the electro-optic terms of Sect. 2.1.12. In this manner, an explicit expression for the first-order electro-optic tensor involving three matrix elements of p [2.184] can be obtained.

The intraband term of (2.214), when replaced into the perturbation theoretical expressions which correspond to Fig. 2.33a, leads to an expression which diverges for $q \to 0$. One has to consider, however, two such terms: one in which $C_F/|q|$ couples two conduction states (electrons) and one in which it couples two valence states (holes). Because of the different charge of electrons (-1) and holes $(+1)$, these two terms cancel and the singularity is lifted. A more careful analysis, carried out to higher order in q, leaves as the balance between the conduction and valence intraband terms a term in χ_s proportional to q which, of course, would be zero by definition in the dipole approximation $(q=0)$ and thus "forbidden". This term, which is responsible for the so-called forbidden LO-scattering, is strongly resonant near critical points and can, under these conditions, produce higher scattering efficiencies than the allowed deformation potential terms discussed above (but only near resonance!). It can even lead to Raman scattering in the case of LO-phonons for materials with inversion symmetry [2.185]. We evaluate these terms in the following three sections.

a) Heuristic Approach

The calculation of the forbidden LO-scattering requires the evaluation of the terms of Fig. 2.33a to *second order* in q: the first-order terms must vanish as the Fröhlich Hamiltonian only depends on $|q|$. For this calculation, straightforward but tedious k-space integrals must be evaluated in 3 dimensions and, in this process, one runs the risk of losing the physical insight. Hence, we give here a *heuristic* calculation of this effect which yields, in the limit $|\omega_L - \omega_0 - i\Gamma| \gg \omega_v$, the same result as the correct calculation to a numerical factor of the order of one. This calculation is the closest we have been able to come to the "polarizability" calculations of the allowed effects discussed in Sect. 2.1. Its physical meaning is quite transparent.

The essence of this method is shown in Fig. 2.39. We assume $\omega_s \simeq \omega_L$ and a *finite* scattering vector q. The finite nature of this vector raises the minimum gap for valence conduction transitions to

$$\omega_0(q) \simeq \omega_0 + \frac{1}{2m_e}\left(\frac{q}{2}\right)^2 \tag{2.215}$$

for the diagram in Fig. 2.39a, and the same expression with m_e replaced by the hole mass m_h for Fig. 2.39b. It is this difference in effective q-dependent gaps which leads to the forbidden LO-effect. The Stokes Raman susceptibility can thus be written for the parabolic 3-dimensional bands of (2.166) [see (2.49)]

$$\chi'_s = \frac{C_F}{|q|} \frac{\partial^2 \chi}{\partial \omega_0^2}\left[\left(\frac{\partial^2(\omega_0)}{\partial q^2}\right)_c - \left(\frac{\partial^2 \omega_0}{\partial q^2}\right)_v\right] : qq, \tag{2.215a}$$

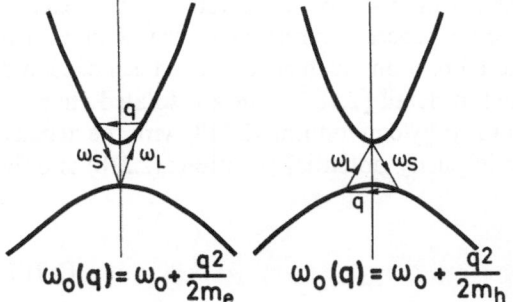

$$\omega_0(q) = \omega_0 + \frac{q^2}{2m_e} \qquad \omega_0(q) = \omega_0 + \frac{q^2}{2m_h}$$

Fig. 2.39. Schematic diagram of the two Fröhlich interaction processes involved in the heuristic calculation of forbidden Raman scattering by LO-phonons

where the subscripts c and v denote the gap shifts of Figs. 2.39a, 2.39b, respectively, i.e.,

$$\left(\frac{\partial^2 \omega_0}{\partial q^2}\right)_{cv} = \frac{1}{2m_{eh}}. \tag{2.216}$$

By replacing (2.216) into (2.215a), we find

$$\begin{aligned}
\chi'_{\underset{\sim}{s}} &= \frac{C_F}{8} \frac{q}{|q|} \cdot \left(\frac{1}{m_e} - \frac{1}{m_h}\right) \cdot q \frac{\partial^2 \chi}{\partial \omega_0^2} \\
&\simeq -\frac{C_F}{8} \frac{q}{|q|} \cdot \left(\frac{1}{m_e} - \frac{1}{m_h}\right) \cdot q \frac{\partial^2 \chi}{\partial \omega^2}.
\end{aligned} \tag{2.217}$$

The second form of (2.217) has been obtained from (2.215a) by dropping terms proportional to $\partial \chi/\partial \omega$ and χ, negligible near resonance. Equation (2.217) represents the Raman susceptibility for forbidden Raman scattering. This susceptibility is linear in $|q|$ and arises as a result of differences between the electron and the hole effective masses. For q along one of the principal directions of the tensor $(m_e^{-1}) - (m_h^{-1})$, this forbidden scattering is *polarized*. Otherwise, depolarized scattering is possible (although weak). For isotropic masses the tensor (2.217) is isotropic and can be written as

$$\chi'_{\underset{\sim}{s}}(|q|) = -\frac{C_F}{8} \frac{m_e - m_h}{M^* \mu^*} \left(\frac{\partial^2 \chi}{\partial \omega^2}\right) |q|, \tag{2.218}$$

where we recall $\mu^{*-1} = m_e^{-1} + m_h^{-1}$ and $M^* = m_e + m_h$. The scattering is completely polarized and isotropic (independent of the direction of q). An ideal configuration to observe this effect for zincblende is for a (100) face with $\hat{e}_L \| \hat{e}_s \| (010)$; in this configuration, no *allowed* scattering is possible (Table 2.2). We point out that (2.2.18) is equal to three times the correct χ_s given in Sect. 2.3.8b in the limit $\omega_L \simeq \omega_s$. It can also be applied, to a numerical factor of the order of unity, to the two-dimensional case of Sect. 2.2.8c. Equation (2.218) leads to "dipole forbidden" scattering only for a perfect crystal. Impurities can contribute a finite q to the scattering process, unrelated to the change in q between the incident and the scattered photon. Such effects have been observed [2.186] and theoretically calculated in detail [2.187]. For an isolated charged impurity, the effect is basically obtained by convoluting (2.218) with the density of phonons and the square of the impurity potential [caution: (2.218) is only valid for small q!, see (2.232)]:

$$\chi_s \propto \int \chi_s(|q|) \left| \frac{4\pi}{q^2 + q_{FT}^2} \right|^2 dq . \tag{2.219}$$

A more detailed treatment of this effect is given in [2.187].

Likewise, the exact cancellation of the conduction and valence terms for $q=0$ can be lifted if the vibrations are localized, provided the corresponding localization lengths of valence and conduction bands are not the same. Such has been found to be the case for superlattices in which the phonons are localized on each layer [2.188]. The valence and conduction wave functions can also be localized but the amount of localization depends on the effective masses and barrier heights. Consequently, within each layer the Fröhlich interactions for electrons and holes do not cancel even for $q=0$ and the LO-scattering becomes allowed. The resonance is then proportional to the first derivative of χ instead of the second. The same thing is found for a resonance near a discrete exciton state (Sect. 2.2.9).

b) 3-d Critical Points (E_0 and $E_0 + \Delta_0$-Edges)

Let us treat the isotropic case. The perturbation theoretical expression for the Raman susceptibility is [2.189]

$$(4\pi)^{-1} \chi'_{s,\alpha\alpha} = \frac{C_F}{|q| V \omega_s^2} P^2 \sum_k \frac{1}{\omega_L - \omega_0 - \frac{1}{2\mu^*} \cdot k^2}$$
$$\cdot \left(\frac{1}{\omega_L - \omega_0 - \omega_{LO} - \frac{1}{2\mu^*}(k - S_e q)^2} - \frac{1}{\omega_L - \omega_0 - \omega_{LO} - \frac{1}{2\mu^*}(k + S_h q)^2} \right) , \tag{2.220}$$

where $S_{e,h} = m_{eh}/M^*$ and the matrix elements of p have been assumed to be independent of k and equal to P. Transforming the sum over q into an integral, we obtain [2.189]

$$(4\pi)^{-1} \chi'_{s,\alpha\alpha} = \frac{4 C_F P^2 \mu^{*2}}{|q| \omega_s \omega_L} [\Xi(q S_e) - \Xi(q S_h)] , \tag{2.221a}$$

where

$$\Xi = \frac{1}{4\pi|q|} \arctan\{iq(2\mu^*)^{-1/2}\omega_{LO}^{-1}[(\omega_L - \omega_0)^{1/2} - (\omega_s - \omega_0)^{1/2}]\}. \qquad (2.221b)$$

By expanding the arctan in (2.221b) to third order in the argument, we find

$$(4\pi)^{-1}\chi'_{s,\alpha\alpha} = \frac{C_F P^2 |q|(m_e - m_h)}{12\pi\omega_L\omega_s\omega_{LO}^3 M^*}(2\mu^*)^{1/2}[(\omega_L - \omega_0)^{1/2} - (\omega_s - \omega_0)^{1/2}]^3. \qquad (2.222)$$

Equation (2.222) is only valid for $|q| \ll (2\mu^*)^{1/2}\omega_{LO}[(\omega_L - \omega_0)^{1/2} - (\omega_s - \omega_0)^{1/2}]^{-1}$. For larger $|q|$'s, (2.221b) goes through a maximum when the magnitude of the argument $\simeq 1$ [compare with (2.231, 244)].

In order to compare (2.222) with (2.218), we first use (2.170) and obtain

$$\chi'_{s,\alpha\alpha} = \frac{m_e - m_h}{M^*\mu^{*4}} \frac{C_F|q|\pi^2}{24P^4\omega_L\omega_s\omega_{LO}^3(4\pi)^2}[\omega_L^2\chi(\omega_L) - \omega_s^2\chi(\omega_s)]^3. \qquad (2.223)$$

In the limit $\omega_{LO} \to 0$, we can rewrite the term in brackets in (2.223) as a function of $\partial\chi/\partial\omega$:

$$\chi'_{s,\alpha\alpha} = \frac{m_e - m_h}{M^*\mu^{*4}} \frac{C_F|q|\pi^2\omega_L^4}{24P^4(4\pi)^2}\left(\frac{\partial\chi}{\partial\omega}\right)^3_{\omega_L - \omega_{LO}/2}$$

$$= \frac{C_F}{24}\frac{m_e - m_h}{M^*\mu^*}\left(\frac{\partial^2\chi}{\partial\omega^2}\right)_{\omega_L - \omega_{LO}/2} \cdot |q|. \qquad (2.224)$$

We thus recover (2.218) except for a factor of 1/3 which stems from the angular integrations which have not been correctly performed in (2.218). Note that whenever $\chi(\omega_L)$ is real (2.224) is pure imaginary as C_F is pure imaginary [see (2.213)]. This means that this type of forbidden scattering does not interfere with allowed scattering.

c) 2-d Critical Points $(E_1, E_1 + \Delta_1)$-Edges

The case of a two-dimensional critical point is treated in detail in [2.74, 190] with particular application to the E_1, $E_1 + \Delta_1$ gaps of germanium and GaAs. We give here the result for a *single* critical point (there are four E_1 gaps along the equivalent {111} directions but their contributions to the Raman suscepti-bility are not equivalent!). For q along one of the "two-dimensional" directions $\bar{x}$ (the other being $\bar{y}$), we have

$$(4\pi)^{-1}\chi'_{s,\bar{y}\bar{y}} = \frac{2C_F|q|}{a_0\omega_s^2}\frac{(S_e - S_h)}{\omega_{LO}^2}J(\alpha,\beta)|\langle c|p_{\bar{y}}|v\rangle|^2, \qquad (2.225)$$

where

$$J(\alpha, \beta) = \left(\frac{\partial}{\partial \beta} + \beta \frac{\partial^2}{\partial \beta^2}\right) \frac{\sqrt{3}}{8\pi} \frac{\ln \beta - \ln \alpha}{\alpha - \beta};$$

$$\alpha = (\omega_L - \omega_1)/\omega_{LO},$$

$$\beta = (\omega_s - \omega_1)/\omega_{LO}.$$

For other directions of the ellipsoid, the contribution to χ_s can be obtained by means of a rotation of the axes [2.190]. Note that, because of the q-dependence, $\underset{\sim}{\chi_s}$ transforms upon rotation like a *fourth rank tensor*.

By transforming the finite differences in (2.225) into a derivative with respect to ω_L (valid for $\omega_{LO} \ll |\omega_L - \omega_0 - i\Gamma|$), we find

$$(4\pi)^{-1}\chi'_{s,\bar{y}\bar{y}} = \frac{\sqrt{3}C_F(S_e - S_h)|\langle c|P_{\bar{y}}|v\rangle|^2}{4a_0\pi\mu\omega_s^2} \frac{q}{(\omega_L - \omega_1)^2}. \qquad (2.225a)$$

For the case of the E_1 critical points, we can rewrite (2.225a) by using (2.185) as:

$$\chi'_{s,\bar{y}\bar{y}} = -\frac{3C_F}{16} \frac{m_e - m_h}{\mu^* M^*} q \left(\frac{\partial^2 \chi}{\partial \omega^2}\right)_{\omega = \omega_L + \omega_{LO}/2}. \qquad (2.226)$$

In (2.226), $\chi_{s,\bar{y}\bar{y}}$ represents the contribution of electrons with k along the [111] direction for $\bar{y}\|1\bar{1}0$. The susceptibility χ which enters in the second derivative of (2.226) has been chosen to be that of *all* the {111} directions, equal to 8/3 that of [111]. Except for a numerical factor of the order of unity, (2.226) is basically the same result as obtained in Sect. 2.2.8a (2.218). We should point out that for this case of an anisotropic mass tensor ($m_\| \gg m_\perp$), forbidden scattering can be observed even for $\hat{e}_L \perp \hat{e}_s$, provided $\hat{e}_L$ and $\hat{e}_s$ do not coincide with the principal axes of the mass tensor. The scattering efficiency for this effect, however, has been estimated to be at most 1/50 of that for $\hat{e}_L \| \hat{e}_s$ [2.74].

d) Electric-Field-Induced Effect

Forbidden LO-scattering can also be induced by a dc electric field E_{dc}. In this case, the effect does not depend on q. A simple estimation of this effect for interband transitions can be made by noting that an electric field E introduces a change in the linear χ by the amount [2.191]

$$\Delta\chi_{ij}(\omega) = \frac{1}{24\omega^2} E \cdot \left(\frac{1}{\mu^*}\right) \cdot E \frac{\partial^3}{\partial \omega^3} [\omega^2 \chi_{ij}(\omega)]. \qquad (2.227)$$

The field E responsible for the effect under discussion is the sum of the Fröhlich field $|C_F|(q/|q|)b_q^+ V^{-1/2}$ [see (2.212), we omit the electron operators as irrele-

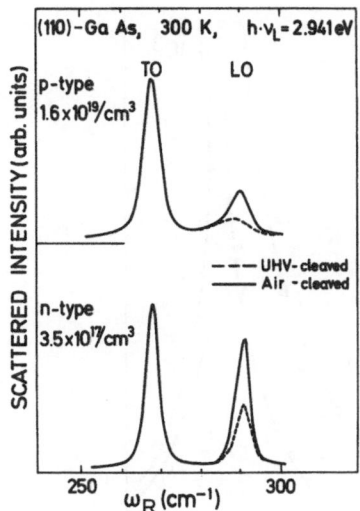

Illustration of surface-field-induced LO-scattering in GaAs; for the cleaved samples no surface field is present and only q-dependent effects (including those induced by impurity scattering) are seen [2.192]

vant for this discussion] and the dc applied field. Hence, we find

$$\chi'_s(\omega) = \frac{|C_F|}{12\omega^2} E_{dc} \cdot \left(\frac{1}{\mu^*}\right) \cdot \frac{q}{|q|} \frac{\partial^3}{\partial\omega^3} [\omega^2 \underset{\sim}{\chi}(\omega)].$$ (2.228)

Note that (2.228), in contrast to (2.224), is real for $\chi(\omega_L)$ real.

A complete calculation of this effect, without the restriction $\omega_{LO} \ll |\omega_L - \omega_0 + i\Gamma|$, has been performed in [2.190]. The result agrees with (2.228) except for an unimportant replacement of 12 by 3π.

Equation (2.228) reveals a few interesting selection rules for the field-induced effect. If μ^* is isotropic (e.g., near the $E_0 + \Delta_0$ gap), the effect only occurs for $E_{dc} \| q$. Surface electric fields in back-scattering are thus ideal candidates for its production. The case of GaAs is particularly interesting (Fig. 2.40). If this material is cleaved well in vacuum, there is no pinning of the Fermi energy at the surface and hence no surface field. If cleaved in air, surface states pin the Fermi energy and, for electron concentrations $\simeq 10^{17}\, cm^{-3}$, surface fields sufficient to enhance the q-induced forbidden LO-scattering result. Note that the selection rules for $\hat{e}_{L,s}$ are the same as in the case of q-induced scattering: for isotropic critical points, the scattering is polarized ($\hat{e} \| \hat{e}_s$).

2.2.9 Resonant Raman Scattering: Effect of Exciton Interaction

We have limited ourselves in the discussion of the quantum theory of resonance Raman scattering to the case of intermediate states which are *uncorrelated* electron-hole excitations (except for the phenomenological treatment of the fluorites in Sect. 2.2.7). The simplest form of correlation between these two

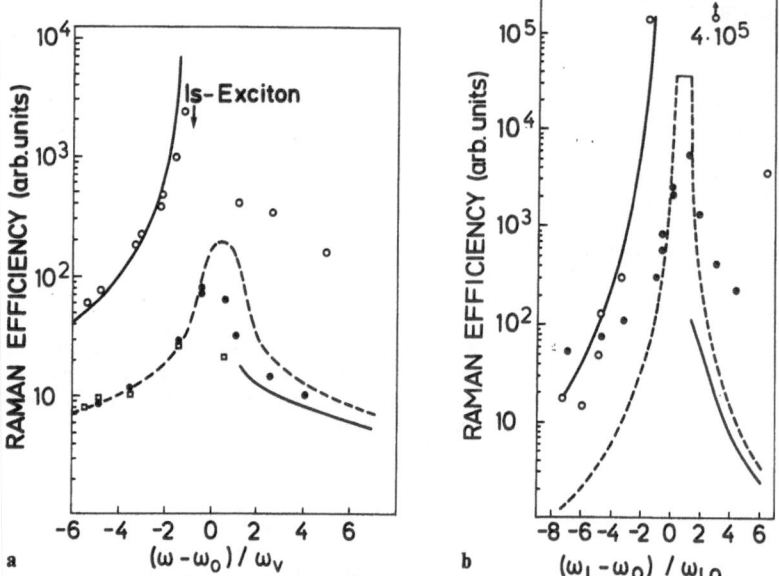

Fig. 2.41a, b. Allowed (**a**) and forbidden (**b**) resonance Raman efficiencies calculated with parameters appropriate to CdS and GaP and also experimental points. The lines are theoretical. The solid lines include exciton interaction; the dashed ones do not [2.195]

particles is the screened Coulomb interaction; it produces bound states (excitons) and modifications of the interband continua [2.48, 193]. As already mentioned, the latter do not modify the behavior of S near resonance in as far as their effect can be taken care of by multiplying the static $\chi(\omega)$ by a phase factor (Sect. 2.2.5).

A number of calculations of allowed and forbidden resonant Raman efficiencies, including discrete excitons, have been published [2.189, 194]. These works involve numerical integrations of gamma-functions whose physical meaning is not very transparent. In order to simplify the treatment, the exciton is often added in ad hoc manner based on the susceptibility expressions such as (2.174, 224). One or more Lorentzian excitonic dispersions are added to the interband terms or, equivalently, experimental spectra including all exciton effects are used for $\chi(\omega)$ in the expression which contains this function [2.52]. A qualitative examination of the published full calculations indicates that this method works well for allowed scattering. For forbidden scattering, however, there is a large cancellation below the lowest exciton of the contributions of this exciton and those of the continuum. In spite of this cancellation, a net *enhancement* of the scattering efficiency near the $n=1$ exciton by about one order of magnitude results in cases such as GaP and CdS [2.189] (Fig. 2.41). Above ω_0 some destructive interference between discrete exciton and continuum occurs and the calculated efficiency is lower than for uncorrelated pairs.

The comparison of the experiments with these predictions is not too satisfactory at the moment: the experimental results (see Fig. 2.41) do not lend support to the asymmetry in the resonance predicted by the theory. Lifetime effects in the continuum probably lift the destructive interference predicted by the theory. Calculations of this effect using a polaron picture for the intermediate state continuum have been performed by *Ferrari* and *Luzzi* [2.195]. These calculations restore, at least qualitatively, the good agreement between theory and experiment which existed before introducing the exciton interaction.

Excitonic effects are observed most clearly in the phenomenon of resonant polariton scattering [Ref. 1.2, Chap. 7]. We present, therefore, before closing this section, the Hamiltonian for the exciton-polariton interaction and discuss some of its features [2.196]. The Hamiltonian for Fröhlich interaction is obtained by taking matrix elements of (2.212) between two excitonic states 1 and 2 of total k equal to k_L and k_s, respectively. Their wave functions are:

$$\psi_1 = \frac{1}{\sqrt{V}} e^{ik_L \cdot R} \phi_1(\varrho)$$

$$\psi_2 = \frac{1}{\sqrt{V}} e^{ik_s \cdot R} \phi_2(\varrho),$$

(2.229)

where $\phi_i(\varrho)$ represent the "envelope function" with respect to the relative coordinate $\varrho = r_e - r_h$ and $R = (M)^{-1}(m_e r_e + m_h r_h)$ the center-of-mass coordinate. The part of $\mathscr{H}_{ev}$ (2.212) which acts on conduction electrons can thus be written as (phonon emission)

$$\langle 2|\mathscr{H}_{ev}|1\rangle_e = \frac{1}{V} \frac{C_F}{|q|} \iint e^{-ir_e \cdot q} \phi_2^*(\varrho)\phi_1(\varrho) e^{i(q_L - q_s)\cdot R} d\varrho dR.$$

(2.230)

By replacing into (2.230) $r_e = R + \frac{m_h}{M^*} \varrho$, we find

$$\langle 2|\mathscr{H}_{ev}|1\rangle_e = \frac{C_F}{|q|} \int e^{-\frac{m_h}{M^*} \varrho \cdot q} \phi_2^*(\varrho)\phi_1(\varrho) d\varrho \quad \text{for} \quad q = q_L - q_s,$$

(2.231)

zero otherwise.

The Hamiltonian (2.212) yields the same expression (2.231) for interaction with the hole component of the exciton, except for a $(-)$ sign and the replacement $m_h \rightarrow -m_e$. We thus obtain for the total Stokes interaction

$$\langle 2|\mathscr{H}_{ev}|1\rangle_F = \frac{C_F}{|q|} \int \left[e^{i\frac{m_e}{M^*}\varrho \cdot q} - e^{i\frac{m_h}{M^*}\varrho \cdot q} \right] \phi_2^*(\varrho)\phi_1(\varrho) d\varrho.$$

(2.232)

A nonvanishing Fröhlich interaction with the exciton thus results only if $m_e \neq m_h$, a fact which we had already encountered in Sect. 2.2.8. Equation

(2.232) can be integrated if we take for $\phi_1(\varrho)=\phi_2(\varrho)$ the envelope function of the ground state of the exciton, which anyhow has the largest contribution to the optical oscillator strength:

$$\phi_1(\varrho)=\frac{1}{\sqrt{\pi a_{ex}^3}}e^{-\varrho/a_{ex}}.\tag{2.232a}$$

We find [2.196]

$$\langle 1|\mathcal{H}_{ev}|1\rangle_F=\frac{C_F}{|q|}\left[\frac{1}{(1+\beta_h^2 q^2)^2}-\frac{1}{(1+\beta_e^2 q^2)^2}\right],\tag{2.232b}$$

where $\beta_{eh}=\dfrac{m_{eh}}{2M^*}a_{ex}$.

The same derivation would apply to deformation potential interaction except that the prefactor $C_F|q|^{-1}$ should be removed and each one of the two terms in (2.232b) must be multiplied by $\bar{\mathcal{E}}_c$ and $\bar{\mathcal{E}}_v$, the deformation potential constants (properly normalized) of conduction and valence band, respectively (2.162):

$$\langle 1|\mathcal{H}_{ev}|1\rangle_{dp}=\frac{\bar{\mathcal{E}}_c}{(1+\beta_h^2 q^2)^2}-\frac{\bar{\mathcal{E}}_v}{(1+\beta_e^2 q^2)^2}.\tag{2.232c}$$

For $q\to0$ (dipole approximation), the interaction of (2.232c) depends only on the difference between $\bar{\mathcal{E}}_c-\bar{\mathcal{E}}_v$, i.e., on $d\omega_0/d\xi$. This is the result we have used so far for uncorrelated electrons. For q finite, however (beyond the dipole approximation), the effect depends on the *separate values* of $\bar{\mathcal{E}}_c$ and $\bar{\mathcal{E}}_v$ if $m_e\neq m_v$. This rather surprising result suggests the possibility of determining absolute deformation potentials (and not only their differences) by measuring resonant Raman or Brillouin effects [2.90]. We note that the modulation of $\chi(\omega)$ by the phonon which results in χ_s takes place, in the present case, through a modulation of the matrix element and not of the energy denominator. Hence, the Raman susceptibility in this case is proportional to $\partial\chi/\partial\omega$ instead of $\partial^2\chi/\partial\omega^2$, see (2.254).

2.2.10 Second Order Raman Scattering by Two Phonons

Some of the basic facts about second-order scattering are discussed in [Ref. 2.1, Chaps. 1 and 3]. The relevant diagrams are given in Fig. 1b–d of that reference. Diagram 1d, a "cascade" process using a photon as intermediate state, is negligible (at least for $\omega_L<\omega_0$): the efficiency should be of the order of the square of the efficiency for first-order scattering, too small to be observed. We are thus left with processes 1b and 1c: 1b is produced by the first-order electron-phonon Hamiltonian discussed so far [$\mathcal{H}_{ev}$ here called $\mathcal{H}_{ev}^{(1)}$], iterated to second-order perturbation theory. Figure 1c of [2.1] is related to the second-order electron-phonon Hamiltonian $\mathscr{H}_{ev}^{(2)}$ in first-order perturbation theory. The theory for the $\mathscr{H}_{ev}^{(2)}$ effect is thus usually isomorphic to that of the first-order scattering discussed above [2.197].

Away from resonances, and strictly speaking for ω_L below all continua, the iterated $\mathscr{H}_{ev}^{(1)}$ effect and that of $\mathscr{H}_{ev}^{(2)}$ can be lumped together into a renormalized $\overline{\mathscr{H}_{ev}^{(2)}}$ and a theory isomorphic to that of the scattering by one phonon can be used. A few remarks concerning $\mathscr{H}_{ev}^{(2)}$ and $\overline{\mathscr{H}_{ev}^{(2)}}$ are in order. The diagonal matrix elements of $\overline{\mathscr{H}_{ev}^{(2)}}$ are (a similar expression holds for non-diagonal ones):

$$\langle i|\overline{\mathscr{H}_{ev}^{(2)}}|i\rangle = \langle i|\mathscr{H}_{ev}^{(2)}|i\rangle + \sum_j \frac{\langle i|\mathscr{H}_{ev}^{(1)}|j\rangle \langle j|\mathscr{H}_{ev}^{(1)}|i\rangle}{\omega_i - \omega_j}. \tag{2.233}$$

For Stokes scattering, the two phonons involved in $\mathscr{H}_{ev}^{(2)}$ must have wave vectors q_1 and q_2 such that $q_1 + q_2 = q$, where q is the scattering vector (Sect. 2.1.13). While q_1 and q_2 can extend to the whole BZ, $q \simeq 0$. For acoustic phonons, $\langle i|\mathscr{H}_{ev}^{(2)}|i\rangle \rightarrow 0$ for $q \rightarrow 0$ as a uniform translation cannot produce energy shifts. The corresponding matrix elements of $\mathscr{H}_{ev}^{(2)}$, however, are not zero. In fact, it is easy to see that within the *rigid ion model*, $\mathscr{H}_{ev}^{(2)}$ is related to the crystal potential U through [2.197]

$$\mathscr{H}_{ev}^{(2)} = \left(\sum_m^{PC} \frac{1}{M_m} \hat{e}_{m2} \cdot \frac{\partial^2 U}{\partial R_m \partial R_m} \cdot \hat{e}_{m1}\right) b_2^\dagger b_1^\dagger \sqrt{\frac{1}{4\omega_{v_1}\omega_{v_2}V^2}}, \tag{2.234}$$

where m are the atoms of the PC and $\hat{e}_m$ the corresponding phonon eigenvectors. For the diamond structure both eigenvectors are equal and, using the orthonormality of the eigenvectors of different phonon bands,

$$\mathscr{H}_{ev}^{(2)} = M^{-1}\left(\frac{\partial^2 U}{\partial R_1^2}\right)\delta_{1,2} b_2^\dagger b_1^\dagger \sqrt{\frac{1}{4\omega_{v_1}\omega_{v_2}V^2}}, \tag{2.234a}$$

where $\delta_{1,2}$ is zero if both phonons belong to *different bands*, one otherwise. Hence, the unrenormalized $\mathscr{H}_{ev}^{(2)}$ only contributes to *overtone*, not to *combination* scattering. Using for U the unscreened Coulomb potential of the atomic nucleus (atomic number Z), we find

$$\frac{\partial^2 U}{\partial R_1^2} = \frac{1}{3}\nabla^2 U \simeq \frac{1}{3} 4\pi Z \, \delta(R). \tag{2.235}$$

By replacing (2.235) into (2.234a), we obtain (for $T \simeq 0\,\mathrm{K}$)

$$\langle i|\mathscr{H}_{ev}^{(2)}|i\rangle = D_{1,i}\frac{1}{3a_0^2\omega_v MV} \simeq \frac{4\pi}{3} Z|\Psi_i(0)|^2 \frac{1}{4\omega_v MV}. \tag{2.236}$$

Equation (2.236) contains the definition of the electron-two-phonon deformation potential D_1 with the normalizing factors with which it is usually given in the literature [2.198]. It shows that $D_{1,i}$ is zero except when $|i\rangle$ is an s-like state (e.g. the conduction band Γ_2, of Ge. For the $\Gamma_{25'}$ valence band, D_0 $(\Gamma_{25'}) = 0$. With the value of $\Psi_i(0)$ obtained from *Herman* and *Skillman's* tables [2.199], one finds for the $4s$ levels of Ge ($\equiv \Gamma_2$, conduction band) $D_1 \simeq -10^6$ eV, independent of the phonon under consideration. Since the *renormalized* $\bar{D}_1$ must be zero for acoustical phonons of $q \to 0$, we conclude that, at least in this case, a strong cancellation between the $\mathscr{H}^{(2)}$ and the $\mathscr{H}^{(1)}$ terms takes place. Such a cancellation is not limited to acoustic phonons with $q \to 0$. For optical phonons, $\bar{D}_1 \simeq 10^3$ eV [2.197] and a strong cancellation must also take place.

It is instructive to evaluate D_1 using pseudopotentials and pseudowave functions. We obtain ([2.169], see also (2.196))

$$D_1(\Gamma_{2'}) = 2\pi^2[8(3 + \gamma^2)v_8 + 2\beta\gamma\sqrt{6}(3v_3 + 11v_{11})]$$

$$\tag{2.237}$$

$$D_1(\Gamma_{25'}) = 2\pi^2[8\beta^2 v_8 + 2\beta\gamma\sqrt{2}(3v_3 - 11v_{11})].$$

For Ge ($v_3 = -3.1$ eV, $v_8 = 0$, $v_{11} = +0.82$ eV), we find $D_1(\Gamma_{2'}) = -12\ D_1(\Gamma_{25'})$ $= -468$, $D_1(\Gamma_{2'}) - D_1(\Gamma_{25'}) = 456$ eV, a factor of five smaller than typical experimental values of $\bar{D}_1$ for the $\Gamma_{25'}$ phonon ($\sim 2 \times 10^3$ eV [2.74]). Hence, the cancellation between $\mathscr{H}_{ev}^{(1)}$ and $\mathscr{H}_{ev}^{(2)}$ terms is not as drastic in the pseudopotential calculation as it is when full wave functions are used. Since (2.237) is independent of the phonons under consideration, a cancellation must take place for TA-phonons of $q \to 0$. This cancellation can be seen explicitly in a recent calculation of the separate $\mathscr{H}_{ev}^{(2)}$ and $\mathscr{H}_{ev}^{(1)}$ terms performed with pseudopotentials for Ge [2.199].

We note that pseudopotential calculations should give for $\bar{\mathscr{H}}_{ev}^{(2)}$ the same results as a calculation using correct wave functions and potentials, as $\bar{\mathscr{H}}_{ev}^{(2)}$ corresponds to an observable energy shift. There is, however, no reason why both types of calculations should give the same values for the separate contributions to $\bar{\mathscr{H}}_{ev}^{(2)}$, $\mathscr{H}_{ev}^{(2)}$ and $\sum (\mathscr{H}^{(1)})^2/\Delta E$. In the pseudopotential calculation, $\bar{D}$ for optical phonons is determined mainly by $\mathscr{H}_{ev}^{(1)}$ terms [2.199].

The Hamiltonian $\mathscr{H}_{ev}^{(2)}$ which results from (2.234, 235) has the full symmetry of the space group of the crystal (Γ_1). It must, therefore, lead to second-order spectra of Γ_1 symmetry in which the pairs of phonons involved have this symmetry. Second-order spectra of $\Gamma_{25'}$ and Γ_{12} symmetry (for Ge) must thus arise from $\mathscr{H}_{ev}^{(1)}$ terms, at least within the context of the rigid-ion model. The scattering efficiency for a perturbation of Γ_1 symmetry is obtained by differentiating the susceptibility with respect to the gap. For the case of the E_0, $E_0 + \Delta_0$ gaps (Sects. 2.2.5 and 6a), one must pay attention to the fact that the density-of-states mass is affected by the change in ω_0 induced by a Γ_1 perturbation according to $m_d^{-1} \sim P^2/\omega_0$. This is automatically taken into account by differentiating (2.181) with C_0'' = constant. We find for E_0 or $E_0 + \Delta_0$

$$\chi_s = \frac{d\chi}{d\omega_0} \frac{\bar{D}_1}{3a_0^2 M} \left[\frac{(1+n_1)(1+n_2)}{\omega_{v_1}\omega_{v_2}} V_c N_{d,12}(\omega_R) \right]^{1/2} V^{-1/2}, \tag{2.238}$$

where $N_{d,12}$ is the density of two-phonon states with $q_{v_1} + q_{v_2} = q$ per PC. $\bar{D}_1$ is to be regarded as an average electron-two-phonon deformation potential including $\mathcal{H}_{ev}^{(1)}$ and $\mathcal{H}_{ev}^{(2)}$ terms. The differential scattering efficiency is obtained, as usual, by replacing (2.238) into

$$\left(\frac{\partial^2 S_s}{\partial \omega_R \partial \Omega} \right) = \frac{\omega_s^4 V}{(4\pi)^2 c^4} |\hat{e}_s \cdot \chi_s \cdot \hat{e}_L|^2 \tag{2.239}$$

[compare with (2.18)]. In terms of the functions F and G of (2.172, 174), (2.239) can be evaluated by introducing the Raman susceptibility

$$(4\pi)^{-1}\chi_s = \frac{C_0''}{24\pi} \frac{\bar{D}_1}{a_0^2\omega_0} \{G(x) + 3F(x) + \tfrac{1}{2}[G(x') + 3F(x')]\}$$

$$\cdot \left[\frac{(1+n_1)(1+n_2)}{\omega_{v_1}\omega_{v_2}M^2} V_c N_{d,12}(\omega_R) \right]^{1/2} V^{-1/2} \tag{2.240}$$

$$x = \frac{\omega_L + \omega_{v_1}/2 + \omega_{v_2}/2}{\omega_0}, \; x' = \frac{\omega_L + \omega_{v_1}/2 + \omega_{v_2}/2}{\omega_0 + \Delta_0},$$

where (2.175) has been used. We note that (2.240) has the same form as (2.175) and thus vanishes for $\omega_L \to 0$. One should not attach much significance to this fact as a constant can always be added to (2.240).

For the component of the second-order spectrum of $\Gamma_{25'}$ symmetry, the $\mathcal{H}^{(2)}$ mechanism yields a susceptibility isomorphic to that of (2.191, 194). It is obtained by means of the replacement [2.197]

$$\frac{d_0(n+1)}{\sqrt{\omega_v}} \to D_{25'} \frac{2(n_1+1)(n_2+1)}{a_0 \sqrt{\omega_{v_1}\omega_{v_2}M^*}} \tag{2.241}$$

and multiplying the resulting expression by the square root of the density of two-phonon states. Equation (2.241) is actually the definition of the deformation potential $D_{25'}$.

The contribution of E_0 and $E_0 + \Delta_0$ to the Γ_{12} component of the second-order Raman spectrum is isomorphic with the $\Gamma_{25'}$ component. To obtain its expression, it suffices to replace $D_{25'}$ by D_{12}. This isomorphism does not hold for the contribution of the E_1, $E_1 + \Delta_1$ edges.

Resonant second-order scattering near the E_1, $E_1 + \Delta_1$ edges has been observed for Ge, Si and a number of III–V compounds. The corresponding expressions can be found in the literature [2.44a]. They depend, for the Γ_1

component of the spectrum, on a deformation potential $\bar{D}_1$ analogous to that defined in (2.236) for the E_0 edge. The $\Gamma_{25'}$ component is isomorphic to (2.201) and depends on two deformation potentials $D^5_{3,0}$ and $D^5_{1,0}$. The Γ_{12} component is isomorphic to the $D^5_{3,0}$ intraband contribution to the $\Gamma_{25'}$ component. It is determined by the deformation potential $D^3_{3,0}$ [2.44a].

We would like to discuss an interesting relationship between the resonant Γ_1 component of the Raman spectrum and the temperature coefficient of the corresponding gap. Let us consider a gap ω_0. In the *overtone* Raman process, two phonons of wave vector q, and $\simeq -q$, are emitted. The Raman effect results from the modulation by the phonons of the ω_0 gap, as given in (2.236). The temperature coefficient of the gap results from the emission of a phonon of wave vector q followed by its absorption. The coupling constants for these processes are the same and hence we conclude that the total efficiency for second-order overtone scattering and the temperature coefficient of the gap are related. Following this argument, the shift in the gap with temperature can be written as

$$\Delta\omega_0 = \frac{6}{3a_0^2 M} \langle (2n+1)\omega_v^{-1}\bar{D}_1 \rangle, \tag{2.242}$$

where the angular brackets represent an average over the whole phonon spectrum and the factor of 6 in the numerator is the number of phonons per unit cell. In the high temperature limit, $2n+1 \simeq 2kT/\omega_v$ and the shift $\Delta\omega_0$ is linear in temperature. Taking for Ge an average phonon frequency equal to half that of the Raman phonon $\langle \omega_v \rangle \simeq 20$ meV, we find for the linear temperature coefficient of the gap

$$\frac{d\omega_0}{d(kT)} = \frac{4}{a_0^2 M \langle \omega_v^2 \rangle} \langle \bar{D}_1 \rangle \simeq 5 \text{ (from experiment)}. \tag{2.243}$$

Equation (2.243) yields $\langle \bar{D}_1 \rangle = 250$ eV, in reasonable agreement with the results obtained from second-order Raman scattering for TA-phonons [2.44a, 74] [the TA phonons give, around room temperature, the main contribution to $d\omega_0/d(kT)$]. We should point out that a pseudopotential calculation seems to overestimate $\bar{D}_1$ for the acoustic modes and underestimate it for the optic modes [2.200]. A more recent erratum to [2.199], however, corrects this insufficiency [2.200].

We have so far discussed two-phonon scattering in the case in which the renormalization of $\mathscr{H}^{(2)}$ is possible. As resonance is approached, one must add ω_L to the denominator of the $\mathscr{H}^{(1)}$ term in (2.233) and this term can become more strongly resonant than $\mathscr{H}^{(2)}$. This effect, however, only occurs for certain phonons, namely, those whose q equals that of the resonant excitation. Thus, a deformation of the two-phonon spectrum from the density of two-phonon states expected in the $\mathscr{H}^{(2)}$ case is observed. Near an indirect gap, the phonons which contribute to the gap absorption are enhanced in the two-phonon

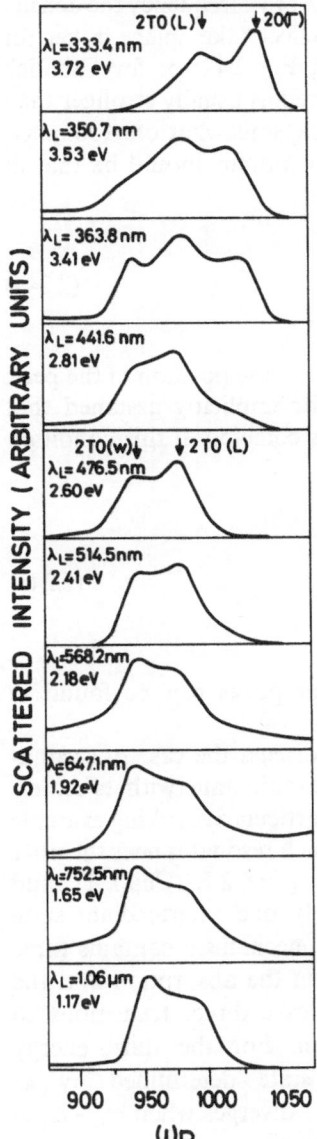

2ⁿᵈORDER RAMAN SCATTERING SILICON

Fig. 2.42. Resonance of the second-order 2 TO Raman spectrum of silicon showing the preferential enhancement of the 2 TO (Γ) phonons as the E_1 gap (~ 3.4 eV) is approached [2.218]

Raman scattering spectrum. Such effects have been observed for Si, GaP [2.171], GaAs [2.201], and the silver halides [2.202] (Sect. 2.3.7). For direct gaps, the phonons with $q = 0$ are enhanced in the two-phonon spectrum as the resonance is approached. For these phonons, the fourth-order perturbation term has *three* equally resonant energy denominators. An example of this effect is shown in Fig. 2.42 for Si: as the lowest direct absorption edge at 3.4 eV is approached, the top of the two-phonon spectrum becomes deformed and one peak at the frequency of two $\Gamma_{25'}$ phonons appears. At this frequency the density of states is zero and hence shows no peak. The effect takes place either for deformation potential coupling (nonpolar modes, Fig. 2.42) or for Fröhlich interaction (Fig. 2.50 below). The latter, when allowed, is usually stronger than the deformation potential effects. At a frequency ω_L somewhat off-resonance, the q for which the scattering efficiency has its maximum should be that at which, see (2.221),

$$\frac{q_{max}^2}{2M^*} \simeq |\omega_L - \omega_0|. \tag{2.244}$$

If the phonon dispersion relation is of the form (2.117), the position of the peak shifts as resonance is approached. We have so far implicitly assumed that $2\omega_v < \Gamma$ so that in-going and out-going resonances coincide. If this condition does not hold, the resonance splits into three:

in-going	for	$\omega_L = \omega_0$
intermediate	for	$\omega_L = \omega_0 + \omega_v$
out-going	for	$\omega_0 = \omega_0 + 2\omega_v$, i.e. $\omega_s = \omega_0$.

$$\tag{2.245}$$

Explicit calculations showing these three resonant peaks can be found in [2.203].

Before closing this section, we want to briefly discuss the case of second-order $\mathscr{H}^{(1)}$-type scattering involving the discrete excitonic state (with its center-of-mass kinetic energy) as an intermediate state. A particularly striking example of this type is the scattering by two $\Gamma_{12'}$ phonons which resonates near the $n = 1$ dipole-forbidden yellow exciton in Cu_2O (see [Ref. 2.1, Chap. 3] and Sect. 2.3.6). In this case there is, for each q, only one intermediate state determined by q-conservation. The $\mathscr{H}^{(1)}$ scattering mechanism contains three energy denominators. Two of them, those involving the absorption and the emission of photons, are nonresonant as in this case direct transitions to the resonating exciton band are dipole-forbidden. For the third energy denominator, there is only one intermediate state determined by k-conservation for each phonon q. Hence, this process diverges when $\omega_L - \omega_v$ is within the continuum unless we take $\Gamma \neq 0$. The scattering is given as a sum of the square of Raman susceptibilities over final, i.e., equivalently, over intermediate states [Ref. 2.1, Eq. (3.86)]

$$\left(\frac{\partial^2 S_s}{\partial \omega \partial \Omega}\right)_s \propto \begin{cases} 0, & \text{for } \omega_L - \omega_v < \omega_{ex}(0) \\ \dfrac{[\omega_L - \omega_v - \omega_{ex}(0)]^{1/2}}{\Gamma}, & \text{for } \omega_L - \omega_v > \omega_{ex}(0). \end{cases} \tag{2.245a}$$

This process can actually be viewed as light absorption by indirect transitions, aided by the absorption of a phonon, followed by re-emission of a photon and emission of a second Γ_{12}, phonon. In this spirit, (2.245a) can be rewritten as

$$\left(\frac{\partial^2 S}{\partial\omega\partial\Omega}\right)_s \propto \alpha(\omega_L)\frac{\Gamma_R}{\Gamma}, \tag{2.246}$$

where α is the absorption coefficient for indirect transitions, Γ_R^{-1} the radiative recombination time and Γ^{-1} the *total* longitudinal recombination time of the intermediate state (indirect exciton). Equation (2.246) represents a situation which could be partially labeled as hot luminescence [Ref. 2.1, Chap. 1]. For $\Gamma_R < \Gamma$, the real excitation to the intermediate state is only partially re-emitted as light. If the phase relationship of the emitted photons to the absorbed ones is determined simply by the two emitted phonons, the phenomenon is labeled Raman scattering. If there are other unspecified or unspecifiable processes determining this relationship, we may partly speak of hot luminescence [Ref. 2.1, Chap. 1]: Γ_R has two components, one which corresponds to "coherent" two-phonon Raman scattering and another due to "incoherent" hot luminescence.

In a two-phonon process of the $\mathcal{H}_{ev}^{(1)}$-type, one can produce first the phonon of the wave vector q_1 and then q_2 or vice versa. The two processes are not equivalent although the final states are the same. The scattering *amplitudes* for these two distinct processes must be added and then squared in order to obtain the scattering efficiency [2.204]; quantum interference can then result if the energies of the intermediate state of these processes are close. It is shown in [2.204] that these quantum interferences are negligible except in a narrow solid angle around the exact back scattering configuration.

2.2.11 Multiphonon Scattering

We have already mentioned that the coupling via the Fröhlich interaction can produce, near resonance, very strong scattering. This is true for forbidden scattering by one LO-phonon and for scattering by 2 LO-phonons near Γ. Electron-phonon coupling via the Fröhlich interaction can actually produce multiphonon scattering near resonance (up to about 10 LO-phonons are observed, see [Ref. 2.1, Fig. 3.20]). Several theories of this scattering have been proposed (for a review see [2.205]). They fall into two categories: those based on Frank-Condon mechanisms [2.185, 206] and those based on some form of perturbation calculation. The latter can actually be subdivided into two categories: those using real [2.8] and those using virtual intermediate states [2.9].

Although they share some aspects in common, these theories lead to partly contradictory results. Their authors claim in all cases partial success when comparing with particular experiments. Actually, the experimental data are not very detailed as no absolute cross sections are usually measured or calculated. The difference between the various theories stems from a partial breakdown of perturbation theory. This is remedied in some of the theories by introducing a broadening of the intermediate states Γ which is produced, in part, by the Fröhlich interaction itself. The results change drastically depending on what fraction of Γ is determined by the Fröhlich interaction. This fraction will usually be a function of ω_L.

In order to illustrate this problem, we can consider the expressions (2.245, 246) for two-phonon scattering. If Γ does not depend on the coupling constant C_F, the scattering efficiency is proportional to $|C_F|^4$, as expected from straightforward perturbation theory. As ω_L becomes much larger than $\omega_{LO} + \omega_{ex}$, the Fröhlich interaction becomes the main scattering mechanism which determines Γ and $\Gamma \propto |C_F|^2$. Thus, the scattering efficiency becomes proportional to the square of the coupling constant or perturbation parameter, a rather unusual result in second-order time-dependent perturbation theory. Although it is often stated that this happens independent of the magnitude of the coupling constant, one has to take this with a grain of salt: as this coupling constant becomes small, Γ is no longer determined by LO-scattering. This proportionality to $|C_F|^2$ also holds in one-phonon processes within the limit of perturbation theory independent of coupling strength [2.205, 207].

This feature is encountered again in the cascade models of multiphonon scattering [2.8]. In [2.8], for instance, *real* electron intermediate states and uncorrelated virtual hole states appear as intermediate levels. These electron states decay by the emission of one, two, or more LO-phonons. After the emission of m such phonons, the probability of emission of a photon is the "branching ratio"

$$S = \frac{\tau_R^{-1}(m)}{\tau_{LO}^{-1}(m)} = \frac{\tau_{LO}(m)}{\tau_R(m)}, \qquad (2.247)$$

where $\tau_R(m)$ is the radiative recombination time for this electronic state and $\tau_{LO}(m)$ the scattering time for the emission of one phonon. This process corresponds to the emission of $m+1$ LO-phonons as, in order to conserve k, the holes must also emit one LO-phonon before recombination. The observation of sharp multiphonon lines at frequencies $m\omega_{LO}(q \simeq 0)$ depends on the existence of a flat dispersion relation near $q = 0$.

The time τ_R which appears in (2.247) has been evaluated in [2.8] for CdS with $\omega_L = \omega_0 + 6\omega_{LO}$. The resulting relative scattering efficiencies reproduce the experimental results well. This treatment also explains the polarized scattering usually observed in the multiphonon case although this depends on the absence of a dephasing scattering mechanism, other than LO-phonon scattering, in the intermediate state. Even for LO-phonon scattering, some depolarization can occur in anisotropic materials such as CdS [2.208a]. In this case, one can calculate a linear polarization ratio for the m-LO process $\varrho_m \simeq (0.85)^m$ [2.208a].

This ratio may be even smaller if other dephasing scattering mechanisms are present (such as impurity scattering). It may be decreased by the presence of a magnetic field B perpendicular to the polarization of the light (Hanle effect). In this case one obtains [2.208b]

$$\varrho_m(H \neq 0) = \varrho_m(H = 0) \frac{1}{1 + \omega_H^2 \tau_i^2}, \qquad (2.248)$$

where $\omega_H = \mu_B g_{ex} B$, g_{ex} is the exciton g-factor, and τ_i the total lifetime of the *real* intermediate states. Equation (2.248) can be used to determine τ_i from the measured dependence of ϱ on H [2.208b].

The models of multiphonon scattering based on real intermediate states (of electrons only or of excitons) are designated as "cascade models". It has been pointed out by *Kochikhin* et al. [2.208b] that there may be an easy way of distinguishing between real and virtual intermediate states in multiphonon processes. For *real* intermediate states, the ratio of the scattering efficiency for the m-LO process S_m to S_{m-1} should be nearly independent of m, as obtained from (2.247) if we assume that the times involved are independent of m. For virtual processes of the m-LO type, however, one has $m!$ topologically equivalent diagrams as one can permute the phonons in all possible ways (this is equivalent to the quantum interference discussed in the previous chapter for $m = 2$). Consequently, $S_m/S_{m-1} = m$. This dependence of S_m/S_{m-1} on m has been observed for ZnTe [2.208b].

A calculation for the case of uncorrelated virtual intermediate states has been given by *Zeyher* [2.9]. It has the drawback of assuming a q-independent coupling constant (thus it strictly applies to deformation potential but not to Fröhlich-type scattering). It illustrates very clearly, however, the effect of the coupling constant C on the lifetime of the intermediate states. For $m \leq 2$, the integrals involved are well behaved in the limit $\Gamma \to 0$ and S_m is proportional to $|C|^{2m}$ as expected from standard perturbation theory. For $m \geq 3$, the integrals diverge when $\Gamma \to 0$ [see (2.245a)] and a weaker dependence on m results from the dependence of Γ on C. In fact, for $m \geq 4$, $S_m \propto |C|^6$ independent of m. We note that on the basis of (2.247), S_m would be independent of $|C|$ provided the lifetime of the intermediate state is determined by LO-scattering exclusively (this, of course, will not be true for small $|C|$).

We show in Fig. 2.43 the scattering efficiency S_m calculated in [2.9] for $\omega_L = \omega_0 + 10\omega_{LO}$ with a non-LO-related broadening $\Gamma_0 = 0.01\omega_{LO}$ and for three values of the coupling parameter D defined as

$$D = |C|^2(1 + n)/8\pi\omega_{LO}^2, \qquad (2.249)$$

where n is the Bose-Einstein factor for the phonons under consideration. The behavior of S_m shown in this figure reproduces qualitatively the experimental results; see Fig. 3.20 in [2.1]. S_1 is very strong and S_m first decreases with increasing m, to increase again strongly for $\omega_L - m\omega_{LO} \simeq \omega_0$. It should become

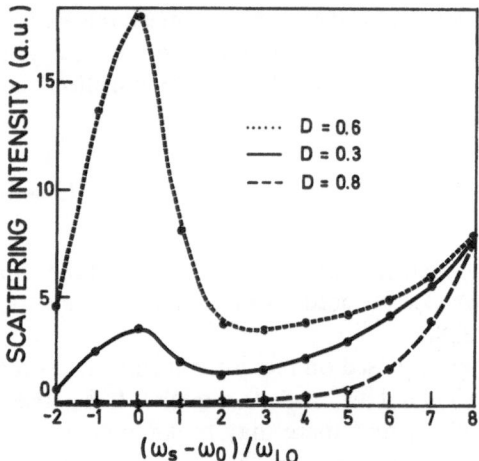

Fig. 2.43. Efficiencies for multiphonon scattering calculated in [2.9] for three values of the coupling parameter D (see text)

clear by now where the difficulties in the theoretical treatment lie: the results critically depend on the broadening parameter Γ and the detailed mechanisms which contribute to it. Different theories often differ only in the assumptions implicitly made for these mechanisms. A generalization of the calculation of [2.9] to the Fröhlich-type of q dependence of $\mathscr{H}_{ev}$, including finite temperature effects, has been recently performed [2.209].

We mentioned in Sect. 2.2.3 that the Frank-Condon terms are not usually relevant to Raman scattering whenever the intermediate states are extended states such as is usually the case for solids and for large molecules. In solids, one may actually have "localized" excitations if the electron and the hole are strongly correlated. Through the Fröhlich interaction, such excitations can couple strongly to the lattice and produce a static displacement of the normal coordinate of the phonons of $q \simeq 0$ (polarons). Such displacement gives rise to Frank-Condon terms, as shown in Sect. 2.2.3. Near resonance, these terms can produce multiphonon scattering which must vanish away from resonance. It is therefore attractive to interpret the observed multiphonon phenomena as the result of Frank-Condon terms. The first such attempt was that of *Williams* and *Smit* [2.206]. These authors fitted the dependence of S_m on m observed for CdS for $m \geq 3$ with a phonon coordinate shift Δ approximately equal to the zero-point vibrational amplitude. A qualitative attempt was made to relate this value of Δ to the ε_{∞} and ε_{rf} of CdS. A more detailed treatment has been given in [2.185] with particular application to the multimodes observed in YbS for $m \leq 5$. In this treatment, one diagonalizes exactly the phonon Hamiltonian plus the Fröhlich interaction to obtain the so-called polaron states by means of the transformation

$$b_q \rightarrow b_q + \frac{C_F^* \cdot \varrho_q^*}{q\omega_{LO}} = b_q + \Delta_q, \tag{2.250}$$

where $\varrho_q(\varrho)$ is the charge density of the exciton (Sect. 2.2.9). Hence, according to (2.250), the phonon coordinates are "displaced" in the excited state by an amount Δ_q which can be easily calculated, as done in Sect. 2.2.9 for the matrix elements of the exciton-phonon interaction in the case of $1s$ excitons (note that in [2.185], it is implicitly assumed that $m_h \gg m_e$). The shift Δ_q is simply the exciton-phonon interaction term (2.231) divided by ω_{LO}.

The linear susceptibility $\chi(\omega)$ can be obtained from (2.157a) for $n_t = n_0$. In the cubic (isotropic) case and at $T \simeq 0$ it is

$$(4\pi)^{-1}\chi(\omega) = |P_0|^2 V^{-1} \sum_{t=0}^{\infty} \frac{\Delta^{2t}}{(\omega_0 + t\omega_{LO} - \omega + i\Gamma)}, \qquad (2.251)$$

where

$$\Delta^2 = \sum_q |\Delta_q|^2. \qquad (2.251a)$$

Equation (2.251) explicitly contains multiphonon sidebands (phonon emission only, as we assumed $T \simeq 0$). Using (2.251) and standard expressions for the overlap integrals of shifted Laguerre polynomials, we find the Raman susceptibility for the nth multiphonon process [2.185]:

$$\chi_s^{(m)}(\omega_L) = \left[\sum_{q,\ldots,qm} |\Delta_{q_1}|^2 \ldots |\Delta_{q_n}|^2 \delta(q_1 + \ldots + q_n) \right]^{1/2}$$
$$\cdot \sum_{p=0}^{m} \binom{m}{p} (-1)^p \chi(\omega_L - p\omega_{LO}). \qquad (2.252)$$

Expression (2.252) enables us to obtain R_s as a function of the experimental linear susceptibility χ, see (2.223). In the limit $\omega_{LO} \ll |\omega_L - \omega_0 - i\Gamma|$, it can be simply transformed into

$$\chi_s^{(m)}(\omega_L) = \left[\sum_{q_1 \ldots q_m} |\Delta_{q_1}|^2 \ldots |\Delta_{q_m}|^2 \delta(q_1 + \ldots + q_m) \right]^{1/2} \left(\frac{\partial^m \chi}{\partial \omega_0^m} \right)_{\omega_L - m\omega_{LO}/2} \omega_{LO}^m. \qquad (2.253)$$

The expected mth derivative of $\chi(\omega)$ thus appears in the expression for the Raman susceptibility for multiphonon processes [compare with (2.218)]. The resulting scattering is polarized for any order m as a result of our having assumed $\chi(\omega)$ to be isotropic. The coefficient in front of the derivative is obtained through a $3 \times (m-1)$-dimensional integration of a product of factors of the form (2.231). This integral is rapidly convergent while the integral which determines Δ in (2.251a) depends strongly on the maximum of q chosen for the integration ($\simeq$ size of Brillouin zone).

We note that (2.253) also gives the susceptibility for forbidden scattering by one LO-phonon provided one keeps the scattering vector in the argument of the Kronecker-δ [$\delta(q_1 - q)$]. The result is the same as that found by treating the exciton-phonon Hamiltonian (2.232b) in perturbation theory:

$$\chi_s = \frac{m_e - m_h}{2M} a_{ex}^2 C_F \left(\frac{\partial \chi}{\partial \omega}\right) q. \tag{2.254}$$

In the discussion above, we have solved exactly the sum of the exciton, phonon and exciton-phonon Hamiltonians under the assumption of only a 1s exciton state. When excited exciton states are included, quasibound exciton LO-phonon states may result [2.210]. In a more accurate treatment, these states must be included as intermediate states for multiphonon LO-scattering. This has been done recently by *Jain* and *Jayanthi* [2.211].

2.2.12 Brillouin Scattering

As discussed in Sect. 2.1.14, the cross section for Brillouin scattering with ω_L not too close to an excitonic resonance is obtained from the elasto-optic constants, see (2.100). In germanium-zincblende and fluorite-type materials, there are three independent elasto-optic constants p_{11}, p_{12}, and p_{44}. For the purpose of their theoretical evaluation, one considers the "irreducible components" of the elasto-optic tensor $p_{11} + 2p_{12}$ (the effect of a hydrostatic strain), $p_{11} - p_{12}$ ([100] shear) and p_{44} ([111] shear).

As already mentioned, the theory of p_{44} in the neighborhood of a given critical point is, to a numerical constant, basically the same as that of the Raman polarizability a. Near E_0, $E_0 + \Delta_0$, for instance, one obtains an expression similar to (2.194) [2.169]:

$$(p_{44})_{E_0} = -\frac{\sqrt{3}}{8} \left(\frac{\varepsilon_0}{\varepsilon}\right)^2 C_0'' \frac{d}{\omega_0} \left\{ -G(x) + \frac{4\omega_0}{\Delta_0} \left[F(x) - \left(\frac{\omega_0}{\omega_0 + \Delta_0}\right)^{3/2} F(x') \right] \right\}, \tag{2.255}$$

where the deformation potential d is defined in (2.197). We mentioned in Sect. 2.2.6a that there is some uncertainty concerning the value of the constant $C_0'' \simeq 2$. Regardless of this uncertainty, this constant should have the same value for p_{44} as for a since the theories of the two parameters are completely isomorphic. It is thus possible to determine C_0'' from the experimental dispersion of p_{44} and the value of d found in independent experiments and to use this C_0'' to calculate a with (2.194).

Near E_0, $E_0 + \Delta_0$, the elasto-optic constant $p_{11} - p_{12}$ also has the same form as (2.253) [2.169]:

$$(p_{11} - p_{12})_{E_0} = -\frac{3}{4} \left(\frac{\varepsilon_0}{\varepsilon}\right)^2 C_0'' \frac{b}{\omega_0} \left\{ -G(x) + \frac{4\omega_0}{\Delta_0} \left[F(x) - \left(\frac{\omega_0}{\omega_0 + \Delta_0}\right)^{3/2} F(x') \right] \right\}, \tag{2.256}$$

where b is the deformation potential (in energy units) for the splitting of the Γ_8 valence band by a [100] shear. Now, however, C_0'' need not be *exactly* the same as in (2.255) because the angular integrations to be performed are not the same. Nevertheless the ratio of the two C_0'', is usually close to unity. The hydrostatic constant $p_{11} + 2p_{12}$ is obtained from an expression isomorphic to (2.240). It is [2.169]

$$(p_{11} + 2p_{12})_{E_0} = -\frac{3a}{2\omega_0}\left(\frac{\varepsilon_0}{\varepsilon}\right)^2 C_0''\{G(x) + 3F(x) + \tfrac{1}{2}[G(x') + 3F(x')]\}, \qquad (2.257)$$

where $a = d\omega_0/d\ln 3a_0$ is the hydrostatic deformation potential of the ω_0 gap. These expressions must be suitably modified to apply then to the E_0 gap of diamond ($\Gamma_{25'} \to \Gamma_{15}$ instead of $\Gamma_{25'} \to \Gamma_{2'}$). The resulting expressions are given in [2.176].

For the E_1, $E_1 + \Delta_1$ gaps, a similar situation obtains. The contribution to p_{44} is given by (see (2.203), [2.169])

$$(p_{44})_{E_1} = -\frac{1}{4\sqrt{3}}\left(\frac{\varepsilon_0}{\varepsilon}\right)^2\left[-d_1^5\frac{d\chi(\omega)}{d\omega_1} + 4\sqrt{2}d_3^5\frac{\chi^{(E_1)} - \chi^{(E_1+\Delta_1)}}{\Delta_1}\right] \qquad (2.258)$$

and similarly for $p_{11} - p_{12}$ and $p_{11} + 2p_{12}$,

$$(p_{11} - p_{12}) = -\sqrt{6}\left(\frac{\varepsilon_0}{\varepsilon}\right)^2 d_3^3\left[\frac{\chi^{(E_1)} - \chi^{(E_1+\Delta_1)}}{\Delta_1}\right] \qquad (2.258a)$$

$$(p_{11} + 2p_{12})_{E_1} = -3\left(\frac{\varepsilon_0}{\varepsilon}\right)^2 \mathscr{E}_1\frac{d\chi}{d\omega_1}, \qquad (2.259)$$

where d_1^5, d_3^5, d_3^3, and $\mathscr{E}_1$ are the appropriate strain deformation potentials, all in units of energy or frequency. We have given here the contributions of the E_0, $E_0 + \Delta_0$ gaps in terms of the functions F and G and those of E_1, $E_1 + \Delta_1$ in terms of $\chi^{(E_1)}$ and $\chi^{(E_1+\Delta_1)}$. Following the method of Sect. 2.2.5, it poses no problems to go from one type of expression to the other.

2.3 Resonant Scattering by Phonons: Experimental Results

For the quantitative interpretation of resonant light scattering in solids, it is necessary to have some fairly detailed prior knowledge of their band structure and lattice dynamics. Only then can one meaningfully proceed to the next step, i.e., to relate quantitatively the measured scattering intensity to electron-phonon interaction mechanisms. Perhaps the best understood nonmetallic materials are the tetrahedral semiconductors, i.e., those with diamond, zincblende, and wurtzite structure. Examples of resonant Raman and Brillouin scattering have been given in previous sections to illustrate specific points of the general theory [see, for instance, Figs. 2.5 (ZnTe), 2.6 (wurtzite-type ZnS), 2.8 (GaAs), 2.10 (zincblende-type ZnS), 2.11 (Si), 2.13 (ZnTe), 2.18 (a-Si), 2.30 (diamond), 2.31 (GaAs), 2.40 (GaAs), 2.41 (GaP and CdS), 2.42 (Si)]. Other somewhat related families of materials have also been studied in detail; among them we mention the antifluorites Mg_2Si, Mg_2Ge, and Mg_2Sn [2.72] and the layer compounds GaS, GaSe [2.168] (see Fig. 2.32). The rocksalt-type silver halides AgCl and AgBr are also very interesting as they possess an indirect gap which gives rise to an enhancement of two-phonon Raman scattering near resonance [2.202]. Of interest is also the case of Cu_2O, in particular the resonances observed near the dipole-forbidden yellow $1s$ exciton [Ref. 2.1, Chap. 3], the two-phonon resonances related to indirect transitions and already described in (2.245a) and the well-resolved in- out-going resonances of the $\Gamma_{12'}$-phonons around the $1s$ exciton [2.212]. In this case, the $1s$ exciton, dipole forbidden, is extremely sharp ($\Gamma_{1s} \sim 0.2\,cm^{-1}$ at 6 K [2.212]) and so are the corresponding resonances. An interesting fact results: the width of out-going resonances is determined by the width of the phonon (0.24 cm^{-1} at low temperature) [2.212]. This material also shows Fröhlich-interaction-induced forbidden scattering for polar Γ_{15}-phonons.

Finally, we will discuss the case of AgCl and AgBr where perhaps the most striking two-phonon resonances at indirect gaps have been observed [2.202].

2.3.1 First-Order Raman Scattering in Germanium-Zincblende: E_0, $E_0 + \Delta_0$ Gaps

a) Allowed Scattering by Phonons

The scattering by free electrons and also by free holes can resonate near the E_0 gap whenever this gap is the lowest. Such is the case for GaAs, InAs, InSb, GaSb, InP. In these materials, the scattering by *free electrons* can also resonate at the $E_0 + \Delta_0$ gap. This effect is easier to observe than the corresponding E_0 resonance as it is free from the strong luminescence associated with the E_0 gap. We have shown in Fig. 2.31 the corresponding resonance of the spin-flip scattering (depolarized!) of n-GaAs. The solid line is a fit with the square of (2.149). Other examples of resonances of free particle scattering and of plasmon-LO-phonon coupled modes will be given in the contributions by Abstreiter and Cardona, Pinczuk, and by Geschwind and Romestain in [1.3]. The resonance near E_0 of the scattering by free holes in ZnTe, a standard material for observing this type of phenomena [2.213, 214], is shown in [Ref. 2.214, Fig. 5]. The resonance of spin-flip scattering near E_0 in InSb is displayed in [Ref. 2.1, Fig. 7.16].

We show in Fig. 2.44 the resonance in the scattering by LO- and TO-phonons measured for GaP near E_0 and $E_0 + \Delta_0$. These gaps are not the lowest in this material, a fact which is rather welcome as it eliminates strong luminescence (which occurs at the lowest gap) and also makes the absorption

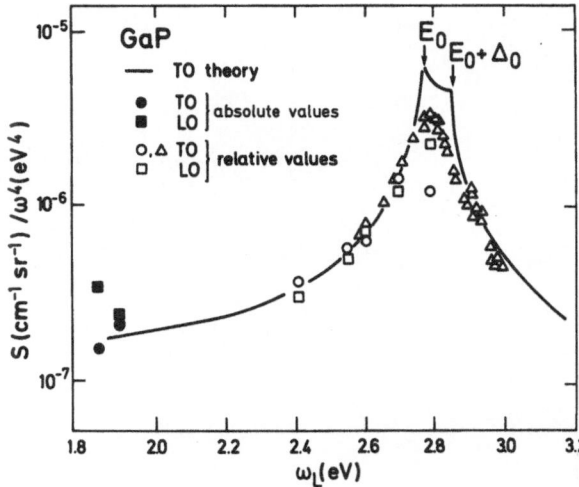

Fig. 2.44. Resonance in the one-phonon Raman scattering of GaP near E_0, $E_0 + \Delta_0$ and fit with (2.194). From [2.49]

correction easier. The agreement between the calculated and the measured curve is rather good although in this case, the spin-orbit splitting Δ_0 is too small to permit an unambiguous determination of the resulting structure. The fit of (2.194) to the observed absolute scattering efficiencies yields the deformation potential d_0 [defined in (2.188)] = 27 eV, in rather good agreement with predictions from pseudopotential ($d_0 \simeq 26$ eV) and LCAO theory ($d_0 \simeq 29$ eV) [2.177].

Similar resonances have been studied for AlSb [2.215], GaAs [2.74] and ZnS, ZnSe, and ZnTe [2.52]. We have already shown the results for ZnTe in Fig. 2.6, where the theoretical fit takes into account some exciton interaction in the phenomenological manner discussed in Sect. 2.2.9. The deformation potential obtained from the fit is $d_0 = 37$ eV, also in reasonable agreement with calculations [2.73, 177]. Below E_0, one sees the difference in LO- and TO-efficiencies ($S_{LO} > S_{TO}$) as required by the Faust-Henry coefficient (Sect. 2.1.12).

In Fig. 2.8 we showed the dispersion of the Raman polarizability a of GaAs obtained from the efficiency S with the expression

$$\frac{dS}{d\Omega}[\text{cm}^{-1}\text{ sterad}^{-1}] = 4.4 \times 10^{-4} \frac{(\omega_s[\text{eV}])^4 (a[\text{Å}^2])^2}{(a_0[\text{Å}])^3 \mu[\text{at}] \omega_j[\text{cm}^{-1}]}. \tag{2.260}$$

Equation (2.260) is written in hybrid units: these units are those most commonly used for each one of the quantities involved; μ, for instance, is the reduced mass of the primitive cell in atomic units. In Fig. 2.44 we notice quite clearly the different character of the E_0 and the $E_0 + \Delta_0$ resonances. The former is sharp, corresponding to a two-band type of phenomenon ($\chi_s \sim \partial\chi/\partial\omega$). The weak $E_0 + \Delta_0$ resonance stems from three-band terms: the electron-phonon Hamiltonian couples the Γ_7 with the Γ_8 states but has no expectation value for Γ_7 (Fig. 2.37). These facts are also evident in (2.194).

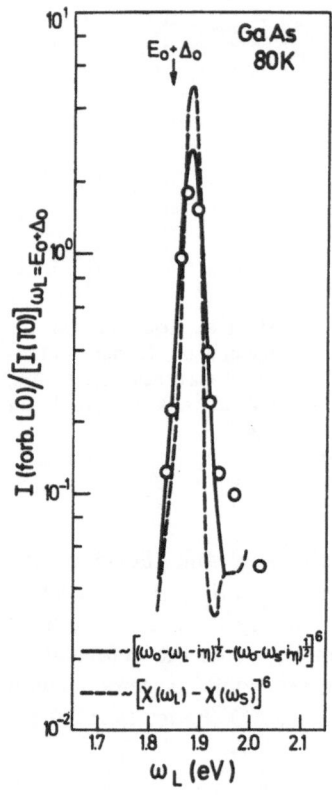

Fig. 2.45. Resonant forbidden LO-scattering observed near the $E_0 + \Delta_0$ gap of GaAs. Two theoretically fitted curves are given [2.74]

The solid line in Fig. 2.8 is a fit to the experimental data using (2.194, 201) with two deformation potentials, d_0 and $d_{3,0}^5$ (the latter will be discussed later). The value of $d_0 = 48$ eV is obtained from the fit, also in agreement with typical values of this deformation potential [2.177].

We note that the fits to the data of Figs. 2.5, 2.8, 2.44 were made simply with (2.194, 195, 201), *without having had to add any additional constants*. Thus, antiresonances of the type described by (2.33a) do not occur (they occur, how-ever, in Brillouin efficiencies for the same materials (Fig. 2.15) and also for the Raman efficiencies of wurtzite-type materials (Fig. 2.6) [Fig. 2.7 of Ref. 2.1]. As we have already mentioned, only the most resonant term of expressions such as (2.194) is, in principle, meaningful; lesser resonant and constant terms strongly depend on how the bands are cut off and on how P depends on ω_s. The empirical realization that (2.194) represents *a* well *without any additional constant is*, however, quite interesting and should be kept in mind. The deformation potentials d_0, so obtained, are basically calculated with (2.196) provided one takes for v_G only the *symmetric* component of the pseudopotential if the structure is zincblende.

The E_0 gap splits into A and B in a wurtzite-type material while $E_0 + \Delta_0$, unsplit, is normally labeled C. Resonances of scattering by Raman active phonons in allowed configurations have been observed in ZnS (wurtzite) (Fig. 2.6), CdS [2.216] and ZnO [2.216, 80]. The latter work is particularly detailed: absolute efficiencies are given (Table 2.8) and electron-phonon deformation potentials derived.

b) Forbidden Scattering by Phonon

Forbidden scattering appears *in polar* materials (zincblende, wurtzite) for ir-active phonons and is always polarized, corresponding to the theory given in Sect. 2.2.8. It strongly resonates both near E_0 and near $E_0 + \Delta_0$; it can be stronger than the allowed deformation potential or electro-optic scattering very near resonance, especially very near the $E_0 + \Delta_0$ edge (the deformation potential scattering is barely resonant in this case).

A *quantitative* evaluation of S in absolute units for forbidden scattering by LO-phonons is seldom found in the literature although it can, in principle, be performed as shown in Sect. 2.2.8. We present data in Fig. 45 for the $E_0 + \Delta_0$ edge of GaAs. These experimental results have been fitted with two alternative theoretical expressions: that of (2.222) based on parabolic bands, and that of (2.223) based on the experimental susceptibility. A quantitative evaluation of the efficiency for forbidden scattering at the peak of the latter (one of such rare evaluations!) [2.74] gives for the ratio S_{LO} (forbidden)/S_{LO} (allowed) 1.7 ± 1, in good agreement with the data of Fig. 2.45. A similar quantitative analysis of the E_0 resonance in GaP is given in [2.217].

2.3.2 First-Order Raman Scattering in Germanium-Zincblende: $E_1, E_1 + \Delta_1$ Gaps

a) Allowed Scattering

The corresponding theoretical expression for the scattering by one phonon via the deformation potential mechanism is given in (2.203). Experiments have been performed for Ge [2.44a], Si [2.218], InSb [2.77], GaSb [2.219], InAs [2.220], GaAs [2.74], InP [2.221] and gray tin [2.222]. All the fits performed to the experimental data with (2.203) show that $|d_{3,0}^5| \gtrsim |d_{1,0}^5|$ and therefore the three-band term involving the finite difference ratio $(\chi^{(E_1)} - \chi^{(E_1 + \Delta_1)})\Delta_1^{-1}$ often dominates, especially in materials with small spin-orbit splitting Δ_1 (e.g., Si, Ge, etc.). In materials with large Δ_1 (e.g., InSb, GaSb), there is a noticeable contribution of the two-band terms represented in (2.203) by $\partial\chi/\partial\omega_1$, especially at low temperatures [Ref. 2.1, Fig. 3.16].

The resonant behavior of first-order Raman scattering by phonons in germanium near E_1 (2.1) eV and $E_1 + \Delta_1$ (2.3 eV) is shown in [Ref. 2.1, Fig. 2.9]. If the two-band terms were dominant, one would expect sharp peaks at $\omega_L \simeq E_1 + \omega_{TO}/2$ and $\omega_L \simeq E_1 + \Delta_1 + \omega_{TO}/2$. Instead, one sees a broad hump peaking between E_1 and $E_1 + \Delta_1$. This is the characteristic result of 3-band terms: the two-band terms are completely negligible and one can set $d_{1,0}^5 \simeq 0$ in

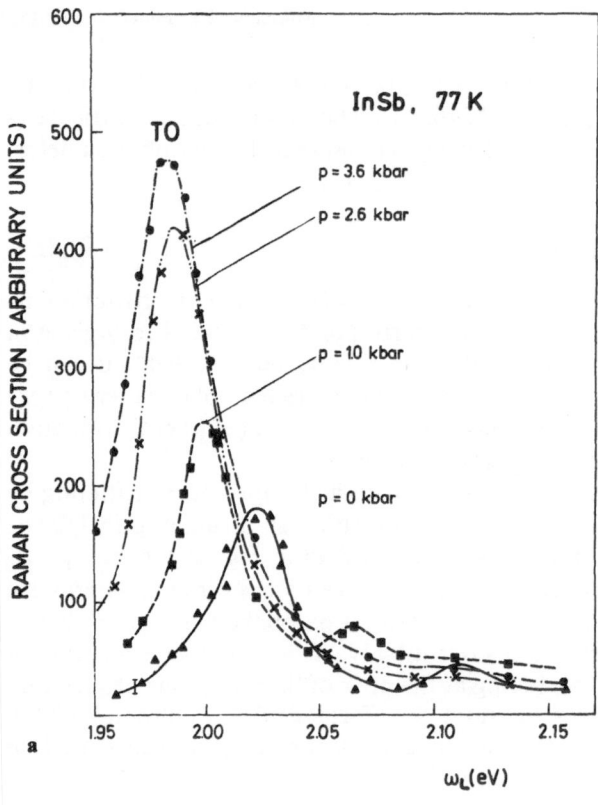

Fig. 2.46. (a) Resonant efficiencies for TO-phonon scattering in InSb as a function of [111] stress in back scattering. The direction of propagation is $[1\bar{1}0]$ while $\hat{e}_L \| \hat{e}_s \| [11\bar{2}]$. (b) Calculated effect of uniaxial stress on scattering efficiencies for InSb under the conditions of (a) [2.190]

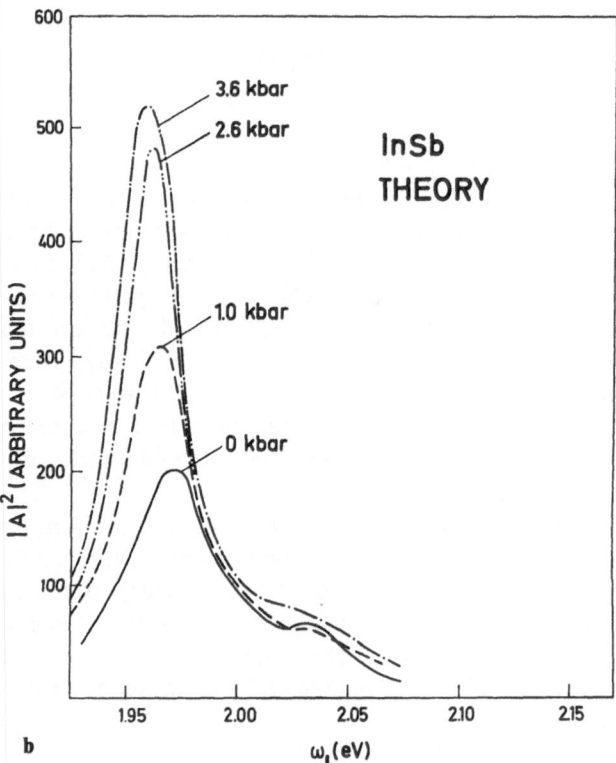

(2.203). A more detailed fit of data obtained under uniaxial stress (see below discussion for InSb) yields for Ge, $d_{1,0}^5 \simeq -0.7\, d_{3,0}^5$.

Figure 2.8 also contains data for the E_1 resonance of GaAs. The fit to these data can also be performed with $d_{1,0}^5 \simeq 0$. We obtain from this fit to the *absolute* efficiencies $d_{3,0}^5 = 37\,\text{eV}$. As we have seen, the three-band terms of (2.203) usually dominate the scattering process. The two-band terms are small as a result of a cancellation [2.190]: a phonon propagating along [111] lifts the energy of the E_1 gap along [111] (singlet) and lowers that of the [$1\bar{1}\bar{1}$], [$\bar{1}1\bar{1}$], [$\bar{1}\bar{1}1$] triplet while keeping the centroid invariant. A detailed calculation [2.190] shows that the contribution of singlet and triplet gaps to χ_s nearly cancel each other [2.190]. If a stress along [111] is applied, the singlet-triplet gap degeneracy is lifted and with it the cancellation just mentioned. The resonance splits into two (singlet and triplet) while gaining considerably in strength. The experimentally observed effect is illustrated in Fig. 2.46a. Figure 2.46b represents a calculation performed by lifting the singlet-triplet degeneracy out splitting the singlet and triplet contributions to a. Figure 2.46b reproduces the increase in strength found experimentally quite well. As a fitting parameter, the ratio $d_{3,0}^5/d_{1,0}^5 \simeq -7$ was determined for InSb.

The values of $d_{3,0}^5$ and $d_{1,0}^5$ obtained from this type of experiment are listed in Table 2.10 and compared with the results of several calculations. In the case of silicon, $\Delta_1 \gtrsim \Gamma$ (= broadening of the intermediate state) and (2.203) reduces

Table 2.10. Deformation potentials $d_{3,0}^5$ and $d_{1,0}^5$ obtained from resonant Raman experiments near E_1 and $E_1 + \Delta_1$, compared with the results of several calculations [2.177] at the L point. The experimental data represent an average over the Λ direction of the BZ

	$d_{1,0}^5$		$d_{3,0}^5$		$d_{3,0}^5 + \dfrac{1}{2\sqrt{2}} d_{1,0}^5$		$d_{3,0}^5/d_{1,0}^5$	
	Theory[a]	Experiment	Theory[a]	Experiment	Theory[a]	Experiment	Theory[a]	Experiment
Ge	−10.7	−30[e]	48.3	45[e]	44.5	34[d]	−2.8	−1.5[b]
Si	−16.4		44.8		39.0	40[c]	−7.5	
Sn	− 8.8		37.6		34.5		−1.7	−3.5[b]
GaAs	−11.8	−22.5[e]	40.3	45[e]	36.1	37[c]	−3.8	−2[b]
InP	−14.5		43.3		38.2	37[d]	−4.0	
InSb	−11.5		38.2		34.1		−2.1	−7[b]
InAs	−11.9	−16[e]	40.3	37[g]	36.1	31[e]	−1.8	−2[f]

[a] [2.177], calculated at the L point with an LCAO method.
[b] [2.190].
[c] [2.53].
[d] S. Onari, R. Trommer, M. Cardona: Solid State Commun. **19**, 1145 (1976), comparison with GaAs.
[e] derived from $d_{3,0}^5/d_{1,0}^5$ and $d_{3,0}^5 + (2\sqrt{2})^{-1} d_{1,0}^5$.
[f] E. Anastassakis, F.H. Pollak: Solid State Commun. **13**, 1755 (1973).
[g] [2.220].

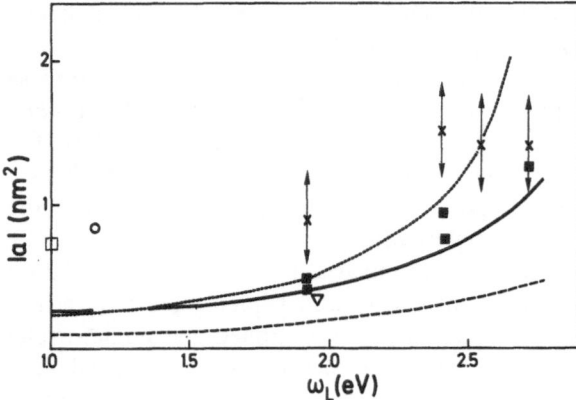

Fig. 2.47. Resonance in the first-order Raman scattering by phonons observed in crystalline Si. The points are experimental, the crosses determined by the Raman-Brillouin method. The solid line fit was made with (2.201) for $\alpha = 1$. The dashed line represents the results of full band structure calculations [2.152]

to (2.201) and only the sum $d_{3,0}^5 + (2\sqrt{2})^{-1} d_{1,0}^5$ can be determined from the resonance data (Fig. 2.47). The fit to the experimental results obtained, in absolute efficiency units using the methods of Sects. 2.1.18b, 2.1.18c, is shown in Fig. 2.47. The fit with (2.201) using for $\chi(\omega)$ the data of [2.223], yields a combined deformation potential $d_{3,0}^5 + (2\sqrt{2})^{-1} d_{1,0}^5 = 27$ eV.

b) Forbidden LO-Scattering

Forbidden LO-scattering, resonant near $E_1 - E_1 + \Delta_1$, has been observed for most III–V semiconductors. For GaAs, a quantitative estimate of the scattering efficiency was made using the expressions of Sect. 2.2.8. This estimate agrees with experimental results [2.74]. For InSb, strong changes in S_s were observed for the forbidden scattering upon application of uniaxial stress along [111] (see Fig. 2.48). In this case, the contributions of singlet and triplet gaps to the Raman polarizability is additive and the sum must be squared to obtain the efficiency. When the singlet gap is split from the triplet upon application of stress, the *square* of χ_s decreases approximately by a factor of 4, in agreement with Fig. 2.48 [2.190].

2.3.3 Second-Order Raman Scattering in Germanium-Zincblende

The macroscopic theory of scattering by two phonons has been discussed in Sect. 2.1.13 and the microscopic treatment is given in Sect. 2.2.10. Examples of second-order spectra are given in Figs. 2.11, 2.42 (Si) and 2.13 (ZnTe). The Γ_1 component of the two-phonon spectra is usually dominant and corresponds mainly to overtone scattering. The Γ_{12} component is negligible. The Raman susceptibility $\partial^2 \chi'/\partial \xi_i \partial \xi_j$ of (2.97) does indeed tend to zero for $\omega_v \to 0$ and can be usually approximated by $\langle \partial^2 \chi'/\partial \xi_i \partial \xi_j \rangle_{\omega_v}, \propto \omega_{v'}$, at least for small ω_v's.

As we have seen in Sect. 2.2.10, away from resonances the second-order scattering is produced by a renormalized second-order electron-phonon Hamiltonian $\bar{\mathcal{H}}_{ev}^{(2)}$. When resonances are approached, terms due to $\mathcal{H}_{ev}^{(1)}$ taken

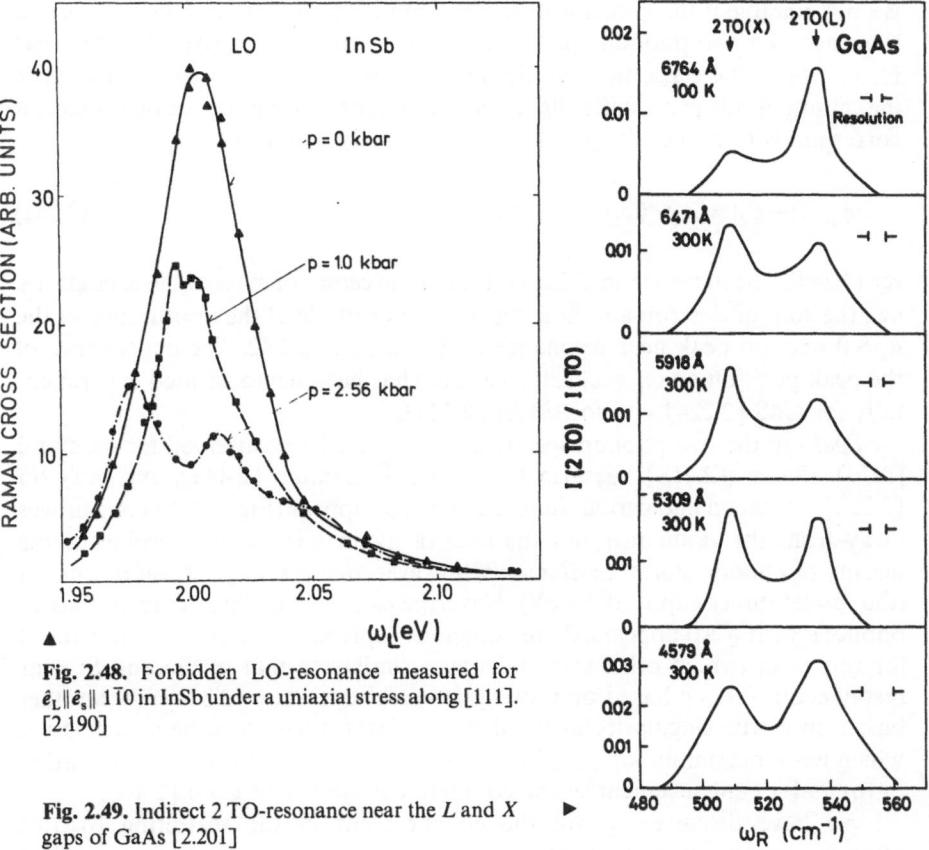

Fig. 2.48. Forbidden LO-resonance measured for $\hat{e}_L \| \hat{e}_s \| 1\bar{1}0$ in InSb under a uniaxial stress along [111]. [2.190]

Fig. 2.49. Indirect 2 TO-resonance near the L and X gaps of GaAs [2.201] ▶

in second order may become dominant; their contribution to χ_s appears in fourth-order perturbation theory and can be strongly resonant. We distinguish two types of $\mathcal{H}_{ev}^{e(1)}$ terms:

i) ω_L *near an indirect gap E_i such that at the gap $\mathbf{k}_c - \mathbf{k}_v = \mathbf{K}$*
In this case, the pairs of phonons with $\mathbf{q}_1 \simeq -\mathbf{q}_2 \simeq \mathbf{K}$ appear in the expressions of χ_s with an energy denominator which vanishes for $\omega_L = \omega_i + \omega_v(\mathbf{K})$ (Stokes). Hence, the second-order spectrum becomes deformed, deviating from the density of states. Such effects can be used to identify the $\mathbf{K}$'s involved if the band structure is not well known but the phonon dispersion is, and vice versa to identify $\mathbf{q}_1 \simeq -\mathbf{q}_2$ if the phonon dispersion is not well known but the band structure is. An example, which helped to identify the relative position of the X and L conduction band minima in GaAs, is shown in Fig. 2.49 [2.201].

ii) ω_L *near an allowed direct gap*
The $\mathcal{H}^{(1)}$ terms have three resonant energy denominators (2.245). The maximum enhancement takes place whenever all denominators are nearly zero. This

is only possible if the q-vector of the phonon is approximately zero. Hence, in this case, the two-phonon spectrum deviates from the density of states near resonance in the sense that for the phonons with $q \simeq 0$, a peak appears. The maximum of this peak shifts slightly as ω_L sweeps through the resonant gap; it corresponds to a phonon q-vector approximately equal to

$$|q_1| \simeq |-q_2| \simeq |2M^*(\omega_L - \omega_0)|^{1/2}, \qquad (2.261)$$

see (2.244). The mass M^* in (2.261) is that of the center of mass of the excitations, i.e., the sum of electron and hole masses. An example of the appearance of the $q_1 \simeq 0$ phonon peak near resonance is shown in Fig. 2.42. The dependence of the peak position on ω_L according to (2.61) has been demonstrated experimentally for CdS [2.224] and for GaAs [2.225].

Peaks in the two-phonon spectra at $q \simeq 0$ have been observed for diamond [2.70], silicon ([2.218], see also Fig. 2.11), germanium [2.44a], and gray tin [2.222]. In silicon and germanium, the peaks disappear (Fig. 2.42) as one moves away from the resonance and the spectra again become isomorphic to the density of phonon states. In diamond, however, one is always far off resonance (the lowest direct gap is at 5.4 eV). Nevertheless, a peak which corresponds to phonons with $q \simeq 0$ appears. Three different explanations have been advanced for this peak (which otherwise looks very similar to that of Ge and Si near resonance!): (a) one based on a two-phonon bound state [2.82a]; (b) another based on a true singularity in the density of states with a dispersion relation which has a maximum for q slightly away from $q = 0$ [2.81]; (c) a combination of (b) and an enhancement of the coupling constant near $q = 0$ [2.40].

We have discussed above the enhancement of the scattering by two phonons near $q = 0$ in the homopolar materials with diamond structure. For the zincblende-type materials, this enhancement is much larger, as a result of the Fröhlich interaction

a) $E_0, E_0 + \Delta_0$ Gaps

The most dramatic effect is perhaps the resonant scattering by 2LO-phonons in polar materials. It has been observed for GaAs [2.74], GaP [2.197], ZnS, ZnSe, ZnTe, [2.52], CdTe [2.226] and the wurtzite-type materials [2.216]. In the case of GaP, *Zeyher* [2.217] made a quantitative estimate of the scattering efficiency and showed that it agrees with experiment [2.197]. In this case, also a mixed TO-LO (Fröhlich) process is observed [Ref. 2.1, Fig. 3.19].

As an illustration of the 2LO-Fröhlich process, we show in Fig. 2.50 the resonance observed near $E_0 + \Delta_0$ for GaAs [2.74]. The resonance of the rest of the two-phonon spectrum (that is, for phonons with $q \neq 0$) has been observed near E_0 (and/or $E_0 + \Delta_0$) for some III–V's (GaP, GaAs), for the Zn-chalcogenides and for the wurtzite-type II–VI's. A peculiarity of the II–VI (both zincblende and wurtzite-type) is that such resonance is observed for $\omega_L \lesssim \omega_0$,

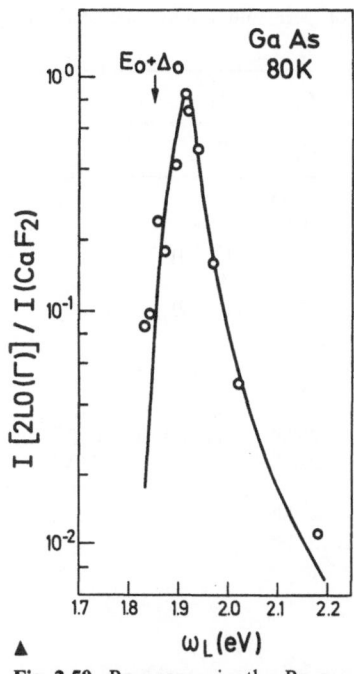

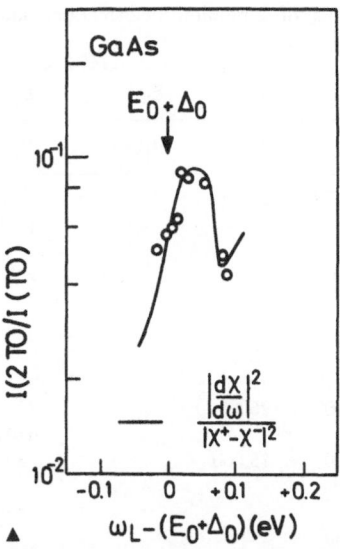

Fig. 2.51. $E_0 + \Delta_0$ resonance for 2 TO-scattering of Γ_1 symmetry normalized to the weakly resonant (see Fig. 2.8) scattering by the TO-phonon in GaAs. The solid line is a theoretical fit [2.74]

Fig. 2.50. Resonance in the Raman scattering by 2 LO (Γ)-phonons near the $E_0 + \Delta_0$ gap in GaAs [2.74]. Plotted is the ratio of 2 TO to TO-scattered intensities versus Raman shift

but above ω_0, the second-order spectrum disappears except for the m-LO(Γ) peaks. The detailed reason for this is unknown to the author.

The resonances of the two-phonon spectra with $q_1 \simeq -q_2 \neq 0$ can, in most cases, be measured and treated as due to $\mathcal{H}_{ev}^{(2)}$ terms. Thus they are isomorphic or nearly isomorphic to the resonances of the scattering by one TO-phonon near E_0. Near $E_0 + \Delta_0$, the latter is weakly resonant (only three-band terms, see Sect. 2.3.1a) while the Γ_1-component of two-phonon spectra can have two-band terms near $E_0 + \Delta_0$ and therefore be more strongly resonant. This is illustrated in Fig. 2.51 for GaAs: a peak in the ratio $I(2\text{TO})/I(\text{TO})$ appears near $E_0 + \Delta_0$.

From the ratio of the second-order scattering intensities or efficiencies to the efficiency for first-order scattering, the renormalized electron-two phonon deformation potentials $\bar{D}$ can be obtained provided d_0 is known. A compilation of existing data is given in Table 2.11. We point out that throughout most of the literature, the density of two-phonon states $N_{d,12}$ of (2.238) in the case of *overtones* appears to be simply the density of the one-phonon states (with the energy scale expanded by a factor of 2) multiplied by the factor $\eta/2$. The 2 in the denominator was meant to represent the fact that the pairs q_1, q_2 and q_2, q_1 are actually the same and should be counted only once (this, of course, can be taken care of in the definition of $\bar{D}$). The constant η is taken to be equal to one for longitudinal phonons and to 2 for transverse phonons in order to take their degeneracy into account.

Table 2.11. Table of 2-phonon deformation potentials of germanium and zincblende-type materials [eV]

	E_0	E_0	E_0	E_1	E_1	E_1
	D_1	D_1	D_{15} or $D_{25'}$	D_1	D_1	$D_{3,0}^5$
	2 TA	2 TO		2 TA	2 TO	
Ge				170[a]	2534[a]	543[a] (2 TO)
Si				80[b]	1220[b]	350[b] (2 TO)
Diamond		4800[c] (E_0')				
GaP	675[d]	1670[d]				
GaAs	170[e]	2600[e]		450[e] 670 (2 LA)[e]	2070[e]	140[e] (TO+TA) 390[e] (LO+LA)
InSb	620[f]	15,000[f]		230[f]	2200[f]	4300[f] (2 TO)
InAs				350[g]	2500[g]	
ZnS	1600[h]	2470[h] (2 TO) 5700 (2 LO)				
ZnSe	545[h]	510[h] (2 LO)	250[h] (TA+LO)	260[h] (TO+LO)		
ZnTe	575[h]	100[h] (2 TO)	340[h] (TA+LO)	100[h] (TO+LO)		

[a] [2.44a].
[b] [2.218].
[c] [2.176].
[d] [2.197].
[e] [2.74].
[f] [2.77].
[g] [2.220].
[h] [2.73].

b) E_1, $E_1 + \Delta_1$ Gaps

Measurements of resonant second-order scattering near E_1, $E_1 + \Delta_1$ have been performed for Ge, [2.44a], Si [2.218] and a large number of III–V compounds [2.74, 220, 221]. In the latter, strongly resonant 2LO(Γ) peaks are seen, no doubt induced by Fröhlich interaction.

The deformation-potential-induced resonance of the renormalized $\mathcal{H}_{ev}^{(2)}$ in germanium is particularly interesting. We saw for the corresponding first-order effect of $\Gamma_{25'}$ symmetry, that the two-band terms were negligible when compared to the three-band terms. This fact is not a requirement of symmetry but results simply from the specific values of the deformation potentials $d_{1,0}^5$ (responsible for two-band terms) and $d_{3,0}^5$ (three-band terms). Surprisingly, the same result is found for the $\Gamma_{25'}$ component of the two-phonon spectrum: the typical three-band resonance with a broad maximum between E_1 and $E_1 + \Delta_1$ appears (see Fig. 2.52).

The Γ_1 components, however, should not produce three-band terms as shown in Sect. 2.2.10. Consequently, two sharp peaks are expected, one at E_1

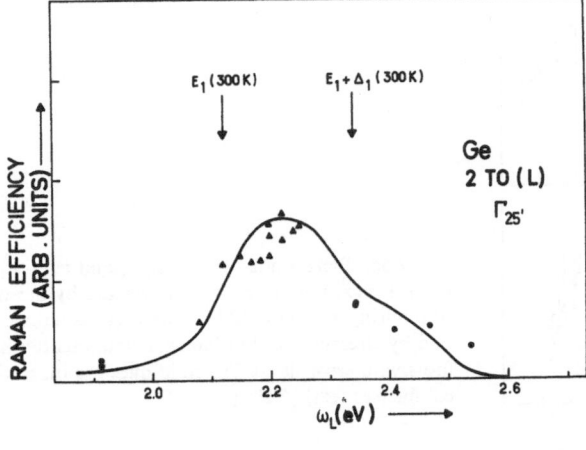

Fig. 2.52. Resonance of the $\Gamma_{25'}$ component [2 TO (L)] of the second-order Raman spectrum of germanium near E_1, $E_1 + \Delta_1$. The resonant shape is characteristic of three-band terms [2.44a]

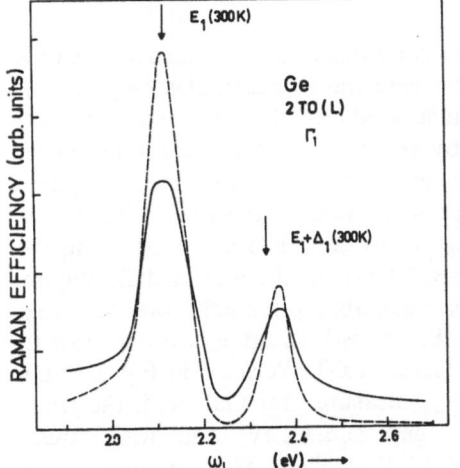

Fig. 2.53. Resonance of the 2 TO (L) peak in the component of the second-order Raman spectrum of germanium near E_1, $E_1 + \Delta_1$. The resonant shape is characteristic of two-band terms [2.44a]. Dashed curve: theory. Solid curve: experiment

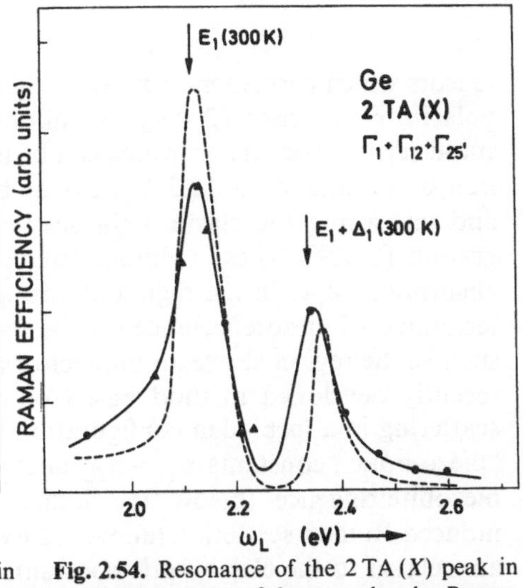

Fig. 2.54. Resonance of the 2 TA (X) peak in the Γ_1 component of the second-order Raman spectrum of germanium near E_1, $E_1 + \Delta_1$. The resonant shape is characteristic of two-band terms [2.44a]. Dashed curve: theory. Points: experiment

and the other at $E_1 + \Delta_1$. This prediction is confirmed in Fig. 2.53 for TO-phonons near L and in Fig. 2.54 for TA-phonons near X.

2.3.4 Elasto-Optic Constants

As mentioned in Sect. 2.1.14, the Brillouin scattering efficiency in cubic materials is determined by the elasto-optic constants p_{ij}. In anisotropic materials, a contribution can also result from the "antisymmetric" elasto-optic

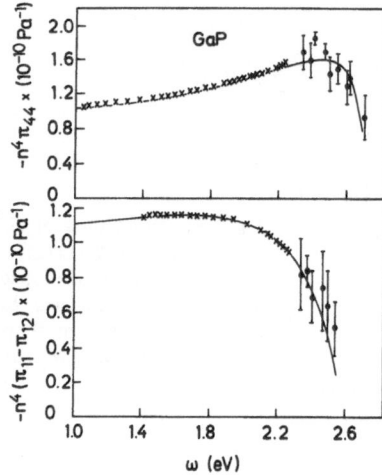

Fig. 2.55. Piezo-optic constants π_{44} and $\pi_{11}-\pi_{12}$ of GaP measured at room temperature by means of birefringence induced by a static stress (crosses) and by stress-induced forbidden Raman scattering (dots with error flags). The solid lines are theoretical fits (see text) [2.229]

tensors which correspond simply to rigid rotations of the standard anisotropic polarizability tensor [2.28a]. We discuss here the standard elasto-optic constants (symmetric tensor) which can be measured directly either by the application of a static stress [2.227, 228] or by setting up acoustic standing waves and measuring the elastic light scattering produced by the resulting phase grating [2.229]. These methods usually work well in the region below the absorption edge. In the region of strong absorption, near a critical point, the technique of piezoreflectance can be used [2.48]. In the intermediate region, such as the region above an indirect or a weak absorption edge, one can use a recently developed method based on the stress-induced first-order Raman scattering in a forbidden configuration [2.152, 153]. We show in Fig. 2.55 the "piezo-optic" constants $\pi_{11}-\pi_{12}$ and π_{44} measured for GaP with the static piezobirefringence (below the indirect gap $\omega_i \simeq 2.2$ eV) and with stress-induced Raman scattering (above 2.2 eV) [2.227]. These piezo-optic constants are related to the elasto-optic constants p_{ij} used so far (2.100) through

$$\chi_{\parallel}-\chi_{\perp} = -\left(\frac{\varepsilon}{\varepsilon_0}\right)^2 \pi_{44} X = -\left(\frac{\varepsilon}{\varepsilon_0}\right)^2 \frac{p_{44}}{c_{44}} X \qquad \text{(for [111] stress)}$$

$$\chi_{\parallel}-\chi_{\perp} = -\left(\frac{\varepsilon}{\varepsilon_0}\right)(\pi_{11}-\pi_{12}) X = -\left(\frac{\varepsilon}{\varepsilon_0}\right)^2 \frac{p_{11}-p_{12}}{c_{11}-c_{12}} X \qquad \text{(for [100] stress)},$$

$$(2.262)$$

where X is the magnitude of the applied uniaxial stress and $(\varepsilon/\varepsilon_0) = n^2$. Hence,

$$p_{44} = c_{44}\pi_{44}$$
$$p_{11} - p_{12} = (c_{11}-c_{12})(\pi_{11}-\pi_{12}). \qquad (2.263)$$

The stiffness constants are, for GaP, $c_{44} = 7.0 \times 10^{10}$ Pa and $c_{11}-c_{12} = 7.9 \times 10^{10}$ Pa [2.230].

Both spectra of Fig. 2.55 contain only the *real* part of π_{ij}: in the photon energy region of these figures (below the *direct* gap), the imaginary part is negligible. Very little experimental information is available on the imaginary parts of elasto-optic constants which, according to (2.111), must also be used for the calculation of Brillouin efficiencies. Recent measurements for silicon in the region 2.4 to 2.7 eV yield $\pi^i_{44} \simeq \pi^i_{11} - \pi^i_{12} \; 3 \times 10^{-14} \, \mathrm{Pa}^{-1}$ [2.228]. The corresponding values of the real parts are, in this region, one to two orders of magnitude larger [2.91]. Around the strong direct gap of Si at $\omega_L \simeq 3.3 \, \mathrm{eV}$, however, real and imaginary parts of π become comparable (see Fig. 72 of [2.48]).

For the theoretical analysis of Fig. 2.55, we use a model composed of the $E_0 - E_0 + \Delta_0$ edges (2.255, 256), the $E_1 - E_1 + \Delta_1$ edges (2.258, 258a) and an additive constant to take into account any other dispersion mechanisms such as the strong E_2 critical points (at 5.3 eV). The need for this constant, labeled C_2 for [100] and C'_2 for [111] stress, appears clearly in Fig. 2.55 since the effect of the E_0 gap has an opposite sign to the background $\pi_{ij}(\omega = 0)$. Such sign reversal of π_{ij}, which corresponds to an antiresonance in the Brillouin efficiency slightly below E_0, can only be obtained with the interference of two dispersion mechanisms. These are related, in our case, to $E_0 - E_0 + \Delta_0$ and either E_2 or E_1, $E_1 + \Delta_1$. We remind the reader that in order to fit the Raman polarizability a of *zincblende-type* materials below E_0, usually no E_1 and E_2 contribution is necessary and hence no antiresonance exists below E_0. The E_2 contribution, in the units of Fig. 2.55 ($-n^4\pi_{44}$ in $10^{-10} \, \mathrm{Pa}^{-1}$), amounts to $C_2 = 1.71$ for [100] stress and $C'_2 = -1.36$ for [111] stress. The contribution of E_0, $E_0 + \Delta_0$ is determined by the deformation potentials b {(2.256), [100] stress} and d {(2.255), [111] stress}. For GaP, $b = -1.8 \, \mathrm{eV}$ and $d = -4.6 \, \mathrm{eV}$. Hence, according to (2.255, 256), the contributions of $E_0 - E_0 + \Delta_0$ to $-n^4\pi_{44}$ and $-n^4(\pi_{11} - \pi_{12})$ are negative *below* E_0. The positive $C_2 = 1.71$ suffices to produce the observed antiresonance, but this is not the case for $C'_2 = -1.36$. In this case ([111]-stress), the antiresonance is produced within our model by interference with the $E_1 - E_1 + \Delta_1$ contribution to π_{44} (2.258): $d^5_1 + 2\sqrt{2}d^5_3$ is positive and large. Thus, $E_1 - E_1 + \Delta_1$ yield a large positive contribution to $-n^4\pi_{44}$.

The antiresonance in p_{44} and $p_{11} - p_{12}$ below E_0 is characteristic of most zincblende-type materials [GaAs, GaSb, Ge, ZnS, ZnSe (Fig. 2.15), ZnTe, CdTe, Inp, GaP (Fig. 2.55)]. For the materials of the family with a very small E_0 gap (InSb, InAs), the E_0 contribution to p_{44} and $p_{11} - p_{12}$ increases enormously as a result of the ω_0-denominator under d and b in (2.255, 256); all other parameters do not depend much upon the material under consideration. Also, the E_1 and E_2 contributions vary much less than the E_0 contribution from material to material since these gaps also vary less than E_0 in *relative* terms (e.g., Ge: $E_2 = 4.5 \, \mathrm{eV}$, $E_0 = 0.8 \, \mathrm{eV}$, InSb: $E_2 = 4.1 \, \mathrm{eV}$, $E_0 = 0.2 \, \mathrm{eV}$). Consequently, for InAs and InSb, the sign of the E_0 contribution dominates even at $\omega = 0$ and no antiresonance takes place [2.48].

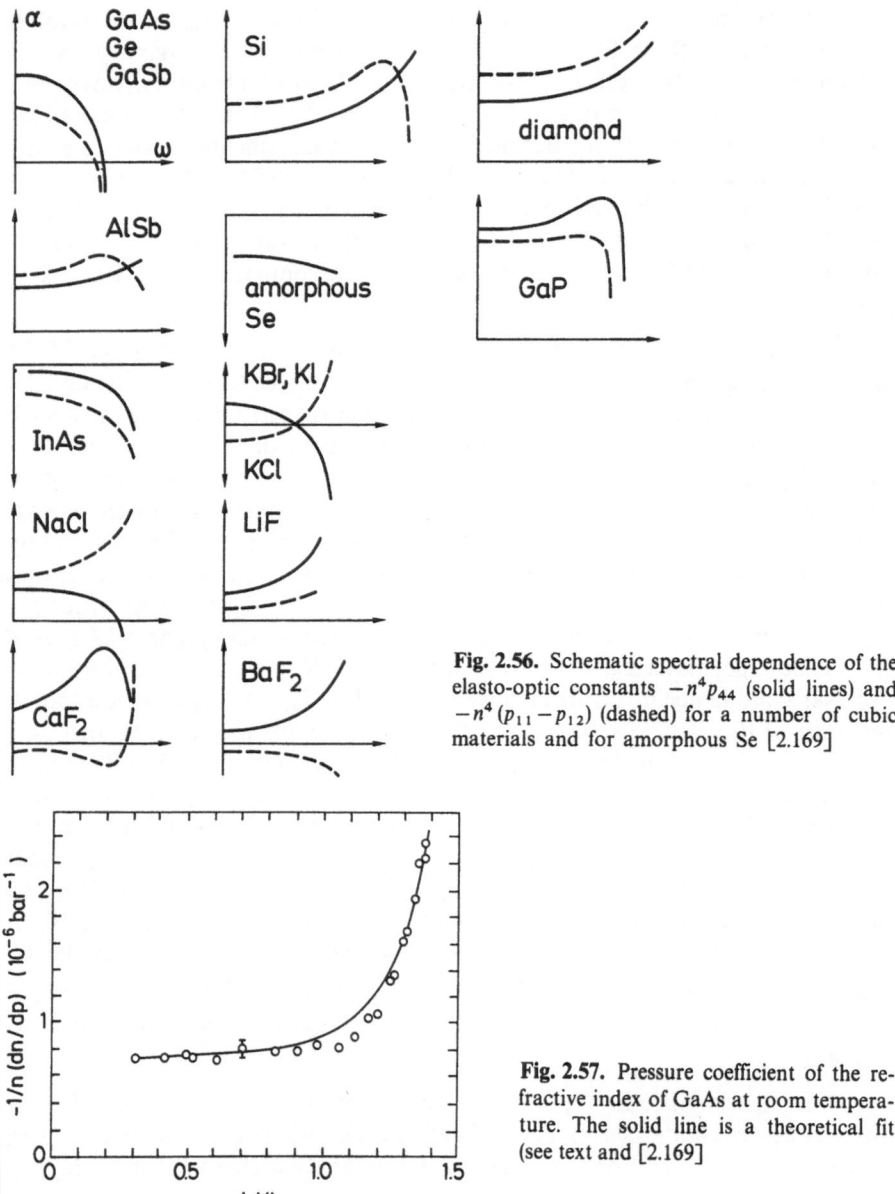

Fig. 2.56. Schematic spectral dependence of the elasto-optic constants $-n^4 p_{44}$ (solid lines) and $-n^4 (p_{11} - p_{12})$ (dashed) for a number of cubic materials and for amorphous Se [2.169]

Fig. 2.57. Pressure coefficient of the refractive index of GaAs at room temperature. The solid line is a theoretical fit (see text and [2.169]

The interference of two or three dispersion mechanisms in the theory of π_{44} and $\pi_{11} - \pi_{12}$ can give rise to a variety of qualitatively different spectral dependences of $-n^4\pi_{44}$ on photon energy. Several typical dependences observed, grouped into similar types, are shown schematically in Fig. 2.56. A compilation of spectral dependences of π_{ij} and/or p_{ij} for zincblende-germanium-type materials is given in [2.91].

The experimental information on the hydrostatic piezo-optical $(\pi_{11} + 2\pi_{12})$ or elasto-optical constant $(2p_{11} + 2p_{12})$ is more scarce. We show in Fig. 2.57, as an example, the hydrostatic pressure coefficient of the refractive index of GaAs which is related to $\pi_{11} + 2\pi_2$ through

$$\frac{1}{n}\frac{dn}{dp} = \frac{1}{2}n^2(\pi_{11} + 2\pi_{12}).$$ (2.263a)

The data in Fig. 2.57 were fitted with (2.257) using the well-known hydrostatic deformation potential a of the E_0 gap and adding a constant to represent the E_2 and E_1 backgrounds. Both the E_0 contribution and the background have the same sign and hence no antiresonance results below E_0.

2.3.5 Multiphonon Scattering

As mentioned in Sect. 2.2.11, scattering by phonon overtones in crystals is characteristic of strongly ir-active phonons: they are not seen in Ge, Si, and the III–V's for $m \gtrsim 3$ (in the III–V's, multimodes up to $m=3$ have been observed only for GaP and possibly for InAs [2.233]). They are, within this family, typical for II–VI compounds of either wurtzite or zincblende structure. They have been observed near E_0 for CdS (to $m=9$ [2.231]), CdSe (to $m=4$ [2.232]), ZnSe (to $m=5$ [2.233]), ZnTe (up to $m=8$ [2.234]), ZnO (to $m=5$ [2.44c, 234]). Unfortunately, few quantitative experiments with efficiencies in absolute units are available, so that a comparison with the theory can be at best only qualitative. It has been conjectured [2.233], however, that the number of overtones seen (m) is roughly proportional to the square of the polaron coupling constant "α" related to $|C_F|^2$ through

$$\text{``}\alpha\text{''}_{\text{eh}} = |C_F|^2\pi(2\mu_{\text{eh}}\omega_{\text{LO}})^{1/2}.$$ (2.263b)

Also, as we have seen in Sect. 2.2.11, the relative intensities depend, for $m \geq 3$, rather critically on the details of the lifetime of the intermediate state. Systematic studies of such dependence are not available. *Klochikhin* et al. [2.208b] have, however, reported multiphonon results for CdS of different intermediate state lifetimes varying from $\tau_i \simeq 3.5 \times 10^{-12}$ to 1.3×10^{-12} s. A slight decrease in the efficiency for 3LO-scattering with decreasing τ_i was reported which could be interpreted with (2.247).

Overtone scattering is also observed for local modes of defects in crystals under resonant conditions. Particularly striking are the results obtained by *Martin* [2.235] for additively colored CsI, which seem due to local modes of I_3^- complexes. These linear complexes have a *symmetric* stretch vibration at 111 cm^{-1}. A symmetric vibration is known to produce overtone scattering for ω_L near resonance via Frank-Condon terms (Sect. 2.2.3). As shown in Fig. 2.58, each one of these overtones is accompanied by a side band of lattice modes

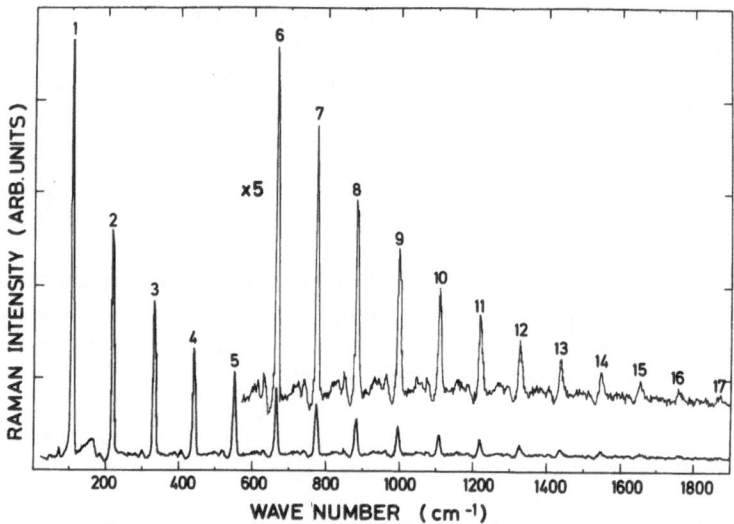

Fig. 2.58. Raman multimode spectra of I_3^- in CsI measured at 8 K for $\omega_L = 2.7$ eV. Overtones of $\omega_v = 111$ cm^{-1} up to $m = 17$ are observed. Each overtone has a sideband of standard lattice phonons [2.237]

roughly equal in shape to the density of one-phonon states, the q-selection rule being broken by the presence of the impurity (Sect. 2.1.16).

Martin [2.235] showed that the line width of the sharp overtones of Fig. 2.58 was determined by anharmonic decay into two lattice modes. Under these conditions, one has to ask whether the observed overtones are really true Raman lines with a final state corresponding to an $m\omega_v$ vibrational excitation of the local mode, or rather, correspond to a final state of the impurity in the vibrational ground state, the $m\omega_v$ vibrational energy being lost in the intermediate state through anharmonic interaction with lattice modes. In the latter case, the phenomenon could be referred to as *hot luminescence*, the various phonons involved in the anharmonic process not being specified. An operational criterion for distinguishing between true Raman scattering and the type of hot luminescence just mentioned was also advanced by *Martin*: in the former case, the line width should be roughly proportional to m while in the latter it should *decrease* with increasing m being proportional to $(m_{max} - m)$, where m_{max} is the highest overtone observed. Application of this criterion shows that, in the case of Fig. 2.58, one is dealing with true Raman scattering and not with hot luminescence.

2.3.6 Cuprous Oxide (Cu$_2$O)

Cu$_2$O is a rather interesting material with very sharp edge excitons. The crystal structure, with 2 formula units per unit cell, has the full cubic point group (O_h).

There are, therefore, 15 optical phonons at the Γ-point. Using the method of Sect. 2.1.10, one finds their symmetries to be [2.236]: $2 \times \Gamma_{15}$ (ir-active), $\Gamma_{25'}$ (the only Raman-active modes), Γ_{25}, $\Gamma_{12'}$, $\Gamma_{2'}$ (silent). The electronic band structure has also been investigated [2.237]. The valence bands correspond to an admixture of the $3d$ wave functions of copper and the $2p$ of oxygen. The top of the valence band has $\Gamma_{25'}$ orbital symmetry and possesses both Cu($3d$) and O($2p$) character. It splits under spin-orbit coupling into Γ_7^+ and Γ_8^+, the former being the uppermost valence band state. Thus, the spin-orbit splitting is *negative* (inverted spin multiplet), a result of the negative spin-orbit splitting of $\Gamma_{25'}$ d-like states [2.238] which is dominant (a similar result obtains for zincblende structure CuCl) [2.239]. The lowest conduction band is Cu($4s$)-like and has Γ_1 orbital symmetry at the center of the BZ. Hence, the lowest direct gap $\Gamma_{25'} \to \Gamma_1$ occurs between states of the same parity and is thus dipole-forbidden (quadrupole allowed), in sharp contrast with most materials discussed so far. It is this unusual feature which makes Cu_2O such an interesting material.

The so-called yellow exciton series is derived from the $\Gamma_7^+ \to \Gamma_1$ energy gap. The binding energy of the $n=1$ exciton is large, approximately 0.1 eV. This series is a classical example of the direct forbidden exciton [2.240]: the admixture of states with $k_e = - k_h \neq 0$ makes the excitons weakly dipole allowed. The allowed exciton states must have a p-like envelope function and, therefore, only the s-like $n=1$ state remains dipole forbidden (but quadrupole allowed). A second "forbidden" exciton series (the green series) is associated with the $\Gamma_8^+ \to \Gamma_1$ edge. The first allowed edge is $\Gamma_{25'}$ (split into Γ_7^+ and Γ_8^+) $\to \Gamma_{12'}$: it gives rise to the dipole allowed blue and violet exciton series (see, for instance, [2.241]).

a) First-Order Raman Spectrum

As discussed in [Ref. 2.1, Chap. 3], some of the Raman-forbidden Γ-phonons of Cu_2O, namely those of odd parity (Γ_{15}, Γ_{25}, $\Gamma_{12'}$, $\Gamma_{2'}$), produce resonant scattering at the forbidden $1s$ yellow exciton. This resonance has been attributed to a dipole-quadrupole mechanism: the incident photon virtually excites the even parity $1s$ exciton through quadrupole interaction, the phonon scatters it into another *odd* parity state from which it decays into the *even* ground state through emission of a photon. In this case, the phenomenon would show an in-going resonance. An out-going resonance is obtained by permuting the first and the last of those steps. The phonons involved in such processes must obviously be odd. Such resonant scattering has been observed for *all* odd phonons at $86\,\mathrm{cm}^{-1}$ (Γ_{25}), $109\,\mathrm{cm}^{-1}$ ($\Gamma_{12'}$), $153\,\mathrm{cm}^{-1}$ (Γ_{15}), $350\,\mathrm{cm}^{-1}$ ($\Gamma_{2'}$), and 640–$660\,\mathrm{cm}^{-1}$ (Γ_{15} TO and LO) [2.242, 246, 247].

Quadrupole transitions are known to be anisotropic with respect to the polarization direction even for cubic materials. Hence, the D–Q (dipole-quadrupole) resonances just mentioned should have rather complex polarization properties represented, near resonances, by nonsymmetric tensors. These tensors are given in [2.243] where it is shown that they give good explanations for the observed polarization properties of the Γ_{15}, Γ_{25}, $\Gamma_{12'}$, and $\Gamma_{2'}$ phonons. An interesting point is also the contribution of the Fröhlich interaction to the D–Q scattering by the polar Γ_{15} phonons. In Sect. 2.1.12, we saw that the interband Fröhlich interaction (also Sect. 2.3.1a) adds an electro-optic contribution to the Raman susceptibility for LO-phonons. This contribution vanishes, in the dipole-dipole case treated in Sect. 2.1.12, for materials with inversion symmetry since the first-order electro-optic tensor is zero. In the dipole quadrupole case we must consider, however, a susceptibility of the form

$$\frac{\partial^2 \chi}{\partial q \partial E} : qE , \tag{2.264}$$

where the q is introduced by the quadrupole operator and E is the field produced by the longitudinal phonon. The differential susceptibility $\partial^2 \chi / \partial q \partial E$, a fourth rank tensor, need not vanish for a material with inversion symmetry. Consequently, we obtain for LO-phonons a "Fröhlich" contribution to the scattering efficiency in the DQ-case which has the same form as the deformation potential contribution: the tensor coefficients for the TO-components of Γ_{15} phonons differ from those of the LO-phonons. For the upper Γ_{15} phonons, the ratio of the corresponding LO to TO efficiencies is $\simeq 5$ [2.55] (it would be one if the effect of (2.264) were not there).

Habinger and *Compaan* have recently reported an interesting phenomenon whose observation is only possible because of the very small width of the 1s quadrupole line of the yellow series ($\sim 0.1\,\mathrm{cm}^{-1}$) [2.243]. These authors observed the in-going and the out-going DQ resonances of the $\Gamma_{12'}$ phonon and found them to have different widths, the second being wider by $\gamma_\mathrm{v} \simeq 0.11\,\mathrm{cm}^{-1}$ (halfwidth), precisely the width of the phonon. The explanation is easily found by looking at the standard expression for such a resonance, isomorphic with that of (2.161). In order to introduce damping in (2.161), we must replace the energy of the electronic intermediate state by $\omega_j + i\Gamma_\mathrm{e}$, where Γ_e is the damping of the electronic excitation ω_j, in our case the 1s exciton. For the in-going resonance, the halfwidth is precisely Γ_e. For the out-going resonance, however, $\omega_\mathrm{s} = \omega_\mathrm{L} - \omega_\mathrm{v}$ and ω_v should *also* be replaced by $\omega_\mathrm{v} + i\gamma_\mathrm{v}$, where γ_v is the damping of the phonon. We thus see that the halfwidth of the out-going resonance is $\Gamma_\mathrm{e} + \gamma_\mathrm{v}$, as shown by the results in [Ref. 2.243, Fig. 1].

We conclude by mentioning that the observation of DQ resonances has proved to be a powerful method for performing "quadrupole spectroscopy", i.e., to observe quadrupole allowed, dipole forbidden excitons [2.244]. By this method, the effect of stress on the 1s, 3s, 4s, and 3d yellow excitons of Cu_2O has recently been investigated [2.245].

b) Higher-Order Raman Spectrum

We have already mentioned in Sect. 2.2.10, see (2.245a, 246), the scattering by $2\Gamma_{12'}$ phonons resonant above the 1s yellow exciton. The effect can be seen in

Fig. 3.14 of [2.1], [2.246]. Another type of two-phonon resonance was reported by *Yu* and *Shen* [2.247]. It involves at least one polar phonon Γ_{15} and a second phonon of either Γ_{15} or $\Gamma_{12'}$ symmetry. While the $2\Gamma_{12'}$ resonance of [2.246] involves only a resonance with the *intermediate state* in which one phonon of wave vector q_1 has been emitted, the present one involves, besides the intermediate state resonance, a resonance with either the initial or the final state ($\omega_L = \omega_{ex}$ or $\omega_s = \omega_{ex}$). The resonant excitonic states are now those for which the optical transitions are dipole allowed ($2p, 3p \ldots$ up to $6p$ have been observed). In these processes like in the $2\Gamma_{12'}$ process, the wave vector of the emitted phonons $q_1 \simeq -q_2$ is determined by the resonance condition in the intermediate state:

$$\omega_L = \omega_{ex} + \frac{q_1^2}{2M^*} + \omega_{v_1}, \tag{2.265}$$

where ω_{ex} is the energy of one of the exciton levels for $K = 0$ and M^* the mass of its center of mass ($M^* = m_e + m_h$). A characteristic of these types of phenomena is the fact that the q vector of the phonons is determined by the laser frequency through the mass M^* (not μ^*!). If the dispersion of the phonons is neglected, (2.265) leads to a nondispersive two-phonon peak in spite of the fact that q_1 changes with $\omega_L - \omega_{ex}$. The phonon dispersion near $q = 0$ can be taken into account through an "effective mass" M_p [see (2.117)]:

$$\omega_{v_1}(q) = \omega_{v_1}(0) + \frac{q^2}{2M_p}. \tag{2.266}$$

Combining (2.266) with (2.265), we find

$$\omega_{v_1} = \omega_{v_1}(0) + (\omega_L - \omega_{ex})\frac{M^*}{M_p}. \tag{2.267}$$

Equation (2.267) can, in principle, also be applied to calculate the shift of the $q_1 \simeq 0$ two-phonon peak in the case of uncorrelated interband transitions (e.g., $E_0 + \Delta_0$ edge of GaAs [2.225], E_0 edge of CdS [2.224]).

For Cu_2O, the $\Gamma_{12'}$ phonons are nondispersive near $q = 0$ ($M_p \simeq \infty$) and hence no dispersion is seen in the $2\Gamma_{12'}$ peak [2.246]. The same thing applies to the $\Gamma_{12'} + \Gamma_{25}$ peak. The peaks involving Γ_{15} phonons (e.g., $\Gamma_{12'} + \Gamma_{15}$, $2\Gamma_{12'} + \Gamma_{15}$ [2.247]) split into two: the splitting is linearly proportional to $(\omega_L - \omega_{ex})$, as predicted by (2.267). This splitting is related to a finite M_p, of opposite sign for the LO- and TO-components of the Γ_{15} phonons.

Perhaps the most spectacular effect is the scattering by $2\Gamma_{12'}$ optical phonons plus one or two acoustic phonons. This scattering, observed for LA, TA, and 2LA acoustic phonons, is strongly dispersive, as expected in view of the linear dispersion of the acoustic phonons for $q \to 0$. The observed dispersion can be fitted well with the known elastic constants of Cu_2O [2.247].

2.3.7 AgCl, AgBr

These materials crystallize in the rocksalt structure. They have indirect absorption edges, from the $L_{3'}$ valence band maximum (mixed p of halogen and $4d$ of silver) to the silver $6s$-like Γ_1 conduction band minimum [Ref. 2.248, Fig. 2.7]. This edge displays well-defined indirect excitons which are excited optically with the help of L-phonons ([2.202] and references therein). Correspondingly, the peaks in the second-order Raman spectrum which are assigned to two phonons at the L point of the BZ resonate strongly near the indirect excitons. The phenomenon is basically the same as the $2\Gamma_{12'}$ two-phonon resonance of the 1s yellow exciton of Cu_2O (Sect. 2.3.6a), except that the phonons are now at L instead of $\simeq \Gamma$. As shown in Fig. 2.59, 2TO(L), 2TO(L) + Acoustic (X) and 2TO(L) + 2 Acoustic (X) phonons are observed. The X-phonons produce "intervalley" scattering between the different {111} exciton branches. Above the 2TO(L) branch, a sideband involving intravalley scattering by LA(Γ)-phonons [2TO(L) + LA(Γ)] is seen. This is shown in Fig. 2.60. The shape of the sideband is determined by the exciton mass of the center of mass M^* [see (2.265), where q_1 is now the q of the intermediate exciton state before acoustic phonon emission, measured from the L point]. Figure 2.60 shows three fits to this sideband for three different values of M^*. From the best one, the value of $M^* = 1.4 \pm 0.2$ [2.202] for the lowest indirect exciton in AgBr is determined.

Resonances of the two-phonon scattering near the indirect edge of AgCl have been recently reported [2.249].

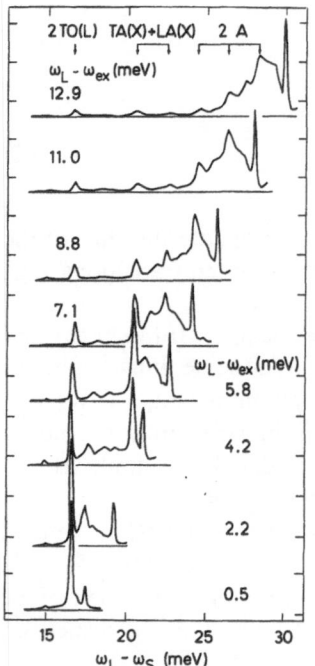

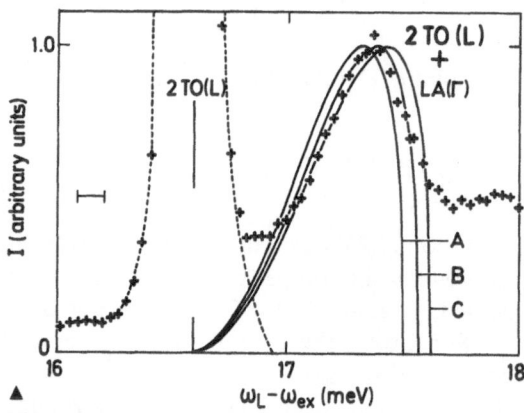

Fig. 2.60. Resonant 2 TO (L) peak and 2 TO (L) + LA(Γ) sideband observed in AgBr for $\omega_L = 2.695$ eV (crosses). The solid lines are theoretical fits for $M^* = 1.2$ (A), 1.4 (B), and 1.6 (C), respectively [2.202]

◄ **Fig. 2.59.** Resonant Raman spectra of AgBr at 1.8 K for different laser energies ω_L labeled by the difference $\omega_L - \omega_{ex}$ (ω_{ex} is the energy of the ground state of the indirect exciton). The following structures have been identified [2.202]: 2 TO (L); 2 TO (L) + TA and LA (X); 2 TO (L) + 2 acoustic phonons at X

References

2.1   M. Cardona (ed.): *Light Scattering in Solids*, Topics Appl. Phys., Vol. 8 (Springer, Berlin, Heidelberg, New York 1975)

2.2   F. Cerdeira, W. Dreybrodt, M. Cardona: Solid State Commun. **10**, 591 (1972)

2.3   P. Y. Yu, Y. R. Shen: Phys. Rev. Lett. **29**, 468 (1972)

2.4   T. W. Hänsch: In *"Dye Lasers"* 2nd ed., ed. by F. P. Schäfer (Springer, Berlin, Heidelberg, New York 1977) p. 194

2.5   A. Compaan, H. R. Cummins: Phys. Rev. Lett. **31**, 41 (1973)

2.6   P. Y. Yu, Y. R. Shen, Y. Petroff, L. Falicov: Phys. Rev. Lett. **30**, 283 (1973)

2.7   R. M. Martin: Phys. Rev. B**4**, 3676 (1971)

2.8   R. M. Martin, C. M. Varma: Phys. Rev. Lett. **26**, 1241 (1971)

2.9   R. Zeyher: Solid State Commun. **16**, 49 (1975)

2.10 W. Brenig, R. Zeyher, J. L. Birman: Phys. Rev. B**6**, 4617 (1972)

2.11 R. G. Ulbrich, C. Weisbuch: Phys. Rev. Lett. **38**, 865 (1977)

2.12 G. B. Wright (ed.): *Light Scattering Spectra of Solids* (Springer, Berlin, Heidelberg, New York 1969)

2.13 M. Balkanski (ed.): *Light Scattering in Solids* (Flammarion Sciences, Paris 1971)

2.14 M. Balkanski, R. C. C. Leite, S. P. S. Porto (eds.): *Light Scattering in Solids* (Flammarion, Paris 1975)

2.15 W. F. Murphy (ed.): *Proc. 7th Intern. Conf. on Raman Spectroscopy* (North-Holland, Amsterdam 1980)

2.16 *Proc. 6th Intern. Conf. on Raman Spectroscopy* (Heyden, London 1978)

2.17 W. Hayes, R. Loudon: *Scattering of Light by Crystals* (Wiley, New York 1978)

2.18 A. Weber (ed.): *Raman Spectroscopy of Gases and Liquids*, Topics Current Phys., Vol. 11 (Springer, Berlin, Heidelberg, New York 1979)

2.19 A. Anderson (ed.): *The Raman Effect*, Vols. 1, 2 (M. Dekker, New York 1973)

2.20 H. Bilz, W. Kress: *Phonon Dispersion Relations in Insulators* (Springer, Berlin, Heidelberg, New York 1979)

2.21 J. Birman: In *Encyclopedia of Physics*, Vol. XXV/2b (Springer, Berlin, Heidelberg, New York 1974)

2.22 L. Ley, M. Cardona: *Photoemission in Solids II*, Topics Appl. Phys., Vol. 27 (Springer, Berlin, Heidelberg, New York 1979) p. 159

2.23 M. Balkanski (ed.): *Lattice Dynamics* (Flammarion, Paris 1978)

2.24 M. Agranovich, J. L. Birman (eds.): *The Theory of Light Scattering in Solids* (Nauka, Moscow 1976)

2.25 J. L. Birman, H. Z. Cummins, K. K. Rebane (eds.): *Light Scattering in Solids* (Plenum, New York 1979)

2.26a W. Richter: In *Springer Tracts in Modern Physics*, Vol. 78 (Springer, Berlin, Heidelberg, New York 1976) p. 121

2.26b J. Raman Spectroscopy, Vol. **10** (1981)

2.27 L. D. Landau, E. M. Lifschitz: *Classical Field Theory* (Addison-Wesley, Reading, Mass. 1962) p. 199

2.28a D. F. Nelson, M. Lax: Phys. Rev. Lett. **24**, 379 (1970)

2.28b D. F. Nelson, P. D. Lazay, M. Lax: Phys. Rev. B**6**, 3109 (1972); D. F. Nelson, P. D. Lazay: Phys. Rev. B**16**, 4659 (1977); J. Opt. Soc. Am. **67**, 1599 (1977); M. H. Grimsditch, A. K. Ramdas: Phys. Rev. B**22**, 4094 (1980)

2.29 R. Zeyher, H. Bilz, M. Cardona: Solid State Commun. **19**, 57 (1976)

2.30 D. Bermejo, M. Cardona: J. Non-Cryst. Solids **32**, 421 (1979)

2.31 G. Placzek: Rayleigh and Raman Scattering VCRL Transl. 526L, available from National Technical Inf. Serv. US Dept. of Commerce, Springfield Va., USA
      Handbuch der Radiologie, Vol. 6, Pt. 2, ed. by E. Marx (Akad. Verlagsgesellschaft, Leipzig 1934) p. 205

2.32 D. L. Rousseau, J. M. Friedman, P. F. Williams: In [2.18], p. 242]

2.33 A. Einstein: Ann. Phys. **38**, 1275 (1910)

2.34   L. D. Landau, E. M. Lifschitz: *Statistical Physics* (Addison Wesley, Reading, Mass. 1958) p. 350
2.35   R. W. Wehner, R. Klein: Physica **62**, 162 (1972)
2.36   E. B. Wilson, Jr., J. C. Decius, P. C. Cross: *Molecular Vibrations: The Theory of Infrared and Raman Spectra* (McGraw-Hill, New York 1955)
2.37   C. Kittel: *Quantum Theory of Solids* (Wiley, New York 1963)
2.38   W. Wettling, N. G. Cottam, J. R. Sandercock: J. Phys. C**8**, 211 (1975)
2.39   M. Wolkenstein: Compt. Rend. Acad. Sci. URSS **32**, 185 (1941)
2.40   S. Go, H. Bilz, M. Cardona: Phys. Rev. Lett. **34**, 580 (1975)
2.41   D. Bermejo, R. Escribano, J. M. Orza: J. Raman Spectrosc. **6**, 38 (1977)
2.42   B. L. Sowers, M. W. Williams, R. N. Hamm, E. T. Arakawa: J. Chem. Phys. **57**, 167 (1972)
2.43   [Ref. 2.18, Fig. 6.14]
2.44a   M. A. Renucci, J. B. Renucci, R. Zeyher, M. Cardona: Phys. Rev. B**10**, 4309 (1974)
2.44b   D. G. Fouche, R. K. Chang: Appl. Phys. Lett. **18**, 5791 (1971)
2.44c   R. H. Callender, S. S. Sussman, M. Selders, R. K. Chang: Phys. Rev. B**7**, 3788 (1973)
2.45   H. A. Hyatt, J. M. Cherlow, W. R. Fenner, S. P. S. Porto: J. Opt. Soc. Am. **63**, 1604 (1973)
2.46a   [Ref. 2.19, p. 130]
2.46b   R. A. Loudon: J. Phys. (Paris) **26**, 677 (1965)
2.46c   R. A. Cowley: [Ref. 2.19, Vol. 1, p. 135]
2.47   C. Kittel: *Introduction to Solid State Physics*, 3rd edition (Wiley, New York 1967)
2.48   M. Cardona: *Modulation Spectroscopy* (Academic Press, New York 1969)
2.49   J. M. Calleja, H. Vogt, M. Cardona: Phil. Mag. (to be published)
2.50   R. Zeyher, H. Bilz, M. Cardona: Solid State Commun. **19**, 57 (1976)
2.51   [Ref. 2.13, p. 43]
2.52   R. L. Schmidt, M. Cardona: In *Proc. 13th Intern. Conf. on Physics of Semiconductors*, ed. by F. G. Fumi (Marves, Rome 1976) p. 239
2.53   M. H. Grimsditch, D. Olego, M. Cardona: [Ref. 2.25, p. 2.49]
2.54   M. Grimsditch, M. Cardona, J. M. Calleja, F. Meseguer: J. Raman Spectr. **10**, 77 (1981)
2.55   A. Z. Genack, H. Z. Cummins, M. A. Washington, A. Compaan: Phys. Rev. B**12**, 2478 (1975)
2.56a   M. Matsumoto, H. Kanamari: Indian J. Pure Appl. Phys. **16**, 254 (1978)
2.56b   D. M. Adams, D. C. Newton: Tables for Factor Group and Point Group Analysis (Beckman, London 1970)
2.56c   D. L. Rousseau, R. P. Bauman, S. P. S. Porto: J. Raman Spectr. **10**, 253 (1981)
2.57a   J. L. Shay, J. H. Wernick: *Ternary Chalcopyrite Semiconductors: Growth, Electronic Properties, and Applications* (Pergamon Press, Oxford 1975)
2.57b   E. Anastassakis, M. Cardona: Phys. Stat. Sol. (b) **104**, 589 (1981)
2.58   P. N. Butcher, N. R. Ogg: Proc. Phys. Soc. **86**, 699 (1965) and references therein; [2.34], p. 398
2.59   P. T. Hon, W. L. Faust: Appl. Phys. **1**, 241 (1973)
2.60   See, for instance, C. Carlone, D. Olego, A. Jayaraman, M. Cardona: Phys. Rev. B**22**, 3877 (1980)
2.61   G. Livescu, O. Brafman: Solid State Commun. **35**, 73 (1980)
2.62   L. N. Ovander, N. S. Tyn: Phys. Stat. Sol. (b) **91**, 763 (1979)
2.63   B. A. Weinstein, M. Cardona: Phys. Rev. B**7**, 2545 (1973): Ce
2.64   W. Weber: Phys. Rev. B**15**, 4789 (1977)
2.65   M. Tinkham: *Group Theory and Quantum Mechanics* (McCraw-Hill, New York 1964)
2.66a   G. F. Koster: *Space Groups and their Representations* (Academic Press, New York 1957)
2.66b   B. K. Vainshtein: *Modern Crystallography I*, Springer Ser. in Solid-State Sci., Vol. 15 (Springer, Berlin, Heidelberg, New York 1981)
2.67   M. Krauzman: Compt. Rend. (Paris) **265**B, 1029 (1967); **266**B, 186 (1968); also [Ref. 2.12, p. 109]
2.68   K. H. Rieder, B. A. Weinstein, M. Cardona, H. Bilz: Phys. Rev. B**8**, 4780 (1973)
2.69   P. A. Temple, C. A. Hathaway: Phys. Rev. **7**, 3685 (1973)
2.70   S. A. Solin, A. K. Ramdas: Phys. Rev. B**1**, 1678 (1970)
2.71   N. Krishnamurty, V. Soots: Can. J. Phys. **50**, 849 (1972); **50**, 1350 (1972)
2.72   S. Onari, M. Cardona: Phys. Rev. B**14**, 3520 (1976) and references therein

2.73   R.L.Schmidt, K.Kunc, M.Cardona, H.Bilz: Phys. Rev. B**20**, 3345 (1979): ZnS, ZnSe, ZnTe
2.74   R.Trommer, M.Cardona: Phys. Rev. B**17**, 1865 (1978): GaAs
2.75   P.B.Klein, R.K.Chang: Phys. Rev. B**14**, 2498 (1976); T.Sekine, K.Uchinoura, E.Matsuura: Solid State Commun. **18**, 1337 (1976): GaSb
2.76   R.Carles, N.Saint-Cricq, J.B.Renucci, M.A.Renucci, A.Zwick: Phys. Rev. B**22**, 4804 (1980): InAs
2.77   W.Kiefer, W.Richter, M.Cardona: Phys. Rev. B**12**, 2346 (1975): InSb
2.78   P.H.Borcherds, K.Kunc, G.F.Alfrey, R.L.Hall: J. Phys. C**12**, 4699 (1979): GaP
2.79   G.F.Alfrey, P.H.Borcherds: J. Phys. C**5**, L275 (1975): InP
2.80   J.M.Calleja, M.Cardona: Phys. Rev. B**16**, 3753 (1977): ZnO
2.81   R.Tubino, J.Birman: Phys. Rev. B**15**, 5843 (1977)
2.82a J.Ruvalds, A.Zawadowski: Phys. Rev. B**2**, 1172 (1970)
2.82b D.Bermejo, S.Montero, M.Cardona, A.Muramatsu: Solid State Commun., in press
2.83   A.D.Bruce, R.A.Cowley: J. Phys. C**5**, 595 (1972)
2.84   K.Kunc, H.Bilz: Solid State Commun. **19**, 1027 (1976)
2.85   J.Sandercock: Phys. Rev. Lett. **28**, 237 (1972)
2.86   A.Dervisch, R.Loudon: J. Phys. C**9**, L669 (1976)
2.87   M.Grimsditch, A.Malozemoff, A.Brunsch: Phys. Rev. Lett. **43**, 711 (1979)
2.88a H.G.Olf, A.Peterlin, W.I.Peticolas: J. Polym. Sci. **12**, 359 (1974)
2.88b G.Benedek, K.Fritsch: Phys. Rev. **149**, 647 (1966)
2.88c H.Z.Cummins, P.E.Schoen: In *Laser Handbook*, ed. by F.T.Arecchi, E.O.Schulz-Dubois (North-Holland, Amsterdam 1972) p. 1029
2.89   G.Winterling, E.S.Koteles, M.Cardona: Phys. Rev. Lett. **39**, 1286 (1977)
2.90   P.Vu, Y.Oka, M.Cardona: Phys. Rev. B**24**, 765 (1981)
2.91   For a tabulation of elasto-optic constants in semiconductors, see Landolt Börnstein tables, ed. by O.Madelung, New Series, Group III, Vol. 17a (Springer, Berlin, Heidelberg, New York 1982)
2.92   D.K.Garrod, R.Bray: Phys. Rev. B**6**, 1314 (1972)
2.93   S.Adachi, C.Hamaguchi: Phys. Rev. B**19**, 938 (1979); **21**, 1701 (1980)
2.94   S.Mishra, R.Bray: Phys. Rev. Lett. **39**, 222 (1977)
2.95   R.Loudon, J.R.Sandercock: J. Phys. C**13**, 2609 (1980)
2.96   A.M.Marvin, V.Bortolani, F.Nizzoli, G.Santoro: J. Phys. C**13**, 1607 (1980)
2.97   R.Loudon: J. Phys. C**11**, 2623 (1978)
2.98   W.E.Spear (ed.): *Amorphous and Liquid Semiconductors* (Univ. of Edinburgh, 1977)
2.99   W.Paul, M.Kastner (ed.): *Amorphous and Liquid Semiconductors* (North-Holland, Amsterdam 1980)
2.100 B.K.Chakraverty, D.Kaplan (eds.): Amorphous and Liquid Semiconductors (éditions de physique, Les Ulis 1981)
2.101 W.E.Spear, P.G.LeComber: Solid State Commun. **17**, 1193 (1975)
2.102 D.E.Carlson, C.R.Wronski: In *Amorphous Semiconductors*, ed. by M.H.Brodsky, Topics Appl. Phys., Vol. 36 (Springer, Berlin, Heidelberg, New York 1979)
2.103 I.Shimizu, T.Komatsu, K.Saito, E.Inoue: [Ref. 2.99, p. 773]
2.104 A.J.Snell, K.D.McKenzie, W.E.Spear, P.G.LeComber, A.J.Hughes: Appl. Phys. **24**, 357 (1981)
2.105 M.H.Brodsky, M.Cardona, J.J.Cuomo: Phys. Rev. B**16**, 3556 (1977)
2.106 P.G.LeComber, W.E.Spear: [Ref. 2.102, p. 251]
2.107 A.Madan: [Ref. 2.99, p. 171]
2.108 D.Bermejo, M.Cardona: J. Non-Cryst. Solids **32**, 405 (1979)
2.109 H.Shanks, C.J.Fang, L.Ley, M.Cardona, F.J.Demond, S.Kalbitzer: Phys. Stat. Sol. (b) **100**, 43 (1980)
2.110 J.S.Lannin: Solid State Commun. **12**, 947 (1973)
2.111 J.Jäckle: [Ref. 131a, p. 135]
2.112 D.T.Pierce, W.E.Spicer: Phys. Rev. B**8**, 3017 (1972)
2.113 F.Yndurain, P.N.Sen: Phys. Rev. B**14**, 531 (1976)
2.114 S.C.Shen, C.J.Fang, M.Cardona, L.Genzel: Phys. Rev. B**22**, 2913 (1980)
2.115 R.Tsu: Solid State Commun. **36**, 817 (1980)

2.116   S.Veprek, Z.Iqbal, H.R.Oswald, A.P.Webb: J. Phys. C14, 295 (1980)
2.117   H.Richter, Z.P.Wang, L.Ley: Solid State Commun. 39, 625 (1981)
2.118   D.Bermejo, M.Cardona: J. Non-Cryst. Solids 32, 421 (1979)
2.119   G.Lucovsky: Solid State Commun. 29, 571 (1979)
2.120   J.J.Hauser: J. Non-Cryst. Solids 23, 21 (1977)
2.121   N.Wada, P.J.Gaczi, S.A.Solin: J. Non-Cryst. Solids 35, 543 (1980)
2.122   C.Carini, M.Cutroni, G.Galli, P.Migliardo, F.Wanderlingh: Solid State Commun. 33, 1139 (1980)
2.123   F.L.Galeener, G.Lucovsky, J.C.Mikkelsen, Jr.: Phys. Rev. B22, 3983 (1980)
2.124   R.J.Nemanich, S.A.Solin, G.Lucovski: Solid State Commun. 21, 273 (1977);
        P.M.Bridenbaugh, G.P.Espinosa, J.E.Griffiths, J.C.Phillips, J.P.Remeika: Phys. Rev. B20, 4140 (1979)
2.125   J.C.Phillips, C.Arnold Beevers, S.E.B.Gould: Phys. Rev. B21, 5724 (1980);
        J.C.Phillips, J. Non-Cryst. Solids 35, 1157 (1980)
2.126   P.N.Sen, M.F.Thorpe: Phys. Rev. B15, 4030 (1977)
2.127   R.M.Martin, F.L.Galeener: Phys. Rev. 23, 3071 (1981)
2.128   G.Winterling: Phys. Rev. B12, 2432 (1975)
2.129   P.A.Fleury, K.B.Lyons: Phys. Rev. Lett. 36, 1188 (1976)
2.130   R.J.Nemanich: Phys. Rev. B16, 1655 (1977)
2.131a P.W.Anderson, B.I.Halperin, C.Varma: Phil. Mag. 25, 1 (1972)
2.131b S.Hunklinger, M. von Schickfus: In Amorphous Solids, Low Temperature Properties, ed. by
        W.A.Phillips, Topics Current Phys., Vol. 24 (Springer, Berlin, Heidelberg, New York 1981)
        p.81
2.131c R.H.Stolen, M.A.Bösch: Phys. Rev. Letters, to be published
2.132   T.C.Rich, D.A.Pinnow: Appl. Phys. Lett. 20, 264 (1972)
2.133   N.L.Laberge, V.V.Vasilescu, C.J.Montrose, P.B.Macedo: J. Am. Cer. Soc. 56, 506 (1973)
2.134   J.A.Bucaro, H.D.Dardy: J. Appl. Phys. 45, 2121 (1974)
2.135   Y.Yacoby: Phys. Rev. B18, 2918 (1978)
2.136   M.Chandrasekhar, H.R.Chandrasekhar, M.Grimsditch, M.Cardona: Phys. Rev. B22, 4825 (1980)
2.137   T.Miyanaga, F.Lüthy: Phys. Rev. (in press)
2.138   B.H.Bayramov, V.V.Toporov, S.B.Ubaydullaev, L.Hildish, E.Jahne: Solid State Commun. 37, 963 (1981)
2.139   S.Schicktanz, R.Kaiser, E.Schneider, W.Gläser: Phys. Rev. B22, 2386 (1980)
2.140   M.V.Klein, J.A.Holy, W.S.Williams, Phys. Rev. B17, 1556 (1978)
2.141   F.Bechstedt, R.Enderlein: Phys. Stat. Sol. (b) 83, 239 (1977)
2.142   P.Kleinert, F.Bechstedt: Phys. Stat. Sol. (b) 85, 253 (1978)
2.143   G.Güntherodt, P.Grünberg, E.Anastassakis, H.Bilz, M.Cardona, H.Hackfort, W.Zinn: Phys. Rev. B16, 3504 (1977)
2.144   See, for instance, M.Garbuny: Optical Physics (Academic Press, New York 1965) p.88
2.145   See, for instance, J.W.Nibler, G.V.Knights: [Ref. 2.18, p. 253]
2.146   M.D.Levenson, N.Bloembergen: Phys. Rev. B10, 4447 (1974)
2.147   J.G.Skinner, W.G.Nielsen: J. Opt. Soc. Am. 58, 113 (1968)
2.148   Y.Kato, H.Takumen: J. Opt. Soc. Am. 61, 347 (1971)
2.149   V.S.Gorelik, M.M.Sushchinskii, Sov. Phys. Solid State 12, 1157 (1970)
2.150   M.D.Levenson: IEEE J. QE-10, 110 (1974)
2.151   M.H.Grimsditch, A.K.Ramdas: Phys. Rev. B11, 3139 (1974)
2.152   M.Grimsditch, M.Cardona: Phys. Stat. Sol. (b) 102, 155 (1980)
2.153   M.Chandrasekhar, M.H.Grimsditch, M.Cardona: J. Opt. Soc. Am. 68, 523 (1978)
2.154   A.K.McQuillan, W.R.L.Clemens, B.P.Stoichef: Phys. Rev. A1, 628 (1970)
2.155   E.Anastassakis, E.Burstein: Phys. Rev. B2, 1952 (1970)
2.156   C.Flytzanis: Phys. Rev. B6, 1264 (1972)
2.157   L.R.Swenson, A.A.Maradudin: Solid State Commun. 8, 859 (1970)
2.158   M.Cardona, M.H.Grimsditch, D.Olego: [Ref. 2.25, p. 639]
2.158a S.Montero: Private communication
        see also: M.A.Morrison, P.J.Hay: J. Chem. Phys. 70, 4034 (1979)
        I.G.John, G.B.Bacskay, N.S.Hush: Chem. Phys. 51, 49 (1980)

2.159   A. Weber: [Ref. 2.19, Vol. 2, p. 544]
2.160   See, for instance, R. Zeyher, J. L. Birman, W. Brenig: Phys. Rev. B6, 4613 (1972) and references therein
2.161   B. Bendow: In *Springer Tracts in Modern Physics*, Vol. 82 (Springer, Berlin, Heidelberg, New York 1978) p. 69
2.162   A. C. Albrecht: J. Chem. Phys. 34, 1476 (1961)
2.163   T. P. Martin, L. Genzel: Phys. Stat. Sol. (b) 361, 493 (1974)
2.164   R. Loudon: Adv. Phys. 13, 423 (1964)
2.165   A. Pinczuk, G. Abstreiter, R. Trommer, M. Cardona: Solid State Commun. 30, 29 (1979)
2.166a D. L. Rousseau, J. M. Friedman, P. F. Williams: [Ref. 2.18, p. 204]
2.166b K. Huang: Sci. Sinica 24, 27 (1981)
2.167   M. Mingardi, W. Siebrand: J. Chem. Phys. 62, 1074 (1975)
2.168   J. Reydellet, J. M. Besson: Solid State Commun. 17, 23 (1975)
2.169   M. Cardona: In *Atomic Structure and Properties of Solids*, ed. by E. Burstein (Academic Press, New York 1972) p. 514
2.170   D. Penn: Phys. Rev. 128, 2093 (1962)
2.171   P. B. Klein, H. Masui, J. Song, R. K. Chang: Solid State Commun. 14, 1163 (1974)
2.172   P. Y. Yu, M. Cardona: J. Phys. Chem. Solids 34, 29 (1972)
2.173   W. C. Higginbotham, F. H. Pollak, M. Cardona: *Proc. Intern. Conf. on Physics of Semiconductors* (Nauka, Moskow 1968) p. 57
2.174   V. Heine, R. O. Jones: J. Phys. C2, 719 (1969)
2.175   J. C. Phillips: *Bands and Bonds in Semiconductors* (Academic Press, New York 1973)
2.176   J. M. Calleja, J. Kuhl, M. Cardona: Phys. Rev. B17, 876 (1978)
2.177   P. Vogl, W. Pötz: Phys. Rev. 24, 2025 (1981)
2.178   R. M. Martin: Phys. Rev. B1, 4005 (1970); H. d'Amour, W. Denner, H. Schulz, M. Cardona: Acta cryst. (in press)
2.179   E. O. Kane: Phys. Rev. 178, 1368 (1969)
2.180   E. W. Aslaksen: Phys. Lett. 40A, 137 (1972)
2.181   A. R. Lubinsky, D. E. Ellis, G. S. Painter: Phys. Rev. B6, 3950 (1972)
2.182   R. A. Heaton, C. C. Lin: Phys. Rev. B22, 3629 (1980)
2.183   J. Frandon, B. Lahaye, F. Pradal: Phys. Stat. Sol. (b) 53, 565 (1972)
2.184   P. N. Butcher: *Non-Linear Optical Phenomena* (Ohio-State University, Columbus, Ohio 1955)
2.185   R. Merlin, G. Güntherodt, R. Humphreys, M. Cardona, R. Suryanaranan, F. Holtzberg: Phys. Rev. B17, 4951 (1978)
2.186   D. Olego, M. Cardona: Solid State Commun. 32, 375 (1979)
2.187   A. A. Gogolin, E. I. Rashba: In *Physics of Semiconductors*, ed. by F. G. Fumi (Marves, Rome 1976) p. 284
2.188   P. G. Manuel, G. A. Sai-Halasz, L. L. Chang, C. A. Chang, L. Esaki: Phys. Rev. Lett. 37, 1701 (1976)
2.189   R. Zeyher, C. S. Ting, J. L. Birman: Phys. Rev. B10, 1725 (1974)
2.190   W. Richter, R. Zeyher, M. Cardona: Phys. Rev. B18, 4312 (1978)
2.191   D. E. Aspnes: Surface Sci. 37, 418 (1973)
2.192   H. J. Stolz, G. Abstreiter: Solid State Commun. 36, 857 (1980)
2.193   W. Hanke, L. J. Sham: Phys. Rev. B21, 4656 (1980)
2.194   R. M. Martin: Phys. Rev. B4, 3676 (1971)
2.195   C. A. Ferrari, R. Luzzi: Phys. Rev. B19, 6284 (1979)
2.196   A. I. Anselm, In. A. Firsov: Sov. Phys. JETP 1, 139 (1955)
2.197   B. A. Weinstein, M. Cardona: Phys. Rev. B8, 2795 (1973)
2.198   F. Herman, S. Skillman: *Atomic Structure Calculations* (Prentice Hall, New York 1963)
2.199   P. B. Allen, M. Cardona: Phys. Rev. 23, 1495 (1981)
2.200   P. B. Allen, M. Cardona: Phys. Rev. B, 24, 7479 (E) (1981)
2.201   R. Trommer, M. Cardona: Solid State Commun. 21, 153 (1977)
2.202   J. Windscheif, W. von der Osten: J. Phys. C13, 6299 (1980)
2.203   A. A. Adumalikov, A. A. Klochikhin: Phys. Stat. Sol. (b) 80, 43 (1977)
2.204   E. L. Ivchenko, I. G. Lang, S. T. Pavlov: Phys. Stat. Sol. (b) 85, 81 (1978)

2.205   B.Bendow: [Ref. 2.24, p. 329]

2.206   M.L.Williams, J.Smith: Solid State Commun. **8**, 2009 (1970)

2.207   V.V.Hizhnyakov, A.V.Sherman: Phys. Stat. Sol. (b) **85**, 51 (1978)

2.208a G.L.Bir, E.L.Ivchenko, G.E.Pikus: Sov. Phys. Izvestia **40**, 1866 (1976)

2.208b A.Klochiknin, Ya.Morozenko, V.Travnikov, S.Permogorov: [Ref. 2.25, p. 215]

2.209   F.Trallero Giner, I.G.Lang, S.T.Pavlov: Phys. Stat. Sol. (b) (in press); Sov. Phys. Solid State **22**, 718 (1980)

2.210   J.Hermanson: Phys. Rev. B**2**, 5043 (1970)

2.211   K.P.Jain, C.S.Jayanthi: To be published

2.212   A.Compaan, R.M.Harbiger: Phys. Rev. B**18**, 2907 (1978)

2.213   J.F.Scott: Rep. Prog. Phys. **43**, 951 (1980)

2.214   Y.Oka, M.Cardona: Phys. Rev. B**23**, 4129 (1981)

2.215   F.Cerdeira, W.Dreybrodt, M.Cardona: Proc. Intern. Conf. Phys. Semiconductors, Warsaw 1972 (PAN, Warsaw 1972) p.1142

2.216   C.A.Arguello, D.L.Rousseau, S.P.S.Porto: Phys. Rev. **181**, 1351 (1969)

2.217   R.Zeyher: Phys. Rev. B **9**, 4439 (1974)

2.218   J.B.Renucci, R.N.Tyte, M.Cardona: Phys. Rev. B**11**, 3885 (1975)

2.219   R.L.Farrow, R.K.Chang, R.M.Martin: In *Physics of Semiconductors 1978*, ed. by B.L.H.Wilson (Institute of Physics, London 1979) p. 485

2.220   R.Carles, N.Saint-Cricq, J.B.Renucci, A.Zwick, M.A.Renucci: To be published

2.221   M.Sinyukov, R.Trommer, M.Cardona: Phys. Stat. Sol. (b) **86**, 5631 (1978)

2.222   M.Iliev, M.Sinyukov, M.Cardona: Phys. Rev. B**16**, 5350 (1977)

2.223   H.R.Philipp, E.A.Taft: Phys. Rev. **120**, 37 (1960)

2.224   A.A.Klochikhin, B.S.Razbirin, G.V.Mikhailov: Sov. Phys. JETP Lett. **9**, 327 (1973)

2.225   D.Olego, M.Cardona: Solid State Commun. **39**, 1071 (1981)

2.226   M.Selders, E.Y.Chen, R.K.Chang: Solid State Commun. **12**, 1057 (1973)

2.227   F.Canal, M.Grimsditch, M.Cardona: Solid State Commun. **29**, 523 (1979)

2.228   M.H.Grimsditch, E.Kisela, M.Cardona: Phys. Stat. Sol. a **60**, 135 (1980)

2.229   D.K.Bigelsen: Phys. Rev. B**14**, 2580 (1976)

2.230   B.Gieseke: In *Semiconductors and Semimetals*, Vol. 2, ed. by R.K.Willardson, A.C.Beer (Academic Press, New York 1966) p. 63

2.231   R.C.C.Leite, J.F.Scott, T.C.Damen: Phys. Rev. Lett. **22**, 780 (1969)
        M.V.Klein, S.P.S.Porto: Phys. Rev. Lett. **22**, 782 (1969)

2.232   E.Gross, S.Permogorov, Ya.Morozenko, B.Kharlamov: Phys. Stat. Solidi (b) **59**, 551 (1973)

2.233   J.F.Scott, T.C.Damen, W.T.Silfart, R.C.C.Leite, L.E.Chessman: Opt. Commun. **1**, 397 (1970)

2.234   A.Klochikhin, Ya.Morozenko, S.Permogorov: Sov. Phys. Solid State **20**, 2057 (1978)

2.235   T.P.Martin: Phys. Rev. **13**, 3618 (1976)

2.236   C.Carabatos, B.Prevot: Phys. Stat. Sol. **44**, 701 (1971)

2.237   J.P.Dahl, A.C.Switendick: J. Phys. Chem. Solids **27**, 931 (1966)

2.238   K.Shindo, A.Morita, H.Kamimura: Proc. Phys. Soc. Jpn. **20**, 2054 (1965)

2.239   M.Cardona: Phys. Rev. **129**, 69 (1963)

2.240   R.J.Elliot: Phys. Rev. **108**, 1384 (1957)

2.241   S.Nikitine: In *Optical Properties of Solids*, ed. by S.S.Mitra, S.Nudelman (Plenum Press, New York 1969) p. 197

2.242   A.Compaan, H.Z.Cummins: Phys. Rev. Lett. **31**, 41 (1973)

2.243   R.M.Habinger, A.Compaan: Phys. Rev. B**18**, 2907 (1978)

2.244   M.A.Washington, A.Z.Genak, H.Z.Cummins, R.H.Bruce, A.Compaan, R.A.Forman: Phys. Rev. B**15**, 2145 (1977)

2.245   R.G.Waters, F.H.Pollak, H.Z.Cummins, R.H.Bruce, J.Wicksted: [Ref. 2.25, p. 229]

2.246   P.Y.Yu, Y.R.Shen, Y.Petroff, L.M.Falicov: Phys. Rev. Lett. **30**, 283 (1973)

2.247   P.Y.Yu, Y.R.Shen: Phys. Rev. Lett. **32**, 373 (1974)

2.248   L.Ley, M.Cardona, R.A.Pollak: In *Photoemission in Solids II*, ed. by L.Ley, M.Cardona, Topics Appl. Phys., Vol. 27 (Springer, Berlin, Heidelberg, New York 1979) p. 11

2.249   K.Nakamura, J.Windscheif, W. von der Osten: Solid State Commun. **39**, 381 (1981)

3. Optical Multichannel Detection

R. K. Chang and M. B. Long

With 9 Figures

Many practical spectroscopic applications require a combination of low-light sensitivity and high stray-light rejection. To accomplish this, the conventional technique (hereafter referred to as single-channel detection) has made use of a multistage spectrometer for high stray-light rejection and a photomultiplier which is a quantum-noise-limited detector. The system has proven to be very effective when coupled with the associated photon-counting electronics and a multichannel analyzer for accumulating repeated scans across the spectral region of interest. However, while one particular wavelength is being monitored with the single-channel detection system, the spectral information at all other wavelengths is totally ignored, causing a considerable waste of time in obtaining the same signal-to-noise ratio. To avoid this, schemes have been devised for simultaneous detection over a wide spectral region. The first detectors for simultaneous multichannel spectroscopy were photographic plates. These simple-to-use, affordable photographic detectors provided spectral resolution which was limited only by the spectrograph. Unfortunately, photographic plates could not compete with the photomultiplier in terms of sensitivity, dynamic range, and real-time response, with the result that this approach remained unused for many years. However, the introduction of low-light-level television cameras and linear array solid-state detectors along with multistage spectrographs and minicomputers has led to the development of a new generation of optical multichannel analyzers (OMA) with drastically improved sensitivity, dynamic range, and real-time response.

Although the impact of the modern OMA system on optical spectroscopy is not comparable to that of the laser, most investigators are increasingly aware of the potentials and merits of the OMA in this field. However, many experimentalists are reluctant to change from the existing single-channel system to the OMA because of the difficulties encountered in deciphering the OMA terminology, gathering information on the physical principles of the television and linear array detectors and raising the necessary funds to purchase the OMA system. Therefore, this chapter emphasizes the physical mechanisms of several nearly quantum-noise-limited OMA detectors as seen from the point of view of the spectroscopist. The pros and cons of cost reduction by interfacing an existing computer with the OMA detector are also discussed. Less emphasis is placed on multistage spectrographs needed for large elastic light rejection and broad-band OMA detection. Excellent papers on the historical development of the OMA, analysis of the signal-to-noise ratio advantages of the OMA in comparison to the single-channel detector and reviews of experiments which use modern OMA systems can be found in [3.1–9] and will, therefore, not be included here. However, a few examples of the unique applications of the OMA approach will be briefly described, and a glossary of terminology and abbreviations related to OMA detectors and used throughout the text is presented at the end of the chapter.

3.1 Image Intensifiers and Detectors

Multichannel detectors currently most commonly used for spectroscopic applications can be divided into two classes, depending on the method by which the stored charge pattern is read out. One group (SEC, SIT, and ISIT) uses the vidicon TV camera technology where the charge pattern on the target of the camera is read out by a scanning electron beam, while another group (ISPD) makes use of a solid-state array of photodiodes and CCD shift registers. These recently developed detectors (SEC, SIT, ISIT, and ISPD) which have been shown to have quantum-noise-limited detection capability will be discussed. Some of the older low-light-level television cameras (e.g., image isocon and image orthicon) will be omitted as they are no longer competitive in sensitivity and overall performance. Since both the ISIT and ISPD detectors make use of image intensifiers, the microchannel plate and the single-stage electrostatic image intensifiers will be briefly reviewed.

3.1.1 Microchannel Plate (MCP) Image Intensifiers

Figure 3.1 is a schematic diagram of the MCP intensifier consisting of a photocathode (e.g., S-20 or ERMA, typically 25 mm in diameter) coated on a fiber-optics faceplate, which is mounted in close proximity to a mosaic of electron channel multipliers. These channels are hollow glass fibers 10–40 μm in diameter that have an internal resistive coating. They are arranged in a regular array and are electrically connected in parallel by metal electrodes on each end of the tubes. The front (facing the photocathode) is maintained at a positive potential relative to the photocathode. Because of the small distance between the photocathode and the glass tubes, photoelectrons emerging from the photocathode are accelerated into the channels with little lateral motion, thus preserving the image integrity. The other end of the mosaic faces the phosphor (e.g., P-11 or P-20) and is held positive ($\simeq 1000$ V) relative to the front end, setting up a quasi-uniform voltage gradient along each channel. The accelerated photoelectrons striking the walls of the channel cause secondary emission ($\delta \simeq 1.4$). These secondary electrons, in turn, travel down the channel with an energy typically of about 100 eV and liberate more secondary electrons where they hit the walls of the channels, eventually creating an avalanche of electrons in each tube. The electron transit time from one end to the other is about 1 ns.

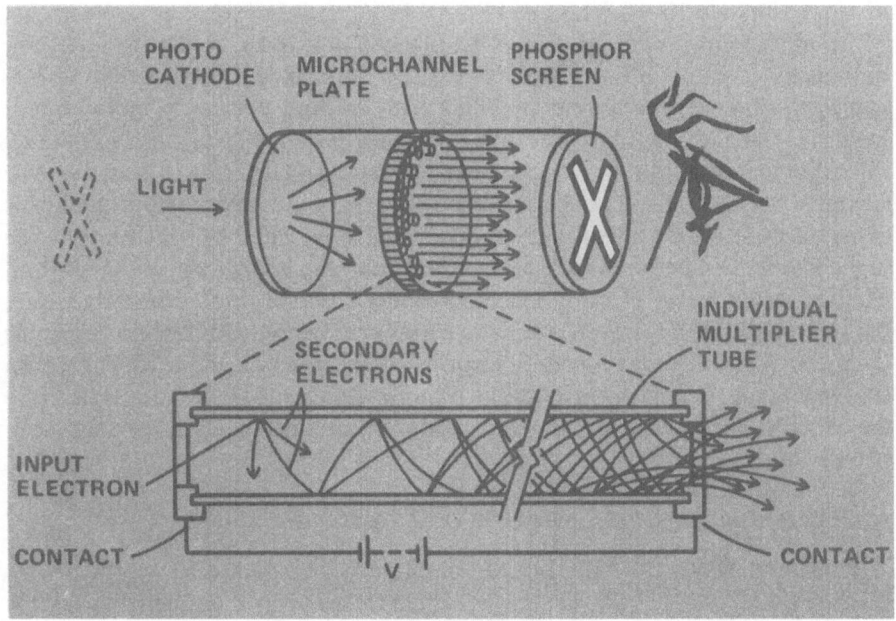

Fig. 3.1. Schematic diagram of the MCP image intensifier. The bottom figure shows one of the many channels where secondary electron emission takes place upon every collision with the inner wall of the tube (from Varian Associates)

The electron gain is about 10^3 to 10^4 and varies, depending on the applied voltage. After leaving the channels and upon striking the phosphor, these electrons produce cathodoluminescence, yielding many output photons in the green. The overall light gain of the MCP intensifier is about 25,000. Registration between the input and output fiber-optics faceplates is reasonably good and is determined by the center-to-center spacing of channels. A spatial resolution of 25 lp/mm at high-light levels and 10 lp/mm at low-light levels is typical. The MCP intensifier can be electronically gated (as fast as 5 ns) by switching the low voltage ($\simeq 300$ V) between the photocathode and the front end of the channels. At present, there appears to be some degradation of the MCP over its lifetime (thousands of hours). An MCP is used as the intensifier section in the ISPD. A review of the MCP can be found in [3.10].

3.1.2 Single-Stage Electrostatic Intensifiers

This device consists of a photocathode (S-20, S-25, or ERMA, typically 25 mm in diameter), an electrostatic lens with radially symmetric and longitudinal electric fields, and a phosphor (P-11 or P-20). If the voltage difference between the photocathode and phosphor is a few thousand volts, the high-energy photoelectrons impinging on the phosphor will produce a maximum light gain of approximately 10–15. Voltages above 15 kV result in excessive dark current, while at the lowest voltages (1–3 kV), the image may rotate and become defocused. Even at the optimum operating voltage, the electrostatic lens is subject to severe aberrations including astigmatism, curvature of the image field, and radial distortion. The latter two distortions can be reduced by coating the photocathode on a curved fiber-optics faceplate which becomes the entrance window. Typical spatial resolution of better than 50 lp/mm can be expected at low-light levels. This device can also be gated by switching off the high electric field between the photocathode and the anode near the phosphor. Magnetically focused electron lenses have much higher resolution even with a flat photocathode. However, magnetic lenses are generally bulky and the fringe magnetic fields can introduce distortion in television cameras which use an electron beam to read out the charge pattern. An electrostatic lens is therefore used for the ISIT. A review of electrostatic image intensifiers can be found in [3.11].

3.1.3 Secondary Electron Conduction (SEC) Cameras

A schematic diagram of the SEC TV camera is shown in Fig. 3.2. A photo-cathode (S-20, typically 25–40 mm in diameter) is deposited on the inner surface of a plano-concave fiber-optics faceplate. Photoelectrons are accelerated toward the target and focused by an electrostatic or magnetic lens. A unique feature of this camera is its target, which consists of a 500 Å thick Al_2O_3

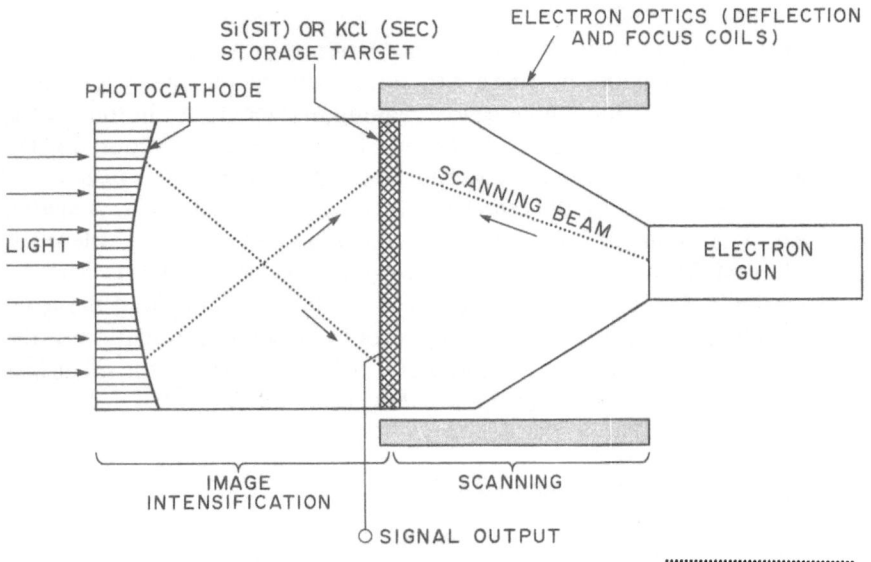

Fig. 3.2. Schematic diagram of the SIT or SEC camera which uses a silicon or KCl storage target, respectively. The photoelectrons are accelerated and impinge on the target, causing a large amount of secondary electron emission. The positive charge distribution stored on the target is brought back to the electron gun potential by the scanning electron beam

membrane followed by a 500 Å layer of Al forming the signal electrode. Deposited on this Al electrode film and facing the reading electron gun is a highly porous layer of KCl with a density of about 0.02 g/cm^3. Photoelectrons with an energy of approximately 8 keV penetrate the Al_2O_3 and Al layers and generate about 100 secondary electrons in the KCl layer. Because of the electric field within the target, these electrons move to the Al signal electrode through the vacuum interstices between the KCl particles, thus the name, Secondary Electron Conduction. Persistence effects are avoided since the electrons do not move in the KCl conduction band. The resultant positive charge pattern on the KCl layer corresponds both spatially and in intensity to the optical image focused onto the faceplate. The scanning electron beam returns the KCl gun cathode potential by depositing electrons on the positively charged areas. This changing current is capacitively coupled to the Al signal plate and consitutes the video signal. Alignment and deflection of the electron beam are accomplished by transverse magnetic fields produced by external coils.

Because of the high resistivity of KCl and its low thermionic emission rate (less than 0.008 electrons/pixel/s), the charge image can be integrated and stored at room temperature for more than five hours without degradation of the image fidelity, provided that the photocathode is turned off or does not saturate the target by photocathode dark emission. The overall gain (typically 15:1) can be varied by the photocathode voltage (3–8 kV) with no perceptible loss in resolution or image rotation. Gating of the SEC detector is possible by rapidly changing its photocathode voltage. This camera has a fast response to optical changes because it does not use a phosphor, which is known to have a luminescence persistence time. Furthermore, the secondary electron conduction takes place in the vacuum and not within the highly resistive KCl. The small amount of lag observed is believed to be caused by the stored charge residing within the KCl layer rather than on its surface. Therefore, in the first scan, more than 90% of the signal is read out, leaving less than 10% for the second scan. About five scans are needed to read out the charge pattern completely and insure linear response. The real-time dynamic range is about 50. For single-pulse spectroscopy, a single or multistage intensifier can be conveniently coupled to the fiber-optics faceplate of the SEC camera, and single photoelectron events can thereby provide sufficient video signal if a properly designed preamp is used. The SEC camera itself is nearly an ideal quantum-noise-limited multichannel detector, although the preamp associated with this camera is quite noisy. Unfortunately, an SEC based OMA system is not commercially available at present, even though Westinghouse developed the SEC tube well over ten years ago. A review of the SEC camera can be found in [3.12].

3.1.4 Silicon Intensified Target (SIT) Detectors

In the SIT detector, the fiber-optics faceplate and the photocathode (which is coated on the concave side of the faceplate) are identical to those in the SEC

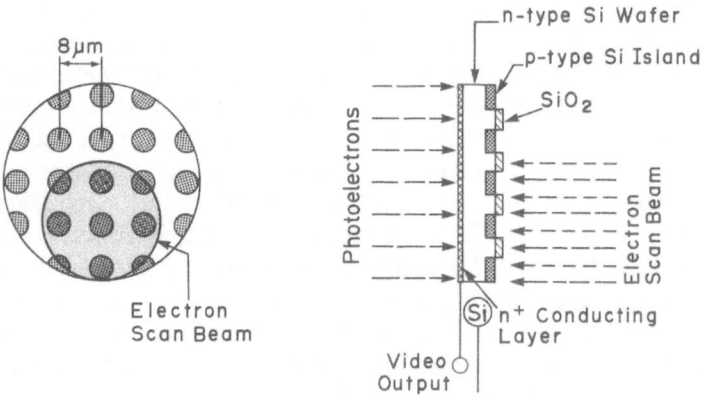

Fig. 3.3. Target structure of the SIT and ISIT cameras. The photoelectrons create electron-hole pairs in the Si wafer which diffuse into the reversed biased p–n diodes and decrease the charge stored by the diode. The electron scan beam restores each diode back to Q_{sat}

camera. However, the target of the SIT (also called the EBS) camera is totally different. It consists of a dense matrix (about 6×10^4 elements) of p-type Si islands (see Fig. 3.3) with a typical center-to-center diode spacing of 8 μm. These diodes are diffused onto one side of an n-type Si wafer 5–50 μm thick. As in the SEC camera, a scanning electron beam is used to read out the signal. The p-type islands facing the electron gun are charged to a negative potential (i.e., the gun potential which is typically − 10 to − 15 V), making them reverse biased with respect to the n-type wafer which is maintained at ground potential. The SiO$_2$ film surrounding the p-type islands isolates the n-type wafer from the electron read beam and, because of the high resistivity of the SiO$_2$, this surface accumulates an electronic charge, causing it to be at a voltage nearly equal to the gun potential. The electron scan beam diameter (typically about 25 μm) is generally larger than the diode spacing which eliminates any need for registration between the beam and the p–n junction diode matrix (Fig. 3.3).

The reverse biased p–n junction will have a depletion width of approximately 5 μm, giving a junction capacitance of approximately 2000 pF/cm^2. The charges stored in these individual storage capacitors can be discharged in two ways: (a) electron-hole pairs may be thermally generated within the depletion region and (b) holes can diffuse into the depletion region from the n-type wafer. Electron-hole pairs can be generated in the n-type wafer either thermally or by the accelerated photoelectrons emitted from the photocathode. The video output signal from each diode is created when the electron read beam returns to a diode and restores the original charge by re-establishing the full value of reverse bias. Since the number of electron-hole pairs produced is proportional to the number of incident photoelectrons, the extent to which the diodes are discharged is proportional to the incoming photoelectron flux.

The photoelectrons striking the nondiode side of the n-type wafer have energies between 3 and 9 keV. Since 3.4 eV of electron kinetic energy are needed to create an electron-hole pair in Si, the 9 keV photoelectrons will produce about 2600 electron-hole pairs. If the target is sufficiently thin, about 70 % of the holes will survive recombination in the n-type region and will diffuse to the depletion region of the p–n junction. Therefore, the target gain for photoelectrons is approximately 1820 for a 9 keV photocathode voltage. Assuming a quantum efficiency of 10 % for the photocathode, the overall gain (holes produced per photon) is about 182.

There are several inherent features of the SIT detector which limit its performance. For example, the variation among the p–n junction diodes leads to a fixed-pattern charge image even though the illumination is uniform across the input faceplate. In addition, the sensitivity is greatest near the center of the tube. This roll-off of sensitivity near the perimeter of the SIT camera is introduced by the electrostatic lens, causing the photoelectrons near the perimeter of the target to possess less kinetic energy than those near the center. The ultimate resolution of the SIT is limited by the combination of the discrete nature of the p-type islands and the lateral diffusion of the holes in the n-type wafer before entering the depletion region.

The problem of lag is an important consideration in attempting to achieve quantum-noise-limited detection. Unlike the SEC camera, where one scan of the electron beam erases about 90 % of the positive charge pattern stored on the KCl target, the lag for the Si target is considerable. For example, as much as 40 % of the charge pattern may still be stored on the target after one electron beam scan. To completely recharge the p-type regions, the target must be scanned many times. This charge retention affects the minimum detectable signal because of the preamplifier noise introduced by multiple read frames and is particularly limiting for single-pulse spectroscopy. The preamp noise of the PAR OMA-2 is equivalent to 2500 electron-hole pairs per channel on the Si target or up to two photoelectrons, depending on the accelerating voltage set by the photocathode potential. In practice, we have found that at least ten frames are necessary to read off 97 % of the charge signal stored on the target. Therefore, the total noise introduced by the ten successive read frames is equivalent to about $\sqrt{20}$ or 4–5 photoelectron counts.

The causes of lag in the Si target are not fully understood. However, the most important parameters affecting lag are the following: (a) the effective junction capacitance of the target; (b) the secondary emission characteristics of the target for the low energy read beam; (c) the electron beam current and its electron energy distribution, and (d) the amount of illumination which affects the reverse bias voltage on each diode. The junction capacitance as well as the capacitance of the depletion region under the SiO_2 are functions of the reverse bias voltage on the target. Increasing this voltage reduces the junction capacitance which decreases the ratio of peak video current to the permissible voltage swing of the individual diodes. Furthermore, this limits the dynamic voltage range of the diodes before electron beam bending will result from the transverse (i.e., parallel to the target surface) electric fields. However, since increasing the reverse bias voltage does substantially decrease the capacitance, the charging ability of the electron beam is improved, and the lag is, therefore, decreased for a fixed electron beam current. Consequently, the usable values of the junction capacitance are restricted to a narrow range. The lag caused by the depletion layer in the Si under the SiO_2 is neither negligible nor well understood.

The incomplete erasure of images on the first readout scan also affects the linearity (gamma value) of the detector. In particular, the acceptance of the scan beam electrons by the p-type islands is a function of their discharge level. For example, in one study, two target charge patterns at 100% and 10% of the maximum discharge level were considered [3.13]. For the particular scan used, the 100% signal was reduced to about 10% after the first readout, while the original 10% signal was reduced to only about 8%. The video current levels for these two signals were not anywhere near the proper ratio of 10:1. Consequently, to insure linearity (gamma of unity), many scans have been found to be necessary for all single-pulse work. The scan rate of approximately 30 ms for the entire target is more or less optimal. Faster scan rates cause more lag and decrease the video output. Slower scan rates improve lag but can decrease the video output because of the capacitive coupling of the video output through the n-type conducting layer (Fig. 3.3) [3.14].

Upon cooling the SIT tube to $-40\,°C$, the lag is so large that, for a high illumination level, several minutes of continuous scanning may be necessary to completely erase the scene to an acceptably low level. We believe this increased lag is caused by the uncontrollable charging of the SiO_2 film by the read beam, which becomes more likely at lower temperatures. The film can accumulate enough negative charge to repel the electron beam and prevent it from impinging on the p-type islands, analogous to the control grid in a triode. Modifications of the basic diode array structure can prevent this charging behavior [3.14]. For example, Westinghouse manufactures a target in which electrically isolated conducting islands are placed over each p-type region. This so-called "deep etch metal cap" (DEMC) structure greatly reduces the area of SiO_2 that is exposed to the read beam [3.15]. Therefore, the uncontrollable over-charging of the SiO_2 is decreased. Furthermore, the acceptance of the scan beam by the p-type islands is increased as a result of the larger beam landing area and the porous nature of the metal caps. For this arrangement, the blooming characteristics are also greatly improved.

Cooling the SIT (EBS) to $-40\,°C$ has been proven to greatly increase the signal-to-noise ratio for cw spectroscopy. The thermionic emission of the photocathode decreases as in the case of photomultipliers. Furthermore, the thermal generation of electron-hole pairs within the p–n junctions and the Si wafer is decreased. Consequently, the target can integrate photoelectron-produced charges for more than two hours without any read beam scanning, provided that the integrated charge does not reach a level that would cause saturation of the preamplifier. For such a long integration period, the read beam should not be scanned and, in addition, the filament of the electron gun should be shut off until about 30 s before the beam readout is to be initiated.

The red glow of the filament can be optically guided into the photocathode by the glass envelope of the camera tube and thereby cause noise beyond photocathode thermionic emission.

For cw operation, a selective scan format can be used to increase the upper limit of intense signal that can be detected, i.e., to improve the real-time dynamic range which is set by the preamp saturation level. The read beam can be made to scan those portions of the target where the intensities are known to be large more often than those portions where intensities are weak. Scanning re-establishes the p-type island potential, enabling that portion to receive more holes from the wafer without saturating the preamp. Accumulation is accomplished in the computer memory which, in general, has an upper limit of about 10^9 (i.e., memory dynamic range).

For selective scanning, the computer does the following: (a) heats up the filament for 30 s before scanning, (b) restricts the scan format to the previously determined "bright" region of the target while the number of scans is determined by the lag considerations, (c) A/D converts these video signals and accumulates them in the appropriate memory locations corresponding to that portion of the target, (d) shuts down the filament for target integration, and (e) repeats (a)–(c), depending on whether or not the "weak" regions of the spectrum need to be scanned. During this time, the computer must keep track of the number of times each portion of the detector has been read out. In this way, the overall dynamic range for cw spectroscopy can be significantly enhanced.

For pulsed operation, the SIT or EBS tubes can be gated by applying a voltage pulse to the intensifier focus grid. A 10^4 gating ratio can be achieved. Signals acquired when the gate is on (as short as 10^{-8} s duration) are retained by the target until the read scan, which can be after the integration of many pulses. We have experienced a substantial increase in the amount of background noise as a result of gating when the PAR OMA-2 detector is integrating in the cooled mode. The origin of this increased background is not well understood.

3.1.5 Intensified Silicon Intensified Target (ISIT) Detectors

Most of the ISIT cameras have an ERMA photocathode allowing the detection of wavelengths up to about 910 nm. The main difference between the ISIT and the SIT is the addition of a single-stage electrostatic image intensifier discussed in Sect. 3.1.2. The output fiber-optics plate of the image intensifier is coupled to the input fiber-optics plate of the SIT with index matching oil. The loss of intensity for such coupling is typically 30 % and, thus, much less than the loss with lens coupling. With the additional gain provided by the image intensifier (about 10 to 40 ×), the ISIT definitely has single photoelectron sensitivity (about 5 × above the preamp noise level). The ISIT is therefore more suitable than the SIT for real-time or pulsed spectroscopy of extremely low intensity where integration on the target is not desirable.

The ISIT can be electronically gated for durations as short as 10 ns by applying a voltage on the image focus grid of the intensifier. The ISIT detector is ideal for single-pulsed spectroscopy in the presence of a cw luminous background or long duration luminescence. The nonsynchronous signal can be gated out by a ratio of $10^4 : 1$ while the synchronous signal can be detected by the preamp with good signal-to-noise ratio since each photoelectron corresponds to about $5 \times$ the preamp noise. The single shot dynamic range is still about 750.

The ISIT is generally not cooled. At low temperatures, the phosphor persistence becomes extremely long, causing long retention of the optical image on the phosphor. Since cooling the photocathode also causes cooling of the phosphors in the same image intensifier, the option of lowering the photocathode dark emission rate is not available. For cw work where integration on the target is possible, we believe that integration on the target of the cooled SIT is equal to, if not superior to, the accumulation of single scans in the computer with the uncooled ISIT. At low signal levels and long accumulation times, the ISIT signal-to-noise ratio switches from photoelectron statistics to preamplifier statistics.

3.1.6 Intensified Silicon Photodiode (ISPD) Array Detectors

EG&G-PAR and Tracor Northern have introduced OMA systems designed specifically as multichannel detectors that utilize an intensified self-scanning linear photodiode array (ISPD). Unlike the SEC, SIT, and ISIT detectors discussed earlier, the ISPD does not use a scanning electron beam for readout but instead uses a multiplex switching scheme for periodic readout with an integrated shift register scanning circuit. The ISPD is currently available only as a one-dimensional detector while the SEC, SIT, and ISIT are inherently two-dimensional detectors (the photomultiplier is a zero-dimensional detector). However, the integrated circuit technology presently being used in the one-dimensional ISPD could be used to make a two-dimensional array detector. Future availability of such solid-state detectors will finally provide the spectroscopist with an excellent variety of quantum-noise-limited detectors from which to choose.

The ISPD from EG&G-PAR consists of an MCP which is optically coupled to a self-scanned photodiode array (SPD from Reticon, S-series with 1024 individual diodes). Each diode is 2.5 mm high, 13 µm wide, and separated by 25 µm (Fig. 3.4). This 100 : 1 aspect ratio is convenient for typical applications where a spectrograph is used. Unlike the Digicon [3.16] where the photoelectrons emitted from the photocathode of the image intensifier are directly bombarded into the Si diode array (with a secondary gain of several thousand), in the ISPD, the photoelectrons from the photocathode (e.g., S-20, ERMA, or S-25) impinge on a phosphor at the end of the MCP intensifier. The resultant cathodoluminescence travels through a fiber-optics endplate which is optically

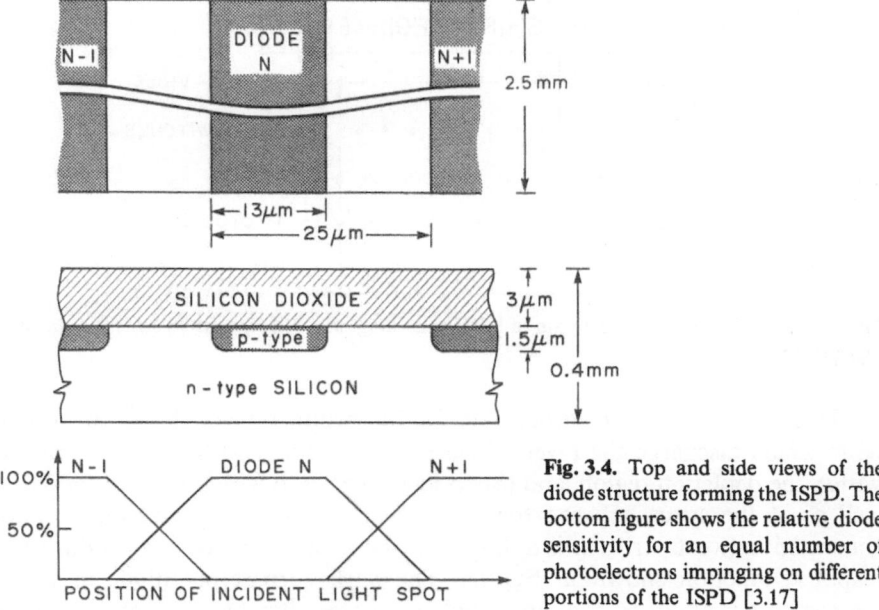

Fig. 3.4. Top and side views of the diode structure forming the ISPD. The bottom figure shows the relative diode sensitivity for an equal number of photoelectrons impinging on different portions of the ISPD [3.17]

coupled to the Si photodiode array. The MCP has sufficient electron gain ($>2 \times 10^4$) to allow the preamp of the SPD to produce one or several counts for one photoelectron. For maximum dynamic range (about 16,000 set by the A/D converter), the MCP gain is best set at one A/D count per photoelectron, which is equal to the rms readout noise of the SPD.

The most unique part of the ISPD is the SPD, its construction, and operation. Each photodiode consists of a diffused p-type Si bar in an n-type Si substrate (Fig. 3.4), and a SiO_2 overlayer covers and protects the entire detector. The electronically driven shift register (see Fig. 3.5) sequentially closes the MOS switches which are associated with each photodiode, completing the circuit from the $+5$ V power supply (common bus in Fig. 3.5) through the reverse biased diode to the input of the charge-sensitive preamplifier (video line in Fig. 3.5). The actual equivalent circuit is more complicated but the essential points are shown in Fig. 3.5. The saturation charge, Q_{sat}, is restored on the junction capacitance when the MOS switch is closed, and the amount of charge needed depends on the discharge state of the diode. The recharging current for a given diode is presented to a common preamplifier, peak detector and A/D converter (14 bit). The output signal obtained from each scan of an N element array is therefore a train of N digital values, each proportional to the amount of charge necessary to recharge the corresponding photodiodes. A 5 MHz clock regulates the shift register and thus the sampling rate. Periodic start pulses from the shift register initiate each complete scan of the N diodes. The time duration between successive start pulses is the integration time of each photodiode (about 16 ms for one scan of a 1024 element array).

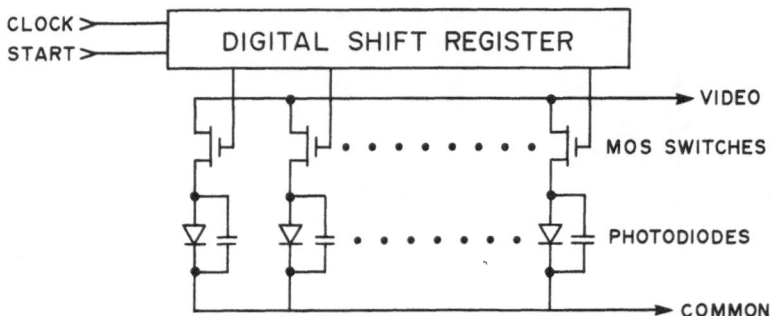

Fig. 3.5. Simplified schematic diagram of ISPD consisting of *p–n* diodes, MOS switches, and shift registers [3.17]

Two mechanisms are responsible for the production of the electron-hole pairs which discharge the reverse biased photodiodes: (a) thermal generation within the depletion region, and (b) optical generation within the *p–n* diode, as well as at the space between the two *p*-type strips. The electron-hole pairs generated between strips will diffuse into the adjacent diodes to produce the response function shown in Fig. 3.4. The dark current per diode at room temperature is about 5600 electron-hole pairs rms after 1 s of integration. This value drops by a factor of 2 for about every 6.7 °C. During the integration interval, the photo-induced electron-hole pairs and the thermally-induced electron-hole pairs remove a proportionate number of charges from the reverse biased diodes. If the thermal electron-hole generation rate is small compared to the optical generation rate, the spatial distribution of the charge deviation from Q_{sat} on the N diode array will mimic the one-dimensional intensity distribution on the phosphor of the MCP intensifier.

The SPD is practically lag free. Less than 0.1 % of the signal remained after the initial readout even when the light level was raised to 5 × the A/D converter saturation level [3.17]. The recharging of each diode lasts for less than 1 μs and does not depend on an electron beam as with the direct beam readout television cameras. In contrast to the latter detectors, the ISPD, because of its very low lag, has linear response (unity gamma) even after one readout scan. Furthermore, the well-separated distance between the photodiodes causes the blooming characteristics of the SPD to be exceptionally good. For a test signal with 1000:1 intensity ratio, even where the more intense beam produced a charge signal 20 × above the A/D converter saturation level, the cross talk was less than 1 % for a spatial separation of 15 diodes [3.17]. In addition, because of the precise location of each photodiode within the SPD, the corresponding wavelength accuracy is excellent, giving an overall wavelength accuracy of 0.025 mm (the diode spacing) multiplied by the linear dispersion of the spectrograph. However, for high resolution spectroscopy, aliasing introduced by the discrete nature of these photodiodes can result which may falsely increase the apparent wavelength separation between peaks and decrease the intensity should the spectral line fall between the *p*-type strips [3.17].

There are several sources of noise in the SPD system: (a) noise is introduced by thermally-generated electron-hole pairs which will completely discharge the diodes in 10 s at room temperature. This thermal shot noise can be greatly reduced by cooling the ISPD to a lower temperature. For example, at $-25\,°C$, less than 550 electron-hole pairs rms are thermally generated during 1 s of integration. Because of this reduced thermal noise upon cooling, the integration time of the diodes can be extended (e.g., 15 s at $-25\,°C$) [3.18]. Upon cooling an SPD to temperatures below $-30\,°C$, a decrease in sensitivity to near IR wavelengths has been noted [3.18]. However, in an ISPD system, the spectral response is determined by the photocathode of the intensifier so that cooling the SPD array will not cause a decreased sensitivity in the red. Even longer integration times (several hours) are possible by cooling to lower temperatures ($-130\,°C$) [3.18]. However, at lower temperatures, the persistence of the phosphor in the MCP intensifier of an ISPD system becomes significantly longer and must be considered in pulsed spectroscopy. (b) Noise from the fixed pattern signal, arising from capacitive coupling of the transients from the clock driver signals onto the video line, is always present even in the absence of any photons or thermal leakage. The fixed pattern signal typically corresponds to about 1 % of the full scale range. Although this signal is more a zero offset than a random noise component, any instability or noise in the clock waveform cannot be accurately subtracted in successive frames. Furthermore, the temperature of the array needs to be held constant, as temperature can change the MOS capacitance and in turn the fixed pattern signal. The fixed pattern noise appears to have a $1/f$ dependence (flicker noise) and is suspected to be related to the $1/f$ noise in the MOS multiplex switches [3.17]. (c) Readout noise constitutes a random noise source arising from the pickup of the clock and start pulses, power supply noise, preamplifier noise, and resetting noise associated with restoring the charge in a given diode. Resetting noise which sets the lower limit to the noise is proportional to the square root of the diode capacitance and temperature and is independent of bandwidth [3.19]. At $25\,°C$, the rms resetting noise per diode is about 1000 electron-hole pairs for the Reticon S-series diode array [3.17]. The preamplifier noise is directly proportional to the capacitance of the array as presented to the preamp. This output capacitance can be fairly high (about 1000 pF) since it is a parallel sum of all MOS switches connected to the video line (Fig. 3.5). The overall noise from all the various sources associated with the SPD is equivalent to about one photoelectron from the photocathode of the MCP. However, the ISPD is not a true single photoelectron detector because of two effects: (a) the secondary emission process in the MCP intensifier is statistical in nature, and (b) it is possible that photoelectrons may be incident between two p-type strips, thus causing about 50 % less electron-hole pairs to be generated than if the photoelectrons hit directly on one strip.

The optoelectronic dynamic range of a single diode is the ratio of the largest signal before diode or preamp saturation to the system rms readout noise. This has been determined to be above 10^4 at $-140\,°C$ (i.e., from about 800 electron-hole pairs rms to about 4.1×10^6 electron-hole pairs per diode, which is about 20 % of the diode saturation) [3.18]. However, the single diode dynamic range is not necessarily the OMA detector dynamic range (i.e., the intrascenic dynamic range), which is generally lower because of the veiling glare phenomenon introduced by the internally reflected light from glass, the Si wafer, and the metallic components. The intrascenic dynamic range of the ISPD, consisting of the MCP and SPD, is reported to be 16,000 : 1 for one readout scan. This ratio is vastly superior to the SIT and ISIT, especially with one readout scan [3.16]. The memory dynamic range is of course much greater, since the computer memory can store up to 2^{31} or 2.1×10^9 counts. However, the $1/f$ noise associated with the fixed pattern signal limits the signal-to-noise ratio improvement that can be realized with the memory accumulation mode to about a factor of 10 [3.16].

Many of the characteristics described earlier for the EG&G-PAR ISPD Model 1420 are also applicable to the Tracor Northern TN-1710 DARSS system. The major difference is in the SPD diode size (50 μm wide by 0.4 mm long for the DARSS system). The computer software of the two companies is somewhat different and needs to be taken into account if one is interested in purchasing an entire OMA system.

3.2 Multistage Spectrographs

Double and triple-stage scanning spectrometers have proven to be ideal for single-channel Raman and fluorescence spectroscopy since their stray-light rejection is many orders of magnitude higher than that of single-stage scanning spectrometers. For these instruments, higher stray-light rejection can be achieved by decreasing the width of the intermediate slits. In doing this, however, the total bandpass of the entire system is reduced, making these spectrometers unsuitable for OMA systems.

When stray-light rejection is not a problem or can be accomplished with the aid of optical filters, it is tempting to convert an existing spectrometer into a spectrograph suitable for OMA detection by opening wide the intermediate slits and removing the exit slit entirely. However, the resulting spectrograph remains less than ideal. In general, spectrometers are less corrected for chromatic aberrations than are spectrographs. The dimension of the flat field region in the focal plane of spectrometers is often quite limited, which limits the spectral coverage having uniform resolution. A specially made fiber-optics plate, concave on one side to match the aberrations and flat on the other for coupling with the flat fiber-optics faceplate of an OMA detector, can extend the spectral coverage with less decrease in resolution to the blue and red side of the central wavelength at which the spectrometer is set. However, the design, fabrication, and finally the alignment of such plates are quite difficult, causing this approach to be unsatisfactory.

SPEX TRIPLEMATE SPECTROGRAPH

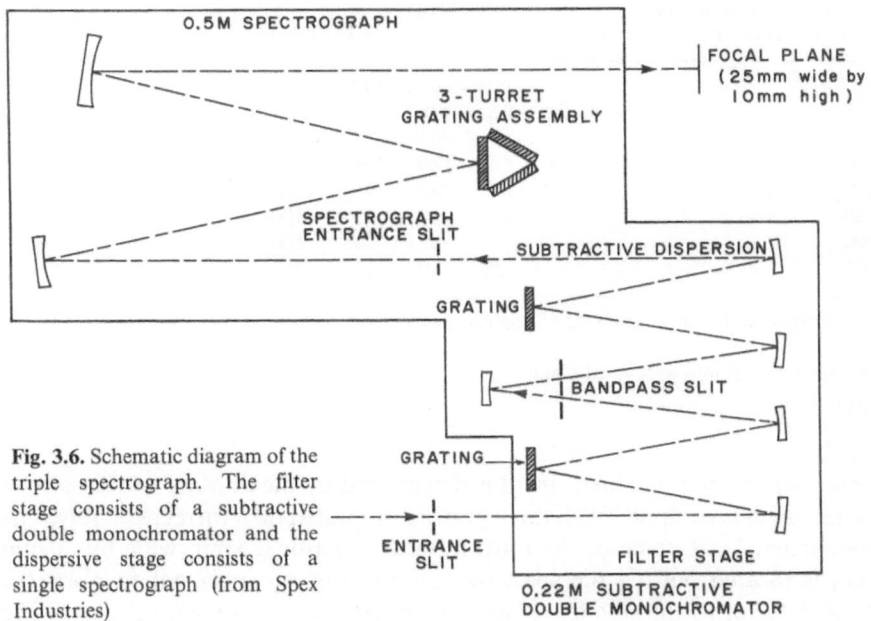

Fig. 3.6. Schematic diagram of the triple spectrograph. The filter stage consists of a subtractive double monochromator and the dispersive stage consists of a single spectrograph (from Spex Industries)

We have designed and constructed a spectrograph (0.98 m focal length, $f/8$, with linear dispersion of 0.5 nm/mm) containing a single Jobin-Yvon $110 \times 110\,\text{mm}^2$ concave holographic grating with 2000 g/mm designed for a single-stage spectrometer [3.20]. The overall chromatic aberration was found to be tolerable when the 12.5 mm diameter SIT detector was placed at the exit plane. The stray-light rejection for such a mirrorless spectrograph was about 10^9 for shifts beyond $500\,\text{cm}^{-1}$. The main advantage of this spectrograph was its low cost, which was essentially that of the off-the-shelf holographic grating. However, an aberration corrected grating (type III) is available from Jobin-Yvon, which instead of focusing on the Rowland circle, is designed to be stigmatic at three wavelengths. The amount of aberration is about $10 \times$ lower than that of classical concave gratings.

Triple-stage spectrographs are now commercially available. Figure 3.6 is a schematic diagram of the Triplemate spectrograph from Spex Industries. The filter stage is a 0.22 m subtractive double monochromator containing two gratings. The first grating disperses the input radiation and the second, which is coupled to the first grating in a subtractive dispersion mode, collapses the dispersed radiation. The net result is zero dispersion at the entrance slit of the third monochromator. The bandpass slit within the filter stage blocks the incident laser radiation but passes a band of the inelastic radiation. The bandpass of the filter stage is dependent on the dispersion of the two $64 \times 64\,\text{mm}^2$ gratings, as well as the slit width (0.5, 1.0, 2.0, and 5.0 mm). The

Table 3.1[a]. Multistage spectrograph coverage and bandpass

Grating [g/mm]	Spectrograph[b] spectral coverage [cm^{-1}]	Filter Stage [c] nominal bandpass [cm^{-1}]			
		Intermediate slit [mm]			
		0.5	1.0	2.0	5.0
150	5257	470	940	1890	4700
300	5257	230	460	930	2300
600	2646	110	230	450	1100
1200	1320	50	100	200	500
1800	870	30	60	120	300

[a] From Spex Industries Preliminary Specifications for 1877 Triple-mate.
[b] At 514.5 nm, 25 mm wide focal plane.
[c] At 514.5 nm.

linear dispersion of the instrument is determined by the third 0.5 m monochromator which is equipped with three gratings mounted on a turret assembly. The unvignetted focal plane of the third monochromator (25 mm wide by 10 mm high) is located outside the spectrograph, making it convenient to place the OMA detector so that the front face is precisely located at the focal plane. The spectral coverage of the triple spectrograph and the nominal bandpass of the filter stages are listed in Table 3.1. The overall stray-light rejection is specified to be 10^{14} at 10 bandpass units from the laser wavelength.

Multistage spectrographs from other manufacturers work on essentially the same principle, incorporating a zero dispersion filter stage and a high dispersion spectrograph which is well corrected for aberrations and vignetting. The stray-light rejection from a triple spectrograph can never equal that from a triple spectrometer. However, the broader bandpass of the former allows OMA systems to be used.

3.3 Computer Control

The volume of data produced by multichannel detectors is potentially very large. In most systems, each pixel is digitized in a format of 8–14 bits and the pixel scan rate can exceed 50 kHz. Consequently, a data rate of 0.7×10^6 data bits/s is not uncommon. This high data rate and the resulting volume of data make it desirable to have multichannel detector systems interfaced to a computer. In addition to collecting the data, the OMA system can be designed in such a way that the computer controls the format in which the data is read out. For example, parameters such as the scan rate, the number of channels scanned and the location of scanned channels must be specified. With computer control, these parameters can be easily changed and complex scanning patterns can be specified. Another important function of the on-line computer system is the analysis and correction of the data. Since the cost of computers with the required performance characteristics has continued to decrease, the computer to control the system may be considerably less expensive than the detector itself.

3.3.1 Computer System Selection

Currently, several commercially available OMA detection systems include a built-in microcomputer for data acquisition and control. Typically, these systems consist of a microprocessor, a memory for parameter and data storage and a read only memory (ROM) which stores the program to control the detector. In addition, various displays, input/output devices and expansion options may be available. The clear advantage of these systems with a built-in computer is their ease of operation. No programming or interfacing is needed and the system can be used immediately. However, the same features that provide ease of operation can lead to limitations. Expansion is often limited to those options offered by the manufacturer. Also, since the sales volume of these specialized computers is relatively low, they tend to be considerably more expensive than general-purpose computers with similar capabilities. In addition, the microprocessor is often isolated from the user in the sense that it is not possible to develop programs and perform specialized analysis on the data obtained within the controlling computer. Therefore, it may be necessary to transfer the data to another computer for analysis.

In some commercial OMA detection systems, the detector and detector controller can be purchased without a built-in computer. The controller, which has the analog electronics to control the detector as well as the A/D converter to digitize the data, can be interfaced to a general-purpose computer by means of a parallel input/output port. This approach has several potential advantages. Since many low cost processors are currently available, a somewhat lower system price may be realized. In addition, expandability is not as limited since a large number of compatible computer peripherals are available from different companies. With a general-purpose processor, specialized analysis programs can be developed and the computer can be used to control aspects of the experiment other than those associated with the detector or to perform general computational tasks. The main disadvantage of this approach is the interfacing and programming required. For experiments where the flexibility inherent in a more general processor is not required and where the capabilities provided by the manufacturer are sufficient, the overhead involved in interfacing and programming may not be justified.

Several factors must be considered in choosing a general-purpose computer system to control an OMA detector or in determining whether or not an existing system is appropriate. The most stringent requirement is that the computer must be able to accept the data at the rate the detector supplies it. As

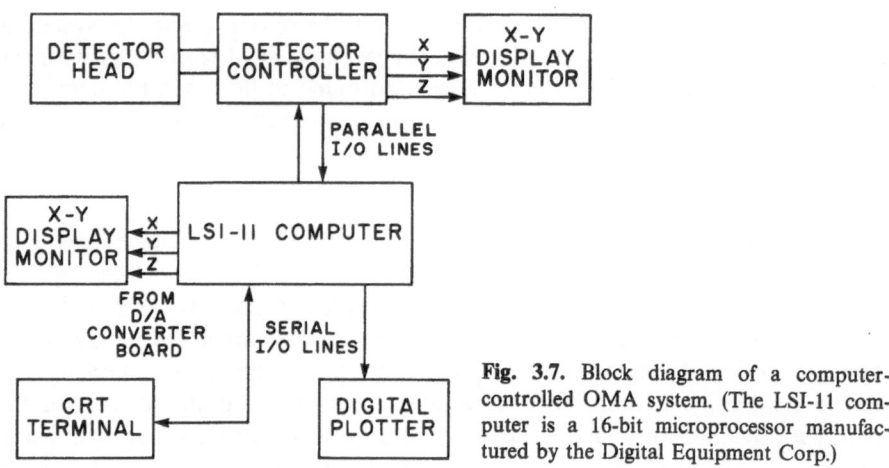

Fig. 3.7. Block diagram of a computer-controlled OMA system. (The LSI-11 computer is a 16-bit microprocessor manufactured by the Digital Equipment Corp.)

mentioned earlier, this rate can be quite high. Therefore, the number of data bits per pixel and the maximum scan rate must be considered when choosing a computer. Both 8-bit and 16-bit processors are currently available that are capable of satisfying the data rate requirements. While the 16-bit processors are somewhat more expensive, the larger data word is helpful in meeting the timing requirements. Another consideration is the amount of memory that the processor can address. For most 8-bit and 16-bit microprocessors, roughly 64,000 8-bit bytes can be addressed. While this is sufficient for many applications, a larger addressing space may be desirable if a large amount of two-dimensional data is being obtained or if the time development of spectral features is to be monitored. A schematic representation of a typical computer configuration used to control a multichannel detector is shown in Fig. 3.7.

3.3.2 Data Analysis and Accumulation

Because of the nature of OMA detectors, certain types of data corrections and analysis are needed. For example, a point-by-point background subtraction is generally required to compensate for the fixed pattern features of the OMA detectors. These features may be due to blemishes on the target in television camera based systems or to the differences in individual diodes in solid-state array detectors. In addition to removing the fixed pattern features, a background subtraction is often useful for reducing the optically generated unwanted signal from the desired spectrum and the thermally generated background. If the scattered background remains constant, a single background run can be subtracted from many different spectra. However, each time the background varies or the detector parameters are changed, a new background must be taken and stored for subsequent subtraction.

Another type of necessary data correction is a detector/optical system response correction. There are often slight spatial differences in the gain of television cameras and intensifier systems. Typically, the gain is greatest at the center of the device and tends to fall off toward the edges. This can give anomalous line shapes and incorrect relative signal intensities. In addition, the throughput of the optical system may not be constant over the entire field of view of the detector (e.g., due to vignetting). To correct for these effects, the signal spectrum can be divided by a "constant signal" spectrum. To obtain this, a broad-band source of radiation can be focused onto the slit of the spectrograph. If the detector response and the optical throughput of the entire system were perfectly uniform, the detected signal in each channel would be the same. If it is not, however, dividing the signal spectrum by this response spectrum will be necessary to correct the data. Note that the background signal should be subtracted from both the data and response signals before the division is done.

3.4 Selected Applications

The applications of OMA systems to optical spectroscopy are extensive and too numerous to cite here individually. A good sampling of applications in different fields can be found in [3.1, 21, 22]. For conventional cw Raman spectroscopy, where the sample is homogeneous, time invariant and unharmed by the incident laser energy, the OMA system simply has the advantage that the spectrum can be obtained N times faster, where N is the number of spectral channels the OMA can monitor simultaneously. With equal monitoring time for the single-channel and multichannel detection schemes, the latter technique is expected to improve the signal-to-noise ratio by $N^{1/2}$. However, we have found that in some cases, this ratio can actually be improved if proper optimization is made in the operation of the SIT based OMA system [3.20]. Laser fluctuation or drift and mechanical vibration or relaxation in the optical system can give rise to time variations in the Raman intensity. In a single-channel system, these variations can cause spurious noise in the detected spectrum as the spectrometer is slowly scanned. On the other hand, when an OMA is used, the entire spectrum is affected equally, and noise is not introduced by time variations in laser intensity or in the optical system.

The unique characteristics of OMA systems become particularly important for experiments in which (a) the sample is spatially inhomogeneous, (b) the signal is temporally evolving, or (c) the laser is pulsed at a low repetition rate. Specific examples of the use of an OMA in each of these cases will be summarized below. Because of limited space, the historical development and a complete list of related publications will not be given.

3.4.1 Spatial Resolution

An *in situ* and nondestructive microanalytical technique based on Raman scattering has been developed for determining the composition within micro-samples which can have spatially inhomogeneous distributions. This technique has proven to be useful in the fields of biology, geology and materials science. The spatial distribution of each chemical component within a heterogeneous sample can be deduced by monitoring the spatial intensity distribution of the Raman radiation uniquely characteristic of each component [3.4, 23–26]. The microprobe was first reported in 1973 [3.8] and by 1975, two groups had independently designed and constructed sensitive and stable detection systems capable of obtaining the Raman signature from micron-sized particles [3.4, 23]. One of these systems made use of a two-dimensional OMA detector [3.27]. A schematic diagram of such a two-dimensional Raman system is shown in Fig. 3.8. A circular area 150–300 μm in diameter is illuminated by directing a laser into the objective annular illuminator (a dark-field illumination device) which surrounds the main objective. To avoid noise associated with laser speckle, the laser beam is continually rotated, and the scattered radiation is collected by the main objective. The image of the sample is first focused onto the concave gratings of the double spectrometer and then onto the front

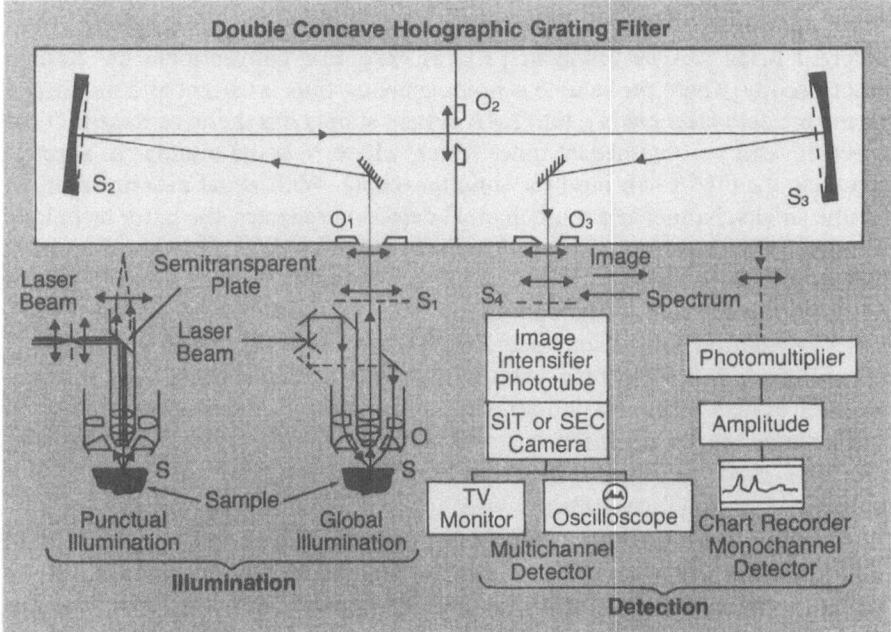

Fig. 3.8. Block diagram for the laser-Raman molecular microprobe. Compositional inhomogeneity of the sample can be detected and displayed on the TV monitor [3.27]

faceplate of the image intensifier of the OMA detector. The spatial resolution is about 1 μm. This system therefore makes full use of the two-dimensional capabilities of the OMA detector. The microprobe system shown in Fig. 3.8 has the capability of changing the microscope objective to a "point illumination" mode in which the double spectrometer is scanned and a photomultiplier is used as the detector. In this mode, the unique Raman signature at any one portion within the sample can be determined by scanning the double spectrometer.

Another way to obtain a two-dimensional Raman distribution, commonly referred to as Ramanography [3.28], is to use an interference filter instead of a spectrometer to isolate the Raman radiation at one wavelength. The sample is uniformly illuminated and the Raman scattered light (selected by the narrow-band interference filter) is imaged onto the face of a two-dimensional OMA detector, such as the SIT [3.29]. To examine Raman radiation at a different wavelength, a different interference filter must be used.

3.4.2 Temporal Evolution

Variations within the sample as a result of externally applied perturbations or naturally occurring changes often make single-channel spectroscopy impractical. This is especially true when the temporal changes are faster than the time necessary to sequentially scan the spectrometer over the entire wavelength region of interest. For example, if the signal-to-noise ratio requires that the counting duration in each wavelength interval be 0.1 s, then the total time to scan over 250 wavelength intervals will be 25 s. However, if an OMA detector is used, the entire 250 wavelength interval can be detected in 0.1 s.

In OMA systems, any temporal evolution within the sample that is shorter than the time required to scan all the desired channels will not be faithfully resolved. For the SEC, SIT, and ISIT, one-dimensional scans of 500 channels can be taken at speeds up to 0.01 s/scan. However, for these fast scanning speeds, the effects of increased lag must be considered when trying to track the temporal evolution. For the ISPD, the shift register sets the rate at which each channel is read out. For a 1024 element linear array, a typical readout rate is 16 μs/channel. For all of these detectors, shorter scan times can be achieved if the wavelength region of interest does not extend across the entire detector (i.e., by scanning fewer channels). The actual size of the image depends on the linear dispersion of the spectrograph. It can be varied, in principle, by using a zoom lens.

We have applied a SIT based OMA system to the study of surface enhanced Raman scattering of molecules adsorbed on electrodes undergoing electrochemical oxidation and reduction cycles [3.30, 31]. Figure 3.9 shows the Raman signal from $Ag(CN)_x^-$ ($x = 1, 2, 3, 4$) over a $300\ cm^{-1}$ interval as the voltage of the electrode is ramped at 50 mV/s over the range from -0.88 to $+0.5$ to -0.88 V. The associated voltammogram is shown in the bottom half of

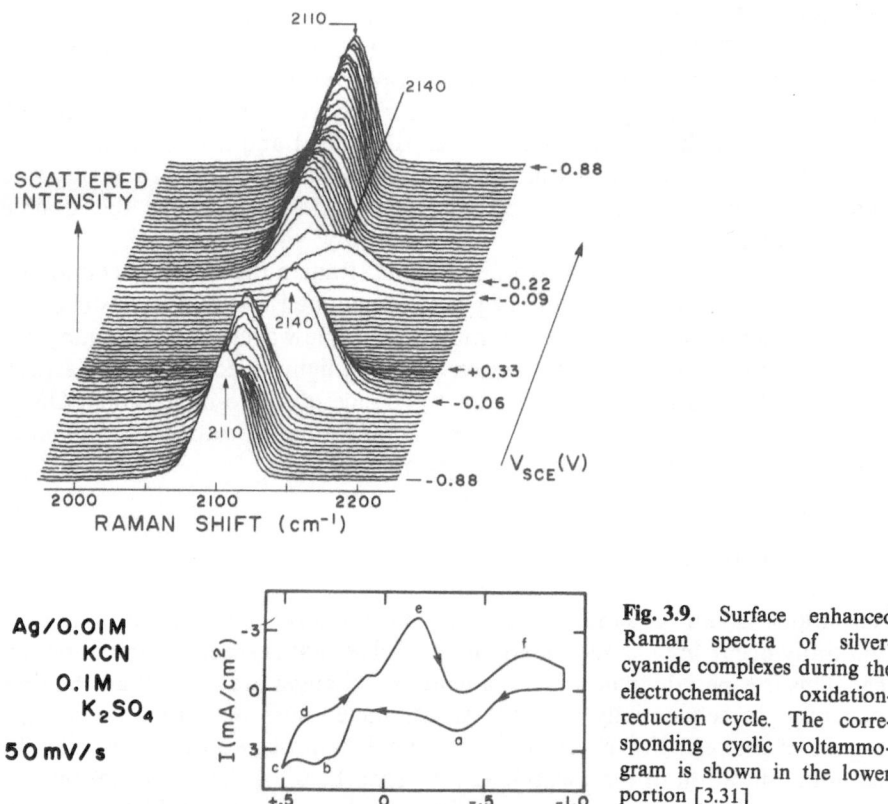

Fig. 3.9. Surface enhanced Raman spectra of silver-cyanide complexes during the electrochemical oxidation-reduction cycle. The corresponding cyclic voltammogram is shown in the lower portion [3.31]

Fig. 3.9 [3.31]. Since the Raman shifts of the various cyanide complexes are different, there is no way to set a single-channel detection system at one specific wavelength and then vary the voltage. Furthermore, the Raman signal is dependent on the voltage scan rate and on the history of the electrode (i.e., whether this is the first electrochemical cycle or a subsequent one), so that it would not have been possible to repeat the voltage cycle at another spectrometer setting within the $300 \, cm^{-1}$ interval. As this example illustrates, the OMA system has a unique place in optical spectroscopy for temporal evolutions in samples which have built-in hysteresis or in which the spectrum is dependent on the history of the sample.

3.4.3 Pulsed Spectroscopy

The OMA system is particularly useful for studying short-lived species, investigating nonlinear optical phenomena that require high laser intensity, or monitoring the discrete Raman signals in the presence of a broad-band fluorescence. The whole spectrum can be monitored after one laser pulse, regardless of its pulse duration (e.g., ps to µs). The spectral information can be stored in the computer for each pulse, many single-pulse spectra can be accumulated in the computer, or many spectra can be integrated on the camera target and then read out after a large number of laser pulses.

Pulsed multichannel spontaneous Raman spectroscopy was first applied in 1970 [3.5]. Since then, ultrashort light spectroscopy [3.20] has been used in many fields. In the ps and µs range, the pulse duration is set by the laser, and the spectral information is read out from the camera after the laser pulse but before the next laser pulse, should target integration not be desirable. The rate at which the camera is ready to receive a new spectrum is dependent on the lag of the camera, which determines how many readout scans are necessary to totally recharge the target of the camera. For the ISPD at low temperatures, the upper limit of the laser pulse rate can be set by the persistence time of the image intensifier phosphor and not by the total time necessary to cycle through N channels. For two-dimensional cameras (SEC, SIT, and ISIT), the laser pulse rate can be speeded up if a rotating mirror is used in conjunction with the spectrograph. The entire spectrum from one laser pulse will be stored on one horizontal strip of the target and the subsequent pulses will be stored on different horizontal strips which are successively displaced in the vertical direction by the rotating mirror which moves the image along the vertical spectrograph entrance slit. The one-dimensional detector (ISPD) does not have such versatility compared to its two-dimensional counterparts.

The OMA detectors cannot follow the temporal development of the optical signal during the short laser pulse as in the case of a fast response photo-multiplier. However, all the OMA detectors can be gated on for durations as short as 10–50 ns either synchronously with the laser pulse or at a preset delay after the laser pulse. Consequently, a luminescence background could be discriminated from the Raman signal by using a "boxcar" approach, i.e., measuring the entire spectrum coincident with the laser pulse and soon after the laser pulse, which would then be appropriately subtracted from the first spectrum in the computer. Another background discrimination technique that can be used with a gated two-dimensional OMA detector is to measure the entire spectrum at the same time (gated synchronously with the laser pulse) but from two different portions of the sample, one portion with laser illumination and another outside the laser focal volume. By imaging the two spectra on two different areas on the OMA detector, the radiation from the nonilluminated part of the sample can be appropriately subtracted from the laser illuminated part of the sample. Placing two spectra onto different areas on the OMA detector is also very useful for wavelength calibration purposes in either a pulsed or cw mode. For example, in CARS spectroscopy, the multiwavelength (broad-band) CARS spectra from the sample gas and from a reference gas have been imaged onto different portions of the SIT detector for the purpose of wavelength calibration [3.32].

3.5 Conclusions

Optical spectroscopists now have multichannel detectors available that are nearly equivalent to having 10^3–10^4 photomultipliers working in parallel. The availability of such detectors, along with high stray-light rejection multistage spectrographs and inexpensive computers, has opened up new capabilities for all types of optical spectroscopy.

New detectors are currently being developed which may prove to be superior to the existing detectors discussed here. Because of their unique capabilities and increased availability, we feel that OMA systems will become even more widely used and will continue to enrich the field of optical spectroscopy.

Acknowledgment. We gratefully acknowledge the Office of Naval Research (Contracts N00014-76-C-0643 and N00014-79-C-0254) and the Gas Research Institute (Basic Research Grant 5080-363-0319) for partial support of this work.

3.6 Glossary

Accumulation: Summing of several scans, channel by channel in memory. Accumulation and integration can be used together for optimal low-light-level sensitivity.

Aliasing: When the spatial frequency of the image is greater than that of the pixels, a Moiré pattern is produced which reduces the monitoring accuracy. The Nyquist criterion requires that the pixel's spatial frequency should be at least twice that of the image to eliminate aliasing.

Blooming: Cross talk between channels caused by charge spreading to adjacent pixels.

Dynamic Range: The limits of this range represent the lowest and highest intensity features that can be monitored in one frame. The upper limit is set by the saturation of the electronics (about 750 counts for the SIT). The noise limit per channel is $\sqrt{Nc_{dark}}$, where N is the number of scans accumulated and c_{dark} is the dark counts per channel per scan.

EBS: Acronym, Electron Bombardment Silicon (tradename used by Westinghouse). Functionally equivalent to the SIT.

EDN: Equivalent Dark Noise, light intensity [photons/s or W] which produces a signal equal to the dark noise.

ERMA: Acronym, Extended Red Multialkaline Photocathode.

Fixed Pattern Noise: Nonrandom noise produced by pixel-to-pixel variations and dark current.

Frame: One scan of the entire target of a detector.

Gain: Ratio of the change in the optical signal to the change in electronic output. The gain of the vidicon target is 1.

Integration: Lengthening of exposure time to allow signal to build up before the target is read. Integration and accumulation can be used together for optimal low-light-level sensitivity.

Intrascenic Dynamic Range: Ratio of the most intense and least intense spectral features that can be *simultaneously* detected within a single readout regardless of the integration time between consecutive readouts. The difference between the individual diode dynamic range and that of the linear array is a result of "veiling glare."

ISIT: Acronym, Intensified Silicon Intensified Target television camera (trademark of RCA).

ISPD: Acronym, Intensified Silicon Photodiode Detector.

Lag: Phenomenon associated with electron readout beam image devices (vacuum devices), in which a complete readout of the signal stored on the target is not achieved in a single scan.

Light Gain: Photon output from the phosphors divided by photon input at the photocathode.

lp/mm: Linepairs/mm. A white line and an adjacent black line are designated a line pair.

MCP: Acronym, Microchannel Plate.

Memory Dynamic Range: Number of counts that can be stored in a given channel of memory.

Nonflatness: Pixel-to-pixel variations in the dark current and wavelength response. Nonflatness can be computer corrected by comparison with a prestored flat field exposure (from a uniform illumination of the images).

Pixel: Each individual sensing element of an image detector.

Resolution: lp/mm at a given Modulation Transfer Function. Defines how well the detector preserves the image details.

Scan Rate: Rate at which the charge image stored on the target is read out. This is determined by the electronic bandwidth and the efficiency of the readout mechanism. If the readout rate is too fast, signals that should have been picked off in previous scans are left on the target because of target lag. A typical readout rate or scan time for the OMA is 32.8×10^{-3} s per frame.

SEC: Acronym, Secondary Electron Conduction television camera (trademark used by RCA).

Sensitivity: A measure of the smallest optical signal (number of photons) that can be usefully detected. The amplifier sets the sensitivity level, about 2500 photoelectrons/count from the A/D converter in the SIT.

Single Shot Dynamic Range: The range of the largest signal the detector can acquire for a single input pulse divided by the noise associated with the readout noise in the absence of signal input.

SIT: Acronym, Silicon Intensified Target television camera (trademark of RCA).

SPD: Acronym, Self-scanned linear Photodiode array. A large scale integrated circuit fabricated on a single monolithic silicon crystal. It contains a row of photodiode sensors (2.5 mm long), typically on 25 μm centers (aspect ratio of 100:1), along with a scanning circuitry for sequential readout.

Veiling Glare: Stray light in the detector enclosure due to internal reflections. This is particularly bad for image intensified detectors because of the optical reflections from glass and metallic components.

Vidicon: An image detector consisting of a photo-sensitive target and an electron beam assembly to read the charge image on the target. The silicon vidicon has a silicon target with a diode array as its photo-sensitive element.

Tracking Dynamic Range: The range of signals that the detector can follow linearly, without overloading, divided by the rms noise for a single readout in the absence of signal input.

References

3.1   Y.Talmi (ed.): *Multichannel Image Detectors*, ACS Symposium Series 102 (American Chemical Society, Washington, D.C. 1979)
3.2   Y.Talmi: Anal. Chem. **47**, 685A (1975)
3.3   Y.Talmi: Anal. Chem. **47**, 697A (1975)
3.4   M.Delhaye, P.Dhamelincourt: J. Raman Spectrosc. **3**, 33 (1975)
3.5   M.Bridoux, M.Delhaye: Nouv. Rev. d'Opt. Appl. **1**, 23 (1970)
3.6   M.Bridoux, M.Delhaye: In *Advances in Infrared and Raman Spectroscopy*, Vol. 2, ed. by R.J.H.Clark, R.E.Hester (Heyden, London 1976) p. 140
3.7   C.M.Savage, P.D.Maker: Appl. Opt. **4**, 965 (1971)
3.8   T.Hirschfeld: J. Opt. Soc. Am. **63**, 476 (1973)
3.9   OMA Application Notebook, EG&G-Princeton Applied Research Corporation, Princeton, New Jersey
3.10 C.E.Catchpole: In *Photoelectronic Imaging Devices*, Vol. 2, ed. by L.M.Biberman, S.Nudelman (Plenum Press, New York, London 1971) p. 167
3.11 J.D.McGee: In *Photoelectronic Imaging Devices*, Vol. 2, ed. by L.M.Biberman, S.Nudelman (Plenum Press, New York, London 1971) p. 133
3.12 G.W.Goetze, A.B.Laponsky: In *Photoelectronic Imaging Devices*, Vol. 2, ed. by L.M.Biberman, S.Nudelman (Plenum Press, New York 1971) p. 217
3.13 H.L.Felkel, Jr., H.L.Pardue: In *Multichannel Image Detectors*, ACS Symposium Series 102, ed. by Y.Talmi (American Chemical Society, Washington, D.C. 1979) p. 59
3.14 M.H.Crowell, E.F.Labuda: In *Photoelectronic Imaging Devices*, Vol. 2, ed. by L.M.Biberman, S.Nudelman (Plenum Press, New York, London 1971) p. 301
3.15 M.Green, A.B.Laponsky, W.D.Frobenius: Proc. Technical Program, Electro-Optical Systems Design Conf. 1973 (Industrial and Scientific Conference Management, Inc., Chicago 1973) p. 224
3.16 R.G.Tull, J.P.Choisser, E.H.Snow: Appl. Opt. **14**, 1182 (1975)
3.17 Y.Talmi, R.W.Simpson: Appl. Opt. **19**, 1401 (1980)
3.18 S.S.Vogt, R.G.Tull, P.Kelton: Appl. Opt. **17**, 574 (1978)
3.19 R.W.Simpson: Rev. Sci. Instrum. **50**, 730 (1979)
3.20 R.E.Benner, R.Dornhaus, M.B.Long, R.K.Chang: In *Microbeam Analysis*, Proc. 14th Annual Conf. Microbeam Analysis Society, ed. by D.E.Newbury (San Francisco Press, San Francisco 1979) p. 191
3.21 E.D.Schmid, R.S.Krishnan, W.Kiefer, H.W.Schrötter (eds.): *Proc. 6th Intern. Conf. on Raman Spectroscopy* (Heyden, London, Philadelphia, Rheine 1978)

3.22 W.P.Murphy (ed.): *Proc. 7th Intern. Conf. on Raman Spectroscopy* (North-Holland, Amsterdam, New York 1980)

3.23 G.J.Rosasco, E.S.Etz, W.A.Cassatt: Appl. Spectrosc. **29**, 396 (1975)

3.24 M.Delhaye, J.Barbillat, P.Dhamelincourt: In *Analytical Techniques in Environmental Chemistry*, ed. by J.Albaiges (Pergamon Press, Oxford, New York 1980) p. 515

3.25 P.Dhamelincourt: In *Microbeam Analysis*, Proc. 14th Annual Conf. Microbeam Analysis Society, ed. by D.E.Newbury (San Francisco Press, San Francisco 1979) p. 155

3.26 P.Dhamelincourt: In *Lasers in Chemistry*, Proc. Conf. on Lasers in Chemistry of the Royal Institution, London, 1977 (Elsevier, Amsterdam, Oxford, New York 1977) p. 44

3.27 P.Dhamelincourt, F.Wallart, M.Leclercq, A.T.N'Guyen, D.O.Landon: Anal. Chem. **51**, 414A (1979)

3.28 D.L.Hartley: In *Laser Raman Gas Diagnostics*, ed. by M.Lapp, C.M.Penney (Plenum Press, New York, London 1974) p. 311

3.29 B.F.Webber, M.B.Long, R.K.Chang: Appl. Phys. Lett. **35**, 119 (1979)

3.30 R.Dornhaus, M.B.Long, R.E.Benner, R.K.Chang: Surf. Sci. **93**, 240 (1980)

3.31 R.E.Benner, R.Dornhaus, R.K.Chang, B.L.Laube: Surf. Sci. **101**, 341 (1980)

3.32 D.V.Murphy, R.K.Chang: Opt. Lett. **6**, 233 (1981)

4. Coherent and Hyper-Raman Techniques

H. Vogt

With 19 Figures

The invention of the laser has allowed the extension of the concept of Raman scattering in two ways termed coherent and hyper-Raman spectroscopy (CRS and HRS)[1]. CRS utilizes laser-like Raman signals generated by the coherent interaction between the input beams and Raman-active modes. HRS exploits the symmetry selection rules for a three-photon process and provides an access to modes not observable by normal two-photon Raman scattering.

Pioneering works in both fields were published around 1965 [4.1–4]. *Bloembergen* and his associates introduced the general framework for describing the phenomena of nonlinear optics in terms of higher-order susceptibility tensors [4.1]. They also invented methods for measuring Raman gain or loss which are often referred to as stimulated Raman spectroscopy (SRS) [4.2]. *Maker* and *Terhune* initiated the development of coherent antistokes Raman spectroscopy (CARS) as well as Raman induced Kerr effect spectroscopy (RIKES) [4.3]. The same group also reported the first observation of hyper-Raman scattering [4.4].

Since tunable laser sources became available, the field of CRS has expanded very rapidly. On the other hand, the number of studies on HRS has remained comparatively small, essentially because the cross sections or efficiencies of the underlying multiphoton processes are extremely weak, except under special resonance conditions.

The present chapter reviews successful applications of CRS and HRS to solids. We concentrate on results obtained with nanosecond laser pulses in the power range up to several MW. Our description is confined to the (ω, k)-domain. Transient phenomena and time-resolved spectroscopy [4.5] are excluded.

After an introductory survey we discuss the structure of the nonlinear susceptibilities in Sect. 4.1. The emphasis is on the multiphoton resonances exploited by the new Raman techniques. Section 4.2 is devoted to CARS which has become the most popular CRS method. SRS and RIKES are briefly illustrated in Sect. 4.3. In Sect. 4.4, we describe double resonance experiments revealing the interference of different two-photon processes and allowing a quantitative comparison between them. Finally, in Sect. 4.5 we demonstrate the merits of spontaneous HRS in the study of infrared and "silent" modes. We also include some results obtained by resonant and stimulated versions of HRS.

---

1 See Table 4.1 for an explanation of the acronyms used in this chapter.

4.1 Basic Principles

4.1.1 Elementary Description of Coherent and Hyper-Raman Effects

The new feature of coherent Raman spectroscopy (CRS) is the phase relationship between the incident and scattered light as well as the elementary excitation or mode involved. In a classical model, the laser radiation coherently drives the Raman mode and imposes a regular forced vibration on the statistical thermal motion. The driving force F arises from the well-known electron-phonon or electron-mode coupling usually described by the derivative $(\partial \alpha / \partial Q)_0$ of the electronic polarizability α with respect to the normal coordinate Q of the mode. An illustrative expression for F can be derived from the Q-dependent potential energy V of the scattering system within the electric field E of the laser light [4.6, 7]. We have

$$V = -\frac{1}{2}\alpha(Q)E^2 = -\frac{1}{2}\left[\alpha_0 + \left(\frac{\partial \alpha}{\partial Q}\right)_0 Q + \ldots\right]E^2 \qquad (4.1)$$

and

$$F = -\frac{\partial V}{\partial Q} = \frac{1}{2}\left(\frac{\partial \alpha}{\partial Q}\right)_0 E^2 + \ldots. \qquad (4.2)$$

Since the frequencies of E are generally much higher than the mode frequency ω_0, only the difference frequency part of E^2 or F can resonantly pump or "phase" the Raman mode. For this purpose the incident laser radiation has to contain two spectral components of frequencies ω_1 and ω_2 with $|\omega_1 - \omega_2|$ approaching ω_0. To avoid confusion in later discussions, let us define $\omega_1 > \omega_2$ so that $\omega_0 \approx \omega_1 - \omega_2$.

The coherent excitation of the mode by F can be probed by various methods. In the most general case, a third beam of frequency ω_p is used as a probe and Raman signals are detected at the antistokes and Stokes frequencies:

$$\omega_s = \omega_p \pm (\omega_1 - \omega_2). \qquad (4.3)$$

The Raman light emerges from the sample as a laser-like beam. It propagates along a direction determined by the wave vector condition

$$k_s = k_p \pm (k_1 - k_2), \qquad (4.4)$$

where k_s, k_p, k_1, k_2 denote the wave vectors of the signal, the probe, and the two pump waves, respectively. There are only special applications of this general scheme of four-wave mixing (4WM). On the other hand, almost every experiment in CRS can be interpreted as a degenerate form of 4WM, where at least two of the four frequencies ω_s, ω_p, ω_1, ω_2 coincide. To give some examples, let us consider the cases $\omega_p = \omega_1$ and $\omega_p = \omega_2$, assuming either of the two pump waves as also being the probe.

Now (4.3) simplifies to

$$\omega_s = \underline{\underline{\omega_1}} \pm (\omega_1 - \omega_2) = \begin{cases} 2\omega_1 - \omega_2 \approx \omega_1 + \omega_0, \\ \omega_2 \end{cases} \tag{4.5}$$

or

$$\omega_s = \underline{\underline{\omega_2}} \pm (\omega_1 - \omega_2) = \begin{cases} \omega_1 \\ 2\omega_2 - \omega_1 \approx \omega_2 - \omega_0 \end{cases}. \tag{4.6}$$

The Raman signals of frequencies $2\omega_1 - \omega_2$ and $2\omega_2 - \omega_1$ travel as new and separate beams along directions given by $2k_1 - k_2$ and $2k_2 - k_1$, respectively. If $(\omega_1 - \omega_2)$ is tuned through ω_0, a Raman resonance occurs in the intensities of both beams. The Raman signal at $2\omega_2 - \omega_1 \approx \omega_2 - \omega_0$ falls into the Stokes region of the lower frequency pump wave and may be obscured by luminescence. Therefore, most attention is given to the antistokes Raman signal at $2\omega_1 - \omega_2 \approx \omega_1 + \omega_0$. The method of measuring its intensity as function of $(\omega_1 - \omega_2)$ has been labeled by the acronym CARS [4.8].

In the Stokes case of (4.5) and in the antistokes case of (4.6), the mixing process reduces to an energy transfer between the two pump waves. One pump wave is collinear with the signal while the other is collinear with the probe. A more detailed analysis shows that the lower frequency intensity $I(\omega_2)$ is amplified at the expense of the higher frequency intensity $I(\omega_1)$. If $(\omega_1 - \omega_2)$ is swept over ω_0, a Raman resonance appears in the gain of $I(\omega_2)$, as well as in the loss of $I(\omega_1)$ [4.2].

As illustrated by (4.2), the coherent interaction underlying all techniques of CRS depends on the same electron-mode coupling parameter $(\partial\alpha/\partial Q)_0$ as the standard spontaneous Raman scattering. Hence, CRS does not open a principally new access to modes and their coupling to electrons. Nevertheless, for many applications, CRS has been found superior to spontaneous Raman scattering. The outstanding merits result from the high intensity and from the beam-like propagation of the Raman signals. Both properties permit extensive spatial filtering for eliminating any background of luminescence and parasitic light. Thus, Raman spectra of combustions, plasmas, and all kinds of highly fluorescent materials become observable. Under favorable circumstances, modes can be studied with a spectral resolution limited by the linewidths of the laser sources and not by the instrumental profile of spectrometers. We should also stress that CRS can be used for a quantitative comparison of various multiphoton processes. For instance, two-photon absorption coefficients can be measured relative to Raman cross sections.

Unlike CRS, hyper-Raman spectroscopy (HRS) is based on a new electron-mode coupling parameter. We have to replace $(\partial\alpha/\partial Q)_0$ by $(\partial\beta/\partial Q)_0$, where β is the nonlinear or hyper-polarizability relating the second-order electronic dipole moment $p^{(2)}$ to the square of the laser field E. We may write

$$p^{(2)} = \beta E^2 \tag{4.7}$$

and

$$\beta = \beta_0 + \left(\frac{\partial \beta}{\partial Q}\right)_0 Q + \dots \tag{4.8}$$

If only one input wave of frequency ω_1 is used, β_0 describes the generation of second harmonic light of frequency $2\omega_1$. As a tensor of third rank, β_0 vanishes in the presence of inversion symmetry. Therefore, the phenomenon of frequency doubling is confined to noncentrosymmetric systems, at least in the dipole approximation. In the second term of the expansion (4.8), however, the odd parity of β can be compensated for by that of the normal coordinate Q, so that $(\partial\beta/\partial Q)_0 \neq 0$ even if $\beta_0 = 0$. Hence, in every material, elementary excitations or modes of an appropriate symmetry are expected to modulate β and give rise to scattered light. The resulting hyper-Raman lines are found at $2\omega_1 \pm \omega_0$, where ω_0 is the mode frequency. In the quantum picture, hyper-Raman scattering is described as a three-photon process by which two photons of frequency ω_1 are simultaneously annihilated to create one scattered photon of frequency $2\omega_1 \pm \omega_0$.

The symmetry selection rules determined by the form of $(\partial\beta/\partial Q)_0$ represent the most important new feature of HRS. In as far as parity is concerned they are complementary to those of normal two-photon Raman scattering and similar to those of one-photon infrared absorption. Generally speaking, all infrared-active modes are also hyper-Raman-active. Moreover, HRS permits the observation of "silent" modes which are both Raman and infrared forbidden [4.9].

In principle, HRS includes both spontaneous and coherent scattering techniques. Analogously to CRS, we also have CHRS (coherent hyper-Raman spectroscopy). However, the extremely weak efficiencies encountered in nonresonant HRS have prevented a rapid development of this field.

It has become customary to coin abbreviations for the various Raman methods. The most common ones are listed and explained in Table 4.1.

In the last few years, a considerable number of reviews on CRS [4.10–18] and HRS [4.19–21] have been published. *Flytzanis* [4.10] and *Hellwarth* [4.11] have described in detail the theoretical background of CRS, while *Maier* [4.12], *Shen* [4.13], and *Akhmanov* and *Koroteev* [4.14] have given general accounts with emphasis on experiments. The utility of CARS in chemistry and molecular physics has been discussed by *Tolles* et al. [4.15], *Druet* and *Taran* [4.16], as well as *Nibler* and *Knighten* [4.17]. Most recently, *Levenson* and *Song* [4.18] have comprehensively covered theory and practice of CRS. Reviews on spontaneous HRS have been published by *Long* and his associates [4.19–21].

Table 4.1. Acronyms of coherent and hyper Raman techniques

Acronym	Explanation
CRS	Coherent Raman spectroscopy
SRS	Stimulated Raman spectroscopy
CARS	Coherent antistokes Raman spectroscopy
CSRS	Coherent Stokes Raman spectroscopy
ASCS	Active spectroscopy of combinatorial scattering (used in Russian literature)
4 WM	Four-wave mixing
BOXCARS	CARS with box-shaped beam configuration
ARCS	Angular resolved coherent Raman spectroscopy
RIKES	Raman induced Kerr effect spectroscopy
OHD RIKES	Optical heterodyne detected RIKES
PARS	Photoacoustic Raman spectroscopy
HRS	Hyper-Raman spectroscopy
CHRS	Coherent HRS

The present article does not aim at completeness in any sense, but concentrates on the basic ideas and the major advantages for solid state spectroscopy. Throughout this chapter cgs units are used.

4.1.2 Definition and General Structure of the Nonlinear Susceptibilities

Nearly all nonlinear optical effects in solids are described by an expansion of the dielectric polarization P in terms of the electric field E of the incident laser radiation. One writes (Sect. 2.1.17)

$$P = \chi^{(1)} E + \chi^{(2)} E^2 + \chi^{(3)} E^3 + \chi^{(4)} E^4 + \chi^{(5)} E^5 + \dots, \tag{4.9}$$

where $\chi^{(n)}$ represents the dielectric susceptibility tensor of rank $(n+1)$ associated with the nth power of E. Nonlinear phenomena beyond this dipole approximation involve spatial derivatives of E and have been studied rather rarely [4.22] (see also Sect. 2.2.8).

In the absence of any resonance enhancement, the ratio of succeeding terms in (4.9) can be estimated by

$$\frac{P^{(n+1)}}{P^{(n)}} = \frac{\chi^{(n+1)} E^{n+1}}{\chi^{(n)} E^n} \approx \frac{E}{E_M}, \tag{4.10}$$

where E_M is the microscopic electric field binding the electrons responsible for the optical effects [4.1]. In insulators, E_M is in the range of 10^8 V/cm whereas the threshold of dielectric breakdown is reached at an electric field strength E of about 10^6 V/cm corresponding to a laser intensity of several GW/cm^2 [4.23]. Thus, the ratio in (4.10) has an upper limit around 10^{-2} and the expansion (4.9) converges so rapidly that nonlinear optical effects beyond the $\chi^{(5)}$ term become barely detectable.

In the Raman experiments under consideration, the input electric field is composed of only a few nearly harmonic waves. We write

$$E(r, t) = \sum_j E_j(r, t) \tag{4.11}$$

with

$$E_j(r, t) = \tfrac{1}{2}[E(\omega_j, k_j)e^{-i(\omega_j t - k_j r)} + E^*(\omega_j, k_j)e^{i(\omega_j t - k_j r)}], \tag{4.12}$$

where the amplitude $E(\omega_j, k_j)$ is allowed to vary slowly with r and t, i.e., over distances which are large compared to $2\pi/|k_j|$ and $2\pi/\omega_j$, respectively. For pump waves, the index j is either 1 or 2, while the probe wave is indicated by $j = p$. Note that the factor of $\tfrac{1}{2}$ in the rhs of (4.12) is not always used in the literature.

Strictly speaking, it is not correct to insert (4.11, 12) directly into (4.9). The power series (4.9) refers to the (ω, k)-domain and hence constitutes a relationship between the Fourier amplitudes $E(\omega_j, k_j)$ and $P(\omega_j, k_j)$ rather than between $E(r, t)$ and $P(r, t)$. Otherwise, (4.9) has to be replaced by a series of intricate convolution integrals [4.11].

To illustrate the meaning of (4.9) with an example, let us consider nondegenerate antistokes 4WM. In this case, (4.9) states that the Raman signal arises from a nearly harmonic third-order polarization wave with an amplitude of the form

$$P_\alpha^{(3)}(\omega_s, k_s) = D \sum_{\beta, \gamma, \delta = 1}^{3} \chi_{\alpha\beta\gamma\delta}^{(3)}(-\omega_s, \omega_p, \omega_1, -\omega_2) E_\beta(\omega_p, k_p) E_\gamma(\omega_1, k_1) E_\delta^*(\omega_2, k_2) \tag{4.13}$$

The frequencies and wave vectors appearing in this expression have already been defined in Sect. 4.1.1 and are related by the plus sign versions of (4.3, 4). The indices $\alpha, \beta, \gamma, \delta$ refer to the three cartesian coordinate axes. The factor D takes into account degeneracies and is equal to the number of distinguishable permutations of the input amplitudes [4.3]. As the two pump waves and the probe differ from one another, we have $D = 3! = 6$.

In order to calculate the electric field $E_s(r, t)$ of the Raman signal, we have to incorporate the nonlinear source term (4.13) into Maxwell's equations. We obtain [4.1]

$$\nabla \times \nabla \times E_s(r, t) + \frac{\varepsilon(\omega_s)}{c^2} \frac{\partial^2 E_s(r, t)}{\partial t^2} = -\frac{4\pi}{c^2} \frac{\partial^2 P^{(3)}(r, t)}{\partial t^2}$$

$$= \frac{4\pi}{c^2} \omega_s^2 [P^{(3)}(\omega_s, k_s)e^{-i(\omega_s t - k_s r)} + \text{c.c.}], \tag{4.14}$$

where $\varepsilon(\omega_s)$ is the dielectric constant at frequency ω_s and c the velocity of light in vacuum. The abbreviation c.c. stands for complex conjugate. Running wave solutions of (4.14) may be written in the form shown by (4.12), the amplitudes again being slowly varying functions of r and t. While the frequency of $E_s(r, t)$ is identical with ω_s, the wave vector may deviate from k_s by an amount Δk. This phase mismatch results from dispersion leading to a phase velocity of $P^{(3)}(r, t)$, different from that determined by $\varepsilon(\omega_s)$. The Raman intensity critically depends on Δk and reaches a pronounced maximum for $\Delta k = 0$. Practitioners of the field always try to meet the phase matching condition $\Delta k = 0$ as perfectly as possible.

The introduction of complex Fourier amplitudes implies complex susceptibilities, i.e.,

$$\chi^{(n)} = [\chi^{(n)}]' + i[\chi^{(n)}]'' . \tag{4.15}$$

As follows from elementary electromagnetic theory, only the imaginary parts of the even rank tensors $\chi^{(1)}$, $\chi^{(3)}$, $\chi^{(5)}$, etc., are associated with loss or gain. We have [4.24]

$$\langle W \rangle = \left\langle E(r, t) \frac{\partial P(r, t)}{\partial t} \right\rangle , \tag{4.16}$$

where W represents the absorbed or released power per unit volume and the bracket indicates the average over time. A nonzero contribution can only be obtained if the bracket encloses the product of an even number of electric field components. This requirement is met by the odd-order terms in the expansion of P which involve $\chi^{(1)}$, $\chi^{(3)}$, $\chi^{(5)}$, etc.

It is important to realize the impact of inversion symmetry on (4.9). Centrosymmetric systems do not reveal any odd rank tensor properties so that (4.9) reduces to

$$P = \chi^{(1)} E + \chi^{(3)} E^3 + \chi^{(5)} E^5 + \dots . \tag{4.17}$$

$\chi^{(3)}$ and $\chi^{(5)}$ are left for describing all multiphoton effects. As will be outlined in the following section, Raman scattering, including all variants of CRS, may be referred to as $\chi^{(3)}$-phenomena, whereas HRS is controlled by $\chi^{(5)}$.

In noncentrosymmetric systems, we have to take cascade processes into account. They consist of at least two distinct and subsequent steps and may result in the same signal frequencies as the one-step Raman processes described by $\chi^{(3)}$ and $\chi^{(5)}$. For instance, the CARS intensity at $2\omega_1 - \omega_2$ may be superimposed to signals arising from two-wave mixing due to $\chi^{(2)}$ [4.25]. In a first step the frequency ω_1 is doubled, while in a second step the second harmonic at $2\omega_1$ mixes with ω_2 to produce a difference frequency wave at $2\omega_1 - \omega_2$. Moreover, $\chi^{(2)}$ allows the generation of a real polariton at $\omega_1 - \omega_2$ which may beat with ω_1 to again give $(\omega_1 - \omega_2) + \omega_1 = 2\omega_1 - \omega_2$. In the case of HRS it is possible that the hyper-Raman signal at $2\omega_1 - \omega_0$ is obscured by a combination of second harmonic generation and normal Raman scattering of the frequency doubled light [4.26]. We shall always neglect such cascade processes and assume either inversion symmetry or scattering configurations where $\chi^{(2)}$ or $\chi^{(4)}$ cannot become operative.

As tensors of fourth and sixth rank, $\chi^{(3)}$ and $\chi^{(5)}$ consist of 81 and 729 elements, respectively. Point group symmetry reduces these numbers considerably. Tables listing the form of $\chi^{(3)}$ for the 32 crystal classes and isotropic media can be found in [4.10, 18]. Similar tables for $\chi^{(5)}$ are not yet available to the authors knowledge.

Additional symmetry relations can be exploited if the mixing process is nonresonant and does not induce any real excitation of the medium under study. Then thermodynamic arguments [4.1] show that the susceptibilities remain invariant when the tensor indices and the corresponding frequency arguments are interchanged simultaneously, e.g.,

$$\chi^{(3)}_{\alpha\beta\gamma\delta}(-\omega_s, \omega_p, \omega_1, -\omega_2) = \chi^{(3)}_{\alpha\gamma\beta\delta}(-\omega_s, \omega_1, \omega_p, -\omega_2). \tag{4.18}$$

If the medium is not only lossless with respect to the interacting waves, but also lacks dispersion, we may freely permute the indices while preserving the order of the frequencies. The identities thus obtained are commonly referred to as *Kleinman's* symmetry relations [4.27]. They do not hold in particular for the Raman terms in $\chi^{(3)}$ and $\chi^{(5)}$ because of the energy transfer during the Raman process. However, they generally apply to the nonresonant background contributions which have to be taken into account in many mixing experiments and represent a severe problem in CARS.

Perturbation theory provides the general scheme for calculating $\chi^{(3)}$ and $\chi^{(5)}$ from the dipole matrix elements of the material. A formula for $\chi^{(3)}$ derived by the conventional time-dependent perturbation technique was published about twenty years ago by *Armstrong* et al. [4.28]. Their result may be recast into the form

$$\chi^{(3)}_{\alpha\beta\gamma\delta}(-\omega_s, \omega_p, \omega_1, -\omega_2)$$
$$= \frac{LN}{3!\hbar^3} \sum_{\mathscr{P}} \left(\sum_{n,n',n''} \frac{\langle i|M_\alpha|n\rangle \langle n|M_\beta|n'\rangle \langle n'|M_\gamma|n''\rangle \langle n''|M_\delta|i\rangle}{[\omega_{ni}-(\omega_p+\omega_1-\omega_2)][\omega_{n'i}-(\omega_1-\omega_2)][\omega_{n''i}+\omega_2]} \right)_{\mathscr{P}}.$$
$$\tag{4.19}$$

Written in full length this expression consists of $4! = 24$ terms differing in the order of α, β, γ, δ in the numerator and the corresponding frequencies $\omega_s = \omega_p + \omega_1 - \omega_2, \omega_p, \omega_1, \omega_2$ in the denominator. A convenient abbreviation is obtained by introducing the sum over all permutations $\mathscr{P}$ of the pairs $(\alpha, -\omega_s)$, $(\beta, \omega_p), (\gamma, \omega_1)$ and $(\delta, -\omega_2)$ [4.10]. Each term in (4.19) contains a product of four dipole matrix elements linking the initial state $|i\rangle$ to the intermediate ones denoted by $|n\rangle, |n'\rangle$, and $|n''\rangle$. We interpret M as the dipole moment operator of a single molecule or unit cell. Then N stands for the number of molecules or unit cells per unit volume. The factor L incorporates local field corrections.

The denominators of (4.19) display two types of resonances. On the one hand, there is a variety of one-photon resonances occurring when the frequency of an input or output wave approaches a transition frequency, e.g.,

$$\omega_{ni} = \frac{1}{\hbar}(\mathscr{E}_n - \mathscr{E}_i).$$ (4.20)

On the other hand, the second factor in the denominators of (4.19) indicates two-photon resonances characterized by

$$\left.\begin{array}{r} \omega_1 - \omega_2 \\ \omega_p - \omega_2 \\ \omega_1 + \omega_p \end{array}\right\} \approx \omega_{ni}$$ (4.21)

with a different ω_{ni} for each row. While the first and second resonance conditions refer to Raman processes, the third one applies to two-photon absorption (TPA). In a nondegenerate 4WM experiment, it is possible to choose ω_1, ω_2, and ω_p in accordance with (4.21) so that two different Raman resonances and one TPA resonance may simultaneously contribute to the intensity of the signal wave [4.29].

We do not gain much new insight in writing down an expression for $\chi^{(5)}$ analogous to (4.19). Now the sum over the permutations $\mathscr{P}$ consists of $6! = 720$ terms. Products of six dipole matrix elements appear in the numerators, whereas the denominators indicate one, two and three-photon resonances.

Formulas like (4.19) have to be improved in two ways. Firstly, we must introduce statistical factors describing the thermal averaging over the initial state $|i\rangle$. Secondly, we have to include the effect of damping. Both improvements can be achieved if the scheme of perturbation theory is applied to the density matrix in place of the wave functions [4.1]. General expressions for $\chi^{(3)}$ calculated on the basis of the density matrix procedure have been published by *Flytzanis* [4.10] as well as *Bloembergen* et al. [4.30]. As a result of damping, the linewidths in the imaginary part of $\chi^{(3)}$ become finite. Moreover, the number of terms in $\chi^{(3)}$ with different resonance denominators may double from 24 to 48. Additional two-photon resonances may occur, characterized by

$$\left.\begin{array}{r} \omega_1 - \omega_2 \\ \omega_p - \omega_2 \end{array}\right\} \approx \omega_{nm},$$ (4.22)

where both n and m label *excited* states [4.31].

4.1.3 Raman Resonances in $\chi^{(3)}$ and $\chi^{(5)}$

In this section we concentrate on the four experimental techniques summarized in Fig. 4.1, i.e., SRS, CARS, RIKES and spontaneous HRS. We relate the

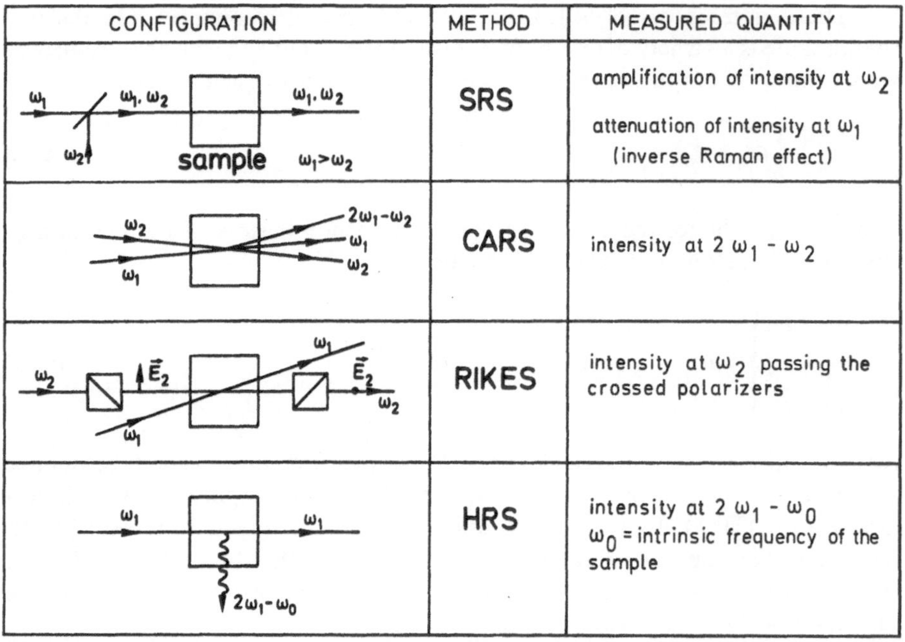

Fig. 4.1. Schematic description of SRS, CARS, RIKES, and spontaneous HRS

measured signal quantities to $\chi^{(3)}$ or $\chi^{(5)}$ and finally to the spontaneous Raman or hyper-Raman scattering cross section. For the sake of simplicity, we always assume the samples to be cubic crystals.

a) SRS

This technique uses two collinear beams of frequencies ω_1 and $\omega_2 < \omega_1$. While traversing the sample, each beam is amplified or attenuated by the other one. Let us consider only the change of $I(\omega_2)$ induced by $I(\omega_1)$. We may write

$$\frac{\partial I(\omega_2)}{\partial z} = gI(\omega_1)I(\omega_2), \tag{4.23}$$

where z indicates the direction of beam propagation and g represents the gain coefficient to be measured [note that $gI(\omega_1)$, as defined in (4.23), equals the g_R defined in (2.129)].

Relation (4.23) can be rigorously derived from (4.13, 14) if we use the ideas of Sect. 4.1.1 and interpret the change of $I(\omega_2)$ as a result of degenerate Stokes 4WM. With $\omega_p = \omega_1$ and $\omega_s = \omega_1 - (\omega_1 - \omega_2) = \omega_2$, the amplitude of the third-order polarization may be written in the form

$$P_\alpha^{(3)}(\omega_2, \boldsymbol{k}_2) = \sum_\delta \left[D \sum_{\beta,\gamma} \chi_{\alpha\beta\gamma\delta}^{(3)}(-\omega_2, \omega_1, -\omega_1, \omega_2) E_\beta(\omega_1, \boldsymbol{k}_1) E_\gamma^*(\omega_1, \boldsymbol{k}_1) \right]$$
$$\cdot E_\delta(\omega_2, \boldsymbol{k}_2). \tag{4.24}$$

Being linear in the electric field at ω_2, $P^{(3)}(\omega_2, k_2)$ looks like a small intensity induced correction to the first-order polarization amplitude $P^{(1)}(\omega_2, k_2)$. Hence, the expression in the square bracket of (4.24) may be understood as in increment of the linear susceptibility $\chi^{(1)}_{\alpha\delta}(-\omega_2, \omega_2)$, i.e.,

$$\Delta\chi^{(1)}_{\alpha\delta}(-\omega_2, \omega_2) = \frac{8\pi}{cn(\omega_1)} D \sum_{\beta,\gamma} \chi^{(3)}_{\alpha\beta\gamma\delta}(-\omega_2, \omega_1, -\omega_1, \omega_2) I_{\beta\gamma}(\omega_1), \tag{4.25}$$

where we have introduced the refractive index $n(\omega_1)$ and the intensity tensor

$$I_{\beta\gamma}(\omega_1) = \frac{cn(\omega_1)}{8\pi} E_\beta(\omega_1, k_1) E_\gamma^*(\omega_1, k_1). \tag{4.26}$$

It is straightforward to transform the imaginary part of (4.25) into an absorption constant or into a gain coefficient in accordance with (4.23). Assuming the waves at ω_1 and ω_2 to be polarized parallel to the β- and α-axis, respectively, we obtain [4.11, 18]

$$g = \frac{32\pi^2\omega_2}{c^2 n(\omega_1) n(\omega_2)} [6\text{Im}\{\chi^{(3)}_{\alpha\beta\beta\alpha}(\omega_2, -\omega_1, \omega_1, -\omega_2)\}]. \tag{4.27}$$

Here we have used $D = 6$ and $\chi^{(3)}(-\omega_2, \omega_1, -\omega_1, \omega_2) = \chi^{(3)*}(\omega_2, -\omega_1, \omega_1, -\omega_2)$ [4.10]. Equation (4.27) is equivalent to (2.130) except for a change in sign which is related to a different choice of the sign of the three frequencies in $\chi^{(3)}$.

Much insight is gained by inserting the perturbation theory result (4.19) into (4.27). Near a Raman resonance, (4.19) reduces to

$$\chi^{(3)}_{\alpha\beta\beta\alpha}(\omega_2, -\omega_1, \omega_1, -\omega_2) = \chi^{NR}_{\alpha\beta\beta\alpha} + \chi^{R}_{\alpha\beta\beta\alpha} \tag{4.28}$$

with

$$\chi^{R}_{\alpha\beta\beta\alpha} = \left(\frac{LN}{6\hbar^3}\right) \frac{1}{[\omega_0 - (\omega_1 - \omega_2)]} \left| \sum_n \frac{\langle i|M_\alpha|n\rangle\langle n|M_\beta|f\rangle}{\omega_{ni} + \omega_2} + \frac{\langle i|M_\beta|n\rangle\langle n|M_\alpha|f\rangle}{\omega_{ni} - \omega_1} \right|^2. \tag{4.29}$$

The Raman part χ^R combines all terms with a resonance denominator of the form $[\omega_0 - (\omega_1 - \omega_2)]$, while the non-Raman part χ^{NR} contains the remaining contributions. $|f\rangle$ denotes the final state of the Raman transition so that $\omega_0 = \omega_{fi}$.

According to (4.29), χ^R turns out to be proportional to the absolute square of the Raman *transition* polarizability $A^{fi}(\omega_1)$ [4.32, 33] if we correct the denominator slightly, i.e., $\omega_{ni} + \omega_2$ to $\omega_{ni} + \omega_1$. Taking into account the factor of $1/2$ in our definition of complex amplitudes [see (4.12)], we obtain

$$\chi^{R}_{\alpha\beta\beta\alpha} = \frac{LN}{24\hbar[\omega_0 - (\omega_1 - \omega_2)]} |A^{fi}_{\alpha\beta}(\omega_1)|^2. \tag{4.30}$$

Equation (4.30), in egs units, is equivalent to (2.133b), which was given in SI units.

A familiar expression for the Raman transition polarizability is provided by Placzek's approximation [4.32]. We have

$$A^{fi}(\omega_1) = \left(\frac{\partial \alpha}{\partial Q}\right)_0 \langle f|Q|i\rangle,$$

(4.31)

where α is the electronic polarizability and Q the normal coordinate of the Raman mode. $(\partial \alpha/\partial Q)_0$ may be referred to as an electron-mode coupling parameter as has been done in Sect. 4.1.1.

At this point we are ready to introduce the spontaneous Raman scattering efficiency S_R defined as a differential cross section per unit volume (see Sect. 2.2.8). S_R is related to $A^{fi}(\omega_1)$ by [4.33]

$$S_R = \frac{1}{V}\frac{d\sigma}{d\Omega} = \left(\frac{\omega_2}{c}\right)^4 LN|A^{fi}_{\alpha\beta}(\omega_1)|^2,$$

(4.32)

where α and β are the polarization directions of the incident and the scattered light. Insertion of (4.32) into (4.30) yields

$$\chi^R_{\alpha\beta\beta\alpha} = \frac{1}{24\hbar}\left(\frac{c}{\omega_2}\right)^4 \frac{S_R}{\omega_0 - (\omega_1 - \omega_2) - i\Gamma},$$

(4.33)

where we have complemented the resonance denominator by an appropriate damping constant. More or less the same result is achieved if (4.19) is replaced by the more elaborate expression based on density matrix calculations.

In most SRS experiments, the non-Raman background does not contribute to loss or gain. Then χ^{NR} may be taken as real and (4.27, 33) lead to [4.11, 12, 34]

$$g = \frac{8\pi^3 c^2}{\hbar n(\omega_1)n(\omega_2)\omega_2^3} S_R \mathcal{F}(\omega_1 - \omega_2),$$

(4.34)

where $\mathcal{F}(\omega)$ is the normalized line-shape function with $\int \mathcal{F}(\omega)d\omega = 1$ [See also (2.130, 133)]. According to (4.33) we have

$$\mathcal{F}(\omega) = \frac{\Gamma}{\pi[(\omega_0 - \omega)^2 + \Gamma^2]}.$$

(4.35)

In applying (4.34), we are not restricted to Lorentzian profiles because (4.35) is only a consequence of the *ad hoc* method by which damping is introduced into $\chi^{(3)}$. We can determine any Raman line shape by measuring g as function of $(\omega_1 - \omega_2)$.

We have taken the route via $\chi^{(3)}$ in order to illustrate the general scheme for describing the techniques of CRS and HRS. As a first step we have related the measured quantity to the underlying susceptibility, [see (4.27)]. As a second step we have simplified the lengthy expression for the susceptibility by extracting the essential resonances, [see (4.33)]. We shall not repeat these steps in the discussion of CARS, RIKES, etc., but only quote the results.

Of course, formula (4.34) can be derived in a much easier way by applying Fermi's golden rule. Spontaneous and stimulated Raman scattering only differ in the initial signal photon number and in the density ϱ_f of final photon states. For the spontaneous process, ϱ_f becomes identical with the well-known spectral density of electromagnetic vacuum oscillators. In the stimulated process, on the other hand, the emission of signal photons is confined to those few modes which are already occupied. Then ϱ_f is determined by the line-shape function $\mathscr{F}$. In fact, the ratio g/S_R can be directly deduced from this difference in ϱ_f [4.2, 33].

As suggested by (4.34), an absolute measurement of g can yield an absolute value of the scattering efficiency S_R ([4.35], see also Sect. 2.1.18d). As far as pulsed lasers are used, the accuracy of such a method is rather limited because the result depends on often unknown laser parameters. In particular, it is influenced by the mode structures and the overlap of the two interacting beams. Both conditions are difficult to control and may even fluctuate within a series of laser shots [4.34].

Nevertheless, formula (4.34) opens an access to extremely narrow Raman lines which can hardly be resolved by conventional Raman spectroscopy because their linewidth is less than about 0.1 cm^{-1} [4.36, 37]. As has been demonstrated by *Owyoung* et al. [4.38], the situation is much improved if cw instead of pulsed lasers can be used. In general, however, the comparatively low power of cw lasers requires a multipass arrangement with more than 50 passes through the sample. Such a technique can be handled in the case of gases confined to appropriate cells, but does not seem to be readily applicable to solids.

b) CARS

This method utilizes the generation of a new and separate beam of frequency $\omega_s = 2\omega_1 - \omega_2$ and wave vector $k_s = 2k_1 - k_2$. In order to derive an expression for the signal intensity, we have to solve the inhomogeneous wave equation (4.14) and adapt the solution to the boundary conditions of the experiment. We obtain [4.14, 18]

$$I_{\alpha\alpha}(\omega_s) \sim \frac{\omega_s^2}{c^4 n^2(\omega_1)n(\omega_2)n(\omega_s)}|\chi_{\alpha\beta\beta\alpha}^{(3)}(-\omega_s,\omega_1,\omega_1,-\omega_2)|^2 I_{\beta\beta}^2(\omega_1)I_{\alpha\alpha}(\omega_2)G^2(\Delta k),$$

$$(4.36)$$

where the intensities are defined according to (4.26). The function $G(\Delta k)$ describes the effect of the phase-mismatch Δk on the signal intensity. It has the form

$$G(\Delta k) \approx \frac{\sin l|\Delta k|}{|\Delta k|} \qquad (4.37)$$

and becomes identical with the sample length l if $\Delta k = 0$.

As in the section on SRS, we have assumed the incident waves at ω_1 and ω_2 to be polarized parallel to the β and α-axis, respectively. Hence, their electric fields are only allowed to be parallel ($\alpha = \beta$) or perpendicular ($\alpha \neq \beta$) to one another. More complex configurations are considered in "polarization active spectroscopy", an ellipsometric variant of CARS [4.14].

In analogy to (4.28, 29), we can decompose $\chi^{(3)}_{\alpha\beta\beta\alpha}(-\omega_s, \omega_1, \omega_1, -\omega_2)$ into a Raman and a background part χ^R and χ^{NR}, respectively. Approximating ω_s and ω_2 in the one-photon resonance denominators by ω_1, we can again express χ^R in terms of the Raman scattering efficiency S_R. According to (4.21), however, we have to take into account two degenerate Raman resonances characterized by $\omega_1 - \omega_2 \approx \omega_0$ and $\omega_p - \omega_2 = \omega_1 - \omega_2 \approx \omega_0$. Therefore, χ^R turns out to be just twice as large as (4.33) [4.18, 39]. On the other hand, we should mention that the factor D in front of $\chi^{(3)}$ [see (4.13)] reduces from 6 to 3 because there are only three distinguishable permutations of the input field amplitudes.

As will be discussed in Sect. 4.2.1, the intensity $I_{\alpha\alpha}(\omega_s)$, measured as a function of $(\omega_1 - \omega_2)$, does not follow the Raman lineshape $\mathscr{F}(\omega_1 - \omega_2)$. The squaring of $\chi^{(3)}$ in (4.36) introduces a crossterm which has a characteristic dispersion shape and deforms the profile of the Raman resonance considerably.

c) RIKES

This technique is similar to SRS and also based on (4.25). It exploits the intensity induced birefringence and dichroism or a mixture of both [4.40]. For the sake of simplicity, let us assume collinear beams with initial amplitudes of the form

$$E(\omega_1, k_1) = \frac{1}{\sqrt{2}}[E(1), E(1), 0]$$

$$E(\omega_2, k_2) = [E(2), 0, 0].$$

(4.38)

Then, (4.25) yields diagonal and nondiagonal increments of the linear suscepti-bility at ω_2. The most important one is

$$\Delta\chi^{(1)}_{21}(-\omega_2, \omega_2) = \frac{24\pi}{cn(\omega_1)}[\chi^{(3)}_{2121}(-\omega_2, \omega_1, -\omega_1, \omega_2) + \chi^{(3)}_{2211}(-\omega_2, \omega_1, -\omega_1, \omega_2)]I(\omega_1),$$

(4.39)

where $I(\omega_1) = I_{11}(\omega_1) + I_{22}(\omega_1)$.

The real and imaginary parts of $\Delta\chi^{(1)}_{21}(-\omega_2, \omega_2)$ describe the Kerr effect birefringence and dichroism (here: anisotropy of the gain coefficient), re-spectively. As well known from crystal optics, either of these phenomena rotates the direction of $E(\omega_2, k_2)$ whereas the presence of both leads to elliptical polarization. In any case, we can observe an intensity $I_{22}(\omega_2)$ polarized perpendicular to the incident one $I_{11}(\omega_2)$. We have

$$I_{22}(\omega_2) = \frac{4\pi^2\omega_2^2}{c^2n^2(\omega_2)}l^2|\Delta\chi^{(1)}_{21}(-\omega_2, \omega_2)|^2 I_{11}(\omega_2),$$

(4.40)

where l is again the crystal length. We can rather easily detect $I_{22}(\omega_2)$ by placing the sample between crossed polarizers as indicated in the third row of Fig. 4.1. Since the signal intensity depends on the absolute square of $\chi^{(3)}$, we expect a similar lineshape as in CARS [4.41].

Finally, let us mention a technique which also belongs to RIKES, but exclusively reveals the *real* part of $\chi^{(3)}$ or χ^R. Using a Jamin-type two-beam interferometer, *Owyoung* and *Peercy* [4.42] have been able to measure the change $\Delta n(\omega_2)$ of the refractive index $n(\omega_2)$ induced by $I(\omega_1)$. Returning to configurations where the two interacting beams are polarized parallel to the α and β-axis, we may write

$$\Delta n(\omega_2) = \frac{96\pi^2}{cn(\omega_1)n(\omega_2)} I(\omega_1) \operatorname{Re}\{\chi^{(3)}_{\alpha\beta\beta\alpha}(\omega_2, -\omega_1, \omega_1, -\omega_2)\}. \tag{4.41}$$

Inserting (4.28, 33), we obtain

$$\Delta n(\omega_2) \sim \operatorname{Re}\{\chi^{NR}_{\alpha\beta\beta\alpha}\} + \frac{S_R}{24\hbar}\left(\frac{c}{\omega_2}\right)^4 \frac{[\omega_0 - (\omega_1 - \omega_2)]^2}{[\omega_0 - (\omega_1 - \omega_2)]^2 + \Gamma^2}. \tag{4.42}$$

Hence, $\Delta n(\omega_2)$, measured as function of $(\omega_1 - \omega_2)$, shows a dispersion-type profile which in general may be interpreted as a Kramers-Kronig transform of $\mathscr{F}(\omega_1 - \omega_2)$ superimposed on a constant background.

d) HRS

Combining (4.27, 34), we can relate the spontaneous Raman scattering efficiency S_R to $\chi^{(3)}$. We have

$$S_R\mathscr{F}(\omega_1 - \omega_2) = \frac{4\hbar}{\pi}\left(\frac{\omega_2}{c}\right)^4 D \operatorname{Im}\{\chi^{(3)}_{\alpha\beta\beta\alpha}(\omega_2, -\omega_1, \omega_1, -\omega_2)\}, \tag{4.43}$$

where we have assumed the incident and the scattered waves to be polarized along α and β, respectively. The product on the left hand side is usually called *spectral* efficiency and defined by (see Sect. 2.2)

$$S_R\mathscr{F}(\omega_1 - \omega_2) = \frac{1}{V}\frac{d^2\sigma}{d\Omega d(\omega_1 - \omega_2)}. \tag{4.44}$$

In the case of HRS, the analogue to (4.43) becomes

$$S_{HR}\mathscr{F}(2\omega_1 - \omega_2) = \frac{32\hbar}{cn(\omega_1)} I(\omega_1)\left(\frac{\omega_2}{c}\right)^2$$

$$\cdot D \operatorname{Im}\{\chi^{(5)}_{\alpha\beta\beta\beta\beta\alpha}(\omega_2, -\omega_1, \omega_1, -\omega_1, \omega_1, -\omega_2)\}. \tag{4.45}$$

As before, ω_2 is the frequency of the scattered light on the Stokes side, i.e., $2\omega_1 - \omega_2 \approx \omega_0$. To keep the geometrical factors simple, we have assumed the incident beam of frequency ω_1 to be linearly polarized parallel to the β-axis, although the four indices of $\chi^{(5)}$ associated with the exciting light allow a large variety of scattering configurations.

Akhmanov and his coworkers [4.14] also used the term hyper-Raman effect to specify a one-photon resonant 4WM process depending on the hyper-Raman efficiency S_{HR}. While scanning a tunable source of frequency ω_2 through an infrared-active transition at ω_0, they detected the intensity at $\omega_s = 2\omega_1 - \omega_2$ generated by the coherent interaction of the beam at ω_2 and another pump of frequency ω_1. In accordance with (4.36), the signal intensity is given by

$$I(\omega_s) \sim |\chi^{(3)}_{\alpha\beta\beta\alpha}(-\omega_s, \omega_1, \omega_1, -\omega_2)|^2 . \tag{4.46}$$

Nevertheless, the method is quite different from CARS because it exploits a one-photon instead of a two-photon resonance. Referring to (4.19), we may extract all terms from $\chi^{(3)}$ which have a resonance denominator of the form $\omega_0 - \omega_2 = \omega_{fi} - \omega_2$. If ω_1 is small compared to electronic transition frequencies, we obtain

$$\chi^{HR}_{\alpha\beta\beta\alpha} = \frac{4LN}{3\hbar} \frac{\langle f|M_\alpha|i\rangle}{(\omega_0 - \omega_2)} [B^{fi}_{\alpha\beta}(\omega_1)]^* . \tag{4.47}$$

Here $B^{fi}(\omega_1)$ denotes the hyper-Raman *transition* polarizability [4.43–45]. It is determined by products of three dipole matrix elements relating $|i\rangle$ to $|f\rangle$ via two intermediate states. In perfect analogy to (4.31), Placzek's approximation yields

$$B^{fi}(\omega_1) = \left(\frac{\partial\beta}{\partial Q}\right)_0 \langle f|Q|i\rangle , \tag{4.48}$$

where Q is the normal coordinate of the mode and $(\partial\beta/\partial Q)_0$ the electron-mode coupling parameter already introduced by (4.8). We also have

$$S_{HR} = \left(\frac{\omega_2}{c}\right)^4 LN|B^{fi}_{\alpha\beta}(\omega_1)|^2 |E(\omega_1, k_1)|^2 . \tag{4.49}$$

Due to the definition of S_{HR}, the intensity is left as a factor on the right hand side, just as in (4.45).

According to (4.47, 49), the absolute square of χ^{HR} is indeed proportional to the hyper-Raman scattering efficiency. It would follow the profile of the infrared absorption if it could be measured independently. In general, however, the presence of a pedestal term to be added to χ^{HR} spoils this feature. Moreover, the single dipole matrix element in (4.47) confines the method to infrared-active transitions and, in contrast to spontaneous HRS, does not allow the study of "silent" modes.

4.2 Coherent Antistokes Raman Spectroscopy (CARS)

4.2.1 Line Shapes

Figure 4.2 shows the energy level scheme usually adopted for describing the CARS process. The dashed lines indicate virtual states while the solid lines denote the initial and final state of the Raman transition. It is clear from the foregoing section that the scheme illustrates a one-step process and should never be interpreted as the time-ordered sequence of Stokes and antistokes Raman scattering.

A typical CARS spectrum is reproduced in Fig. 4.3 [4.39, 46]. It refers to the totally symmetric A_{1g} mode of the CO_3-ion in calcite ($CaCO_3$) at $\omega_0 = 1088\,\mathrm{cm}^{-1}$. The logarithm of the signal intensity has been plotted as a function of $\omega_0 - (\omega_1 - \omega_2)$. The observed contour drastically deviates from the Lorentzian line shape measured by spontaneous Raman spectroscopy. In particular, an additional minimum or antiresonance appears on either side of the maximum. Moreover, the signal intensity does not vanish far from resonance, but forms a constant background usually normalized to unity. All these features can be easily explained by the structure of $\chi^{(3)}$ and its squaring in (4.36). Omitting tensor indices, we may write

$$I(\omega_s) \sim |\chi^{(3)}|^2 = |\chi^{NR} + \chi^R|^2 = |\chi^{NR}|^2 + |\chi^R|^2 + 2\mathrm{Re}\{\chi^R(\chi^{NR})^*\}. \qquad (4.50)$$

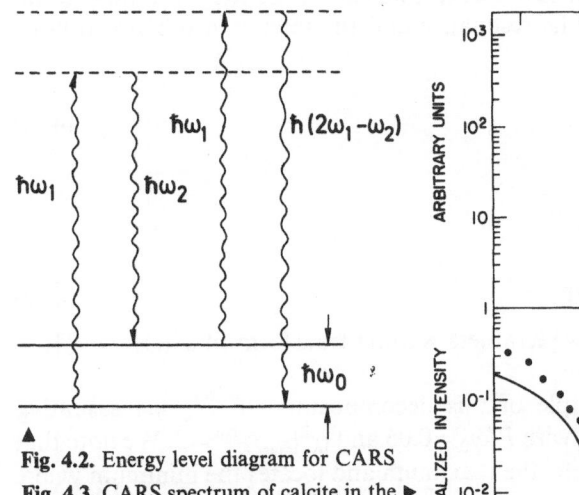

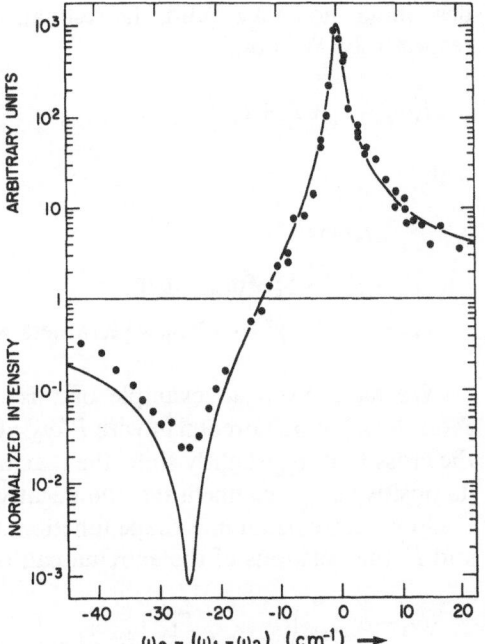

Fig. 4.2. Energy level diagram for CARS

Fig. 4.3. CARS spectrum of calcite in the ▶ frequency range of the internal A_{1g} mode at $\omega_0 = 1088\,\mathrm{cm}^{-1}$. The dots are experimental points. The solid line is a theoretical fit based on (4.50) and (4.33) with $\Gamma = 0.75\,\mathrm{cm}^{-1}$ and $|\chi^R|_{max}/\chi^{NR} = 67$. $\lambda_1 = 530\,\mathrm{nm}$ [4.46]

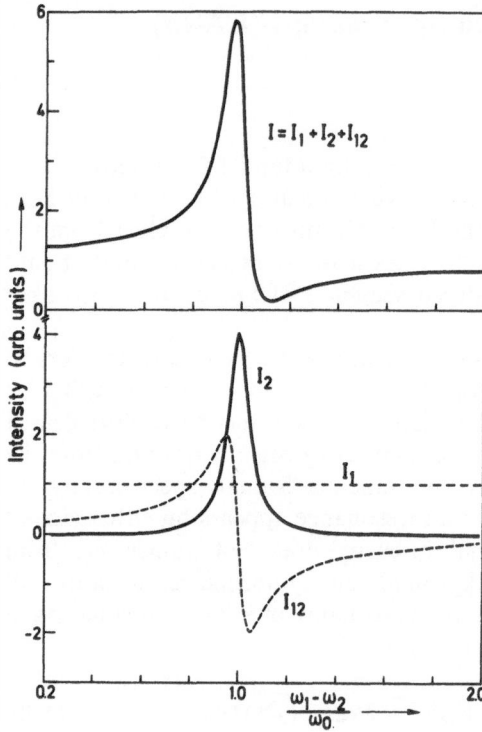

Fig. 4.4. Decomposition of the CARS intensity I into its three components I_1, I_2, and I_{12} according to (4.51, 52). Parameters have been chosen to be $\Gamma/\omega_0 = 0.05$ and $|\chi^R|_{max}/\chi^{NR} = 2$

If χ^{NR} is real and independent of frequency, we can divide $I(\omega_s)$ into three terms describing the background, the resonance and the interference between both, respectively. We obtain

$$I(\omega_s) = I_1 + I_2 + I_{12} \tag{4.51}$$

with

$$
\begin{aligned}
I_1 &\sim (\chi^{NR})^2 \\
I_2 &\sim |\chi^R|^2 \sim S_R^2 \mathscr{F}(\omega_1 - \omega_2) \\
I_{12} &\sim \chi^{NR} \operatorname{Re}\{\chi^R\} \sim \chi^{NR} S_R \times [\text{Kramers-Kronig transform of } \mathscr{F}(\omega_1 - \omega_2)].
\end{aligned}
\tag{4.52}
$$

$\vdots$

We have given an example of this decomposition in Fig. 4.4 assuming $\mathscr{F}(\omega_1 - \omega_2)$ to be Lorentzian with $\Gamma/\omega_0 = 0.05$ and $|\chi^R|_{max}/\chi^{NR} = 2$. We note that the cross term I_{12} slightly shifts the maximum and locates the minimum where its positive slope compensates the negative one of I_2.

For a Lorentzian line-shape function and arbitrary parameters χ^{NR}, S_R, ω_0, and Γ, the positions of the extrema can be easily calculated [4.39]. We find

$$
\left.\begin{aligned}
(\omega_1 - \omega_2)_{min} \\
(\omega_1 - \omega_2)_{max}
\end{aligned}\right\} = \omega_0 + \frac{\Gamma|\chi^R|_{max}}{2\chi^{NR}} \left\{ 1 \pm \left[1 + \left(\frac{2\chi^{NR}}{|\chi^R|_{max}} \right)^2 \right]^{1/2} \right\}
\tag{4.53}
$$

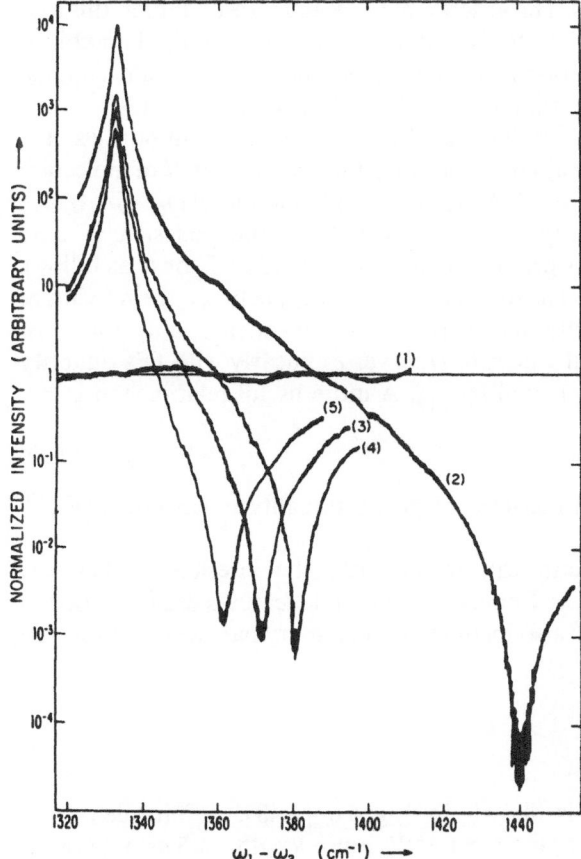

Fig. 4.5. CARS spectrum of diamond. The numbers in parentheses refer to different polarization conditions. $\omega_0 = 1332\,\text{cm}^{-1}$; $\lambda_1 = 545\,\text{nm}$ [4.39]

with

$$\Gamma|\chi^{R}|_{\max} = \frac{1}{12\hbar}\left(\frac{c}{\omega_2}\right)^4 S_R > 0. \tag{4.54}$$

Accordingly, the sign of χ^{NR} decides on which side of the maximum the antiresonance is observed. The value of χ^{NR} follows immediately from the distance $[(\omega_1 - \omega_2)_{\min} - (\omega_1 - \omega_2)_{\max}]$ between minimum and maximum, provided the damping constant Γ and the spontaneous Raman scattering efficiency S_R are known.

Figure 4.5 demonstrates the influence of χ^{NR} on the extrema of $I(\omega_s)$. Several CARS spectra of diamond ($\omega_0 = 1332\,\text{cm}^{-1}$) have been plotted referring to different polarizations of the incident fields and hence to different tensor elements of χ^{NR} [4.39]. While the maximum remains nearly stationary at ω_0, the mimimum shifts considerably. Such behavior is in accordance with (4.53) if $|\chi^{R}|_{\max}$ is large compared to χ^{NR}. As shown by curve (1), the CARS intensity reduces to the constant background as soon as both input beams are polarized parallel to the same cubic axis. The absence of the Raman resonance in this case results from the special form of the Raman tensor or transition polarizability. The F_{2g} symmetry of the phonon under consideration requires all diagonal elements $A_{\alpha\alpha}^{fi}(\omega_1)$ to vanish so that $\chi_{\alpha\alpha\alpha\alpha}^R = 0$ [see (4.31) and Sect. 2.1.9].

If CARS is used to detect small concentrations of gases or impurities, it is essential to know the relationship between $I(\omega_s)$ and the number N of molecules or unit cells per unit volume [4.15]. As I_{12} is an odd function of $\omega_0 - (\omega_1 - \omega_2)$, the integrated CARS intensity, after subtraction of the pedestal, is only determined by I_2 and becomes proportional to the square of S_R or N as follows from the second row of (4.52). The difference $(I_{max} - I_{min})$ between the maximum and minimum CARS intensity shows the same dependence only for large values of N, whereas for small values of N, it varies linearly with this quantity because the line shape is dominated by I_{12}. A more useful relationship is

$$(I_{max})^{1/2} - (I_{min})^{1/2} \sim N \tag{4.55}$$

which holds for all N with a constant of proportionality independent of χ^{NR} [4.15].

Rather complicated spectra can result from the interference between neighboring Raman resonances. Some of the possible features are described in [4.15, 47]. The case where χ^{NR} also includes an imaginary part will be discussed in Sect. 4.4.

4.2.2 Experimental Problems

The classical CARS setup is shown in Fig. 4.6 [4.48]. A nitrogen laser (wavelength: 337.1 nm, peak power: ≈ 1 MW, pulse length: ≈ 5 ns, repetition rate: ≈ 10 Hz) simultaneously pumps two dye lasers. These tunable sources (peak power: $10 - 100$ kW, linewidth: < 1 cm^{-1}) provide the two interacting beams of frequencies ω_1 and ω_2. The nitrogen laser can be replaced by the second harmonic of a Q-switched Nd-YAG laser [4.8]. The better beam quality and more convenient wavelength of 532.4 nm permit the use of this radiation also as a ω_1-beam, so that a dye laser is only needed for tuning ω_2.

Figure 4.7 lists some wave vector configurations which satisfy the phase-matching condition $\Delta k = 0$. Small dispersion often prevents the signal beam from being clearly separated from the input ones, so that the wave vector triangle almost collapses to a straight line and spatial filtering or angular resolution cannot be realized. The situation is much improved by splitting the ω_1-beam into portions of wave vectors k_1' and k_1'' having the same length but different directions. Then the phase-matching condition is generalized to

$$k_s = k_1' + k_1'' - k_2 \tag{4.56}$$

and can be fulfilled with more degrees of freedom than the original one. Three of the new possibilities are shown in Fig. 4.7. They utilize counter-propagating

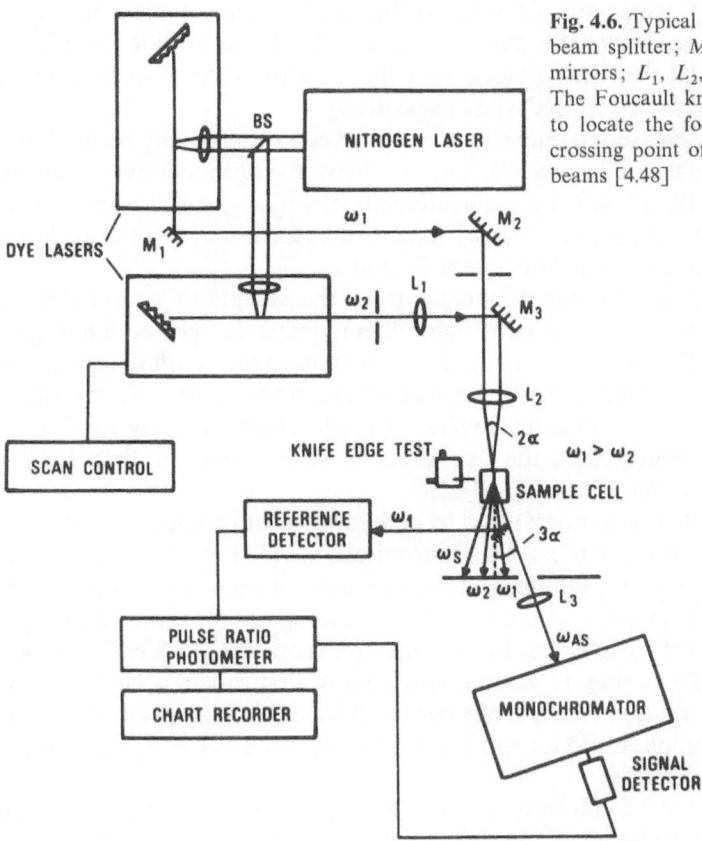

Fig. 4.6. Typical CARS setup. BS is a beam splitter; M_1, M_2, and M_3 are mirrors; L_1, L_2, and L_3 are lenses. The Foucault knife-edge test is used to locate the focal volume and the crossing point of the two interacting beams [4.48]

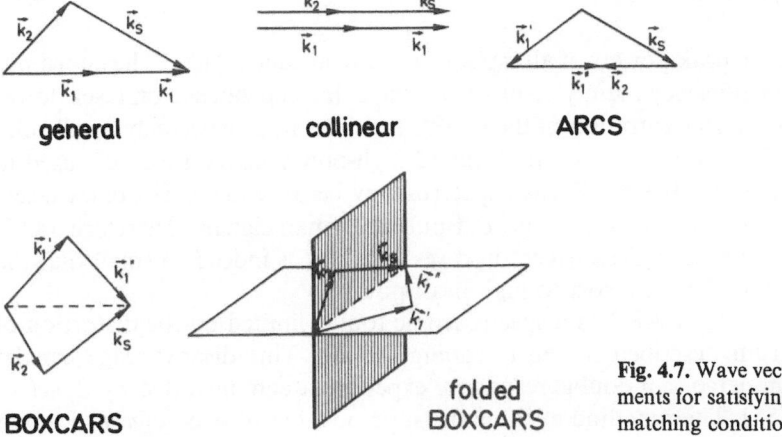

Fig. 4.7. Wave vector arrangements for satisfying the phase-matching condition in CARS

beams [4.49, 50], a box-shaped wave vector configuration [4.51] or even a nonplanar scheme, where the pairs k'_1, k''_1 and k_2, k_s lie in different planes. Inventors have labeled them ARCS (angular resolved CARS spectroscopy), BOXCARS, and folded BOXCARS, respectively.

In order to increase the intensity, the input beams are usually focused into the sample. Special care has to be taken to achieve the optimum overlap of the beam waists [4.18, 39, 48]. In the equipment of Fig. 4.6, a lens with a focal length of about 20 cm simultaneously focuses the radiation of both dye lasers and also provides an angle 2α between k_1 and k_2.

Although the CARS signal emerges from the sample as a well defined coherent beam, the presence of stray light often requires the spectral filtering by a single or double monochromator. If ω_2 is tuned twice as fast as ω_1, the gratings of this instrument can be kept at a fixed position because the signal frequency $\omega_s = 2\omega_1 - \omega_2$ remains constant [4.39]. Otherwise, they have to be scanned synchronously with the dye lasers in such a manner that $I(\omega_s)$ is monitored as function of $(\omega_1 - \omega_2)$ [4.54].

The CARS intensity is normalized by comparison with a reference signal. In the spectrometer of Fig. 4.6, $I(\omega_s)$ is electronically divided by the cube of $I(\omega_1)$ because the simultaneous pumping of the dye lasers suggests $I(\omega_2) \sim I(\omega_1)$ and hence, $I(\omega_s) \sim I^2(\omega_1) I(\omega_2) \sim I^3(\omega_1)$. A better averaging over the laser fluctuations is obtained by referring $I(\omega_s)$ to the nonresonant CARS emission of a material like NaCl having no Raman spectrum of first order [4.39].

In studying liquids and solids, one can easily achieve signal peak powers up to 1 W. Therefore, photodiodes and boxcar integrators are appropriate means for detecting $I(\omega_s)$.

Defining the CARS efficiency S_{CARS} in analogy to S_R, we can estimate the ratio S_{CARS}/S_R on the basis of (4.36, 54). In solids, S_R is typically 10^{-7} cm^{-1} sr^{-1} (see Sect. 2.1.18). With $\Gamma = 1$ cm^{-1}, this value leads to $|\chi^R|_{max} = 2 \times 10^{-13}$ cm^2 dyn^{-1}. Assuming the same order of magnitude for $|\chi^{(3)}|$, we find [4.14, 15, 18]

$$S_{CARS}/S_R \approx 10^6 - 10^9, \tag{4.57}$$

provided the peak powers of all input beams are around 50 kW. The enormous increase in efficiency mainly results from the cubic dependence on laser power and from the concentration of the CARS emission in an extremely small solid angle $d\Omega$. We note, however, that pulsed high-power lasers are rarely used in spontaneous steady-state Raman spectroscopy because the merits of cw lasers outweigh any advantage of large, but pulsed Raman signals. Therefore, (4.57) seems rather meaningless, except in cases where S_R is indeed so small that the spectroscopist has to resort to high peak powers.

The utility of CARS as a spectroscopic tool is limited by the distortion of the spectrum described in the foregoing section. This disadvantage can be partly eliminated in a double resonance experiment demonstrated by Lynch et al. [4.29, 55]. Their method exploits the superposition of two different Raman contributions χ^{R_1} and χ^{R_2} to $\chi^{(3)}$ in a nondegenerate 4 WM process. As follows from (4.21), no less than three independently tunable dye lasers of frequencies $\omega_p \neq \omega_1 \neq \omega_2$ are required. As a first step, $(\omega_p - \omega_2)$ is scanned across the Raman resonance R_1 and fixed at a position where χ^{NR} is cancelled by the real part of χ^{R_1}. At this position, only a small imaginary amount of χ^{R_1} is left and $\chi^{(3)}$ reduces to $ir_1 + \chi^{R_2}$. As a second step, $(\omega_1 - \omega_2)$ is tuned through the Raman resonance R_2 by varying ω_1. Then the signal intensity at $\omega_s = \omega_p + (\omega_1 - \omega_2)$, measured as function of $(\omega_1 - \omega_2)$, shows the undistorted Raman line shape with only a small pedestal due to r_1. Improved versions of this method have been developed by *Song* et al. [4.18, 56].

4.2.3 CARS Spectra of Solids

a) Early Work

Before the introduction of tunable lasers, CARS-type spectra of solids had been obtained by scanning the material resonance instead of the laser frequencies ω_1 and ω_2. *Yablonovitch* et al. [4.57] have exploited the tunability of Raman transitions between the Landau levels of the conduction electrons in n-InSb. In this case, the Raman frequency ω_0 is identical with the cyclotron frequency ω_c (or with $2\omega_c$) and thus can be varied with the applied magnetic field. When ω_c (or $2\omega_c$) is swept through the difference $(\omega_1 - \omega_2)$ of two input CO_2 laser frequencies, the CARS or CSRS intensity shows a resonance entirely dominated by the background contribution I_1 and by the cross term I_{12}. This observation is in accordance with (4.52) if χ^{NR} is large compared to $|\chi^R|$. The difference in magnitude can be quantitatively explained by a semiclassical model relating the existence of both parts of $\chi^{(3)}$ to the nonparabolicity of the conduction band [4.58].

The spin-flip process presents another tunable Raman transition and has been studied in a similar manner [4.59].

b) Centrosymmetric Crystals

The CARS spectra of centrosymmetric crystals have been used to determine χ^{NR} relative to known Raman scattering efficiencies S_R [4.39, 60–62]. By application of (4.53, 54), the ratio S_R/χ^{NR} can be easily obtained from the distance between maximum and minimum. The coherence of the interacting beams also allows two different materials to be combined in a "sandwich" and ratios of the form $S_{R_1}/(\chi^{NR_1} + \chi^{NR_2})$ or $S_{R_2}/(\chi^{NR_1} + \chi^{NR_2})$ to be measured, where the contributions of the two components are distinguished by indices. In this way, unknown Raman efficiencies can be calibrated by comparison to a standard [4.39, 61]. Although the necessity of placing two different samples consecutively in the same position is avoided, special care must be taken to provide an identical beam overlap throughout the "sandwich" (see also Sect. 2.1.18).

For insulators, the order of magnitude of χ^{NR} turns out to be $10^{-15} - 10^{-13}$ cm^2 dyn^{-1}. A knowledge of χ^{NR} is very helpful in evaluating the degree of self-focusing. This effect results from the change Δn of the refractive index induced by a single laser beam. As shown by (4.41), Δn is determined by the real part of $\chi^{(3)}(\omega_1, -\omega_1, \omega_1, -\omega_1)$ which is nearly identical with χ^{NR} in regions of small dispersion. Self-focusing seriously complicates the construction of large high-power laser systems and has to be taken into account in the interpretation of many studies on nonlinear optical phenomena [4.63].

c) Noncentrosymmetric Crystals

As already mentioned in Sect. 9.1.2, the second-order susceptibility $\chi^{(2)}$ of noncentrosymmetric crystals gives rise to cascade processes resulting in the same signal frequency as the one-step process described by $\chi^{(3)}$ [4.25, 64, 65]. Figure 4.8 shows the CARS spectrum of LiNbO$_3$ (point group symmetry C_{3v}) at room temperature [4.64]. The observed line shapes cannot be interpreted in terms of the foregoing formulas. The maxima appear at positions where $(k_1 - k_2)$ and $(\omega_1 - \omega_2)$ coincide with the wave vectors and frequencies of polaritons. Such matching conditions, however, clearly indicate that the polaritons are mainly driven by an infrared electric field generated by the input beams in a second-order mixing process. Hence, the nonlinear force of (4.2) involving the Raman transition polarizability no longer plays the dominant role, but is outweighed by the forces directly acting on the effective charges.

d) D4WM

In recent literature, the term Degenerate Four-Wave Mixing (D4WM) has been used to specify a method which in a sense may be regarded as a borderline case of CARS. The input beams and the signal have equal frequencies and the wave vectors are usually arranged in a cross-like configuration with $k_1' = -k_1''$ and $k_s = -k_2$ [see (4.56)]. The underlying susceptibility is that of CARS in the limit

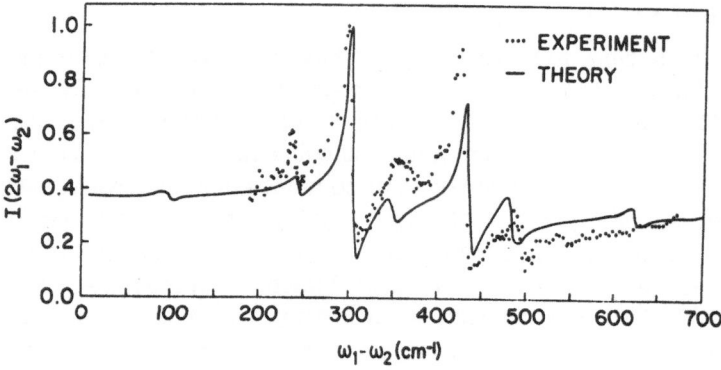

Fig. 4.8. CARS-type spectrum of LiNbO$_3$. Both curves are normalized to unity at their maxima. $\lambda_1 = 577$ nm [4.64]

of $\omega_2 \to \omega_1$, i.e., $\chi^{(3)}(-\omega_1, \omega_1, \omega_1, -\omega_1)$. D4WM has attracted much technological interest because it allows the generation of time-reversed or phase-conjugated wave fronts [4.66].

Of course, no Raman resonances can be studied by tuning ω_1. Refferring to (4.21), however, we note that $\chi^{(3)}(-\omega_1, \omega_1, \omega_1, -\omega_1)$ still includes contributions from TPA resonances characterized by $2\omega_1 \approx \omega_{fi}$. *Maruani* et al. [4.67] have applied D4WM to study the biexciton two-photon transition in CuCl. As the energy of the biexciton is nearly twice the energy of the single exciton, the signal is enhanced not only by a two-photon, but also by a one-photon resonance. It becomes so strong that it can already be detected with moderate laser powers in a nonphase matching configuration with $k'_1 = k''_1$. Moreover, higher-order scattering beams can be observed resulting from cascades via $\chi^{(3)}$ or from one-step processes involving higher-order susceptibilities up to about $\chi^{(9)}$. The measured spectrum is somewhat similar to the line shape encountered in CARS. It also shows an antiresonance caused by a background contribution. The background is interpreted as a continuum formed by the pairs of unrelated elementary excitations into which the biexciton can decay [4.68].

4.3 Raman-Induced Kerr Effect Spectroscopy (RIKES) and Related Techniques

In contrast to CARS, RIKES-type methods do not utilize the generation of new beams but exploit the intensity induced changes of the linear optical constants. Interpreting the gain coefficient as a negative absorption constant per unit power, we may also subsume the technique of SRS under this section.

As already outlined before, SRS has been successfully applied in the analysis of extremely narrow Raman lines. An example is given in Fig. 4.9 showing the line shape of the spin-flip Raman transition in InSb [4.37]. The gain coefficient has been plotted as a function of the magnetic field which tunes the Raman frequency ω_0 through the fixed frequency difference $\omega_1 - \omega_2 = 4.17 \text{ cm}^{-1}$ of two CO lasers. The relative stability of the two lasers is better than 1 MHz so that the spin-flip linewidth of about 200 MHz (0.007 cm^{-1}) can be easily resolved.

It is also promising to use SRS in the study of low frequency modes. Due to the directional nature of the laser beams, the interference of the elastic scattering can be reduced to a greater degree than in conventional Raman spectroscopy [4.69, 70].

Figure 4.10 demonstrates the present state of the art in RIKES [4.71]. The spectrum reveals the sharp peak near the two-phonon cutoff in diamond. This structure is well known from spontaneous Raman spectroscopy [4.72]. The background has been suppressed by a heterodyne technique based on the superposition of the signal field $E_2(\omega_2)$ and a "local oscillator" field E_{LO} with $|E_{LO}| \gg |E_2(\omega_2)|$ (concerning the signal field, see Fig. 4.1 and Sect. 4.1.3). In

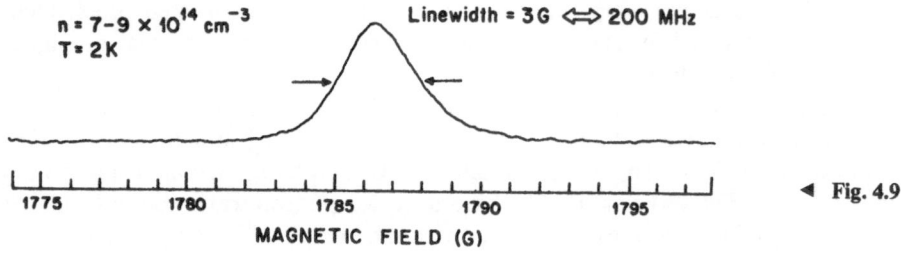

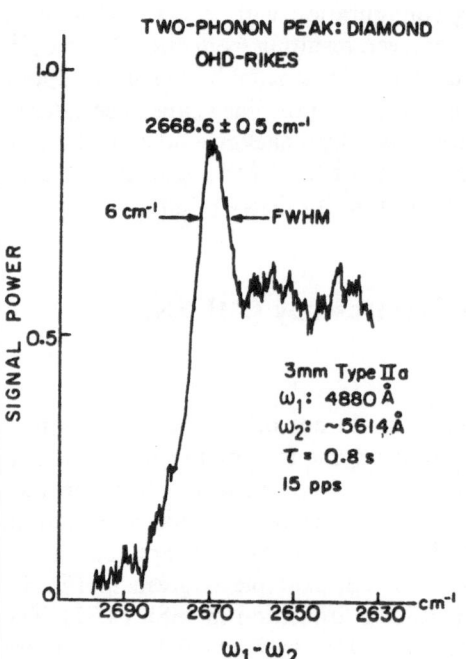

TWO-PHONON PEAK: DIAMOND
OHD-RIKES

2668.6 ± 0 5 cm⁻¹

6 cm⁻¹———←——FWHM

3mm Type Ⅱa
ω_1: 4880 Å
ω_2: ~5614 Å
τ = 0.8 s
15 pps

Fig. 4.9. Gain of the spin-flip Raman transition in n-InSb. The Raman frequency is linearly tuned through the frequency difference $\omega_1 - \omega_2 = 4.17\,\mathrm{cm}^{-1}$ of the input lasers. $\lambda_1 = 5.35\,\mu\mathrm{m}$ [4.37]

Fig. 4.10. Two-phonon cut off in diamond as revealed by RIKES. The background is suppressed by an optical heterodyne (OHD) technique [4.18, 71]

combination with electrical filtering, such a superposition leads to a signal which is proportional to the real part of the product $[E_{LO}^* E_2(\omega_2)]$ and hence to $\chi^{(3)}$. Depending on the phase relationship between E_{LO} and $E_2(\omega_2)$, the observed spectrum displays either the real or the imaginary part of $\chi^{(3)}$ without any distortion by cross terms [4.18, 71].

4.4 Double Resonance Interferences in Four-Wave Mixing

We have stated in (4.21) that $\chi^{(3)}$ includes both Raman and TPA resonance contributions. If the TPA transition can be characterized by a Lorentzian lineshape function with a well-defined transition frequency ω_{fi} and damping

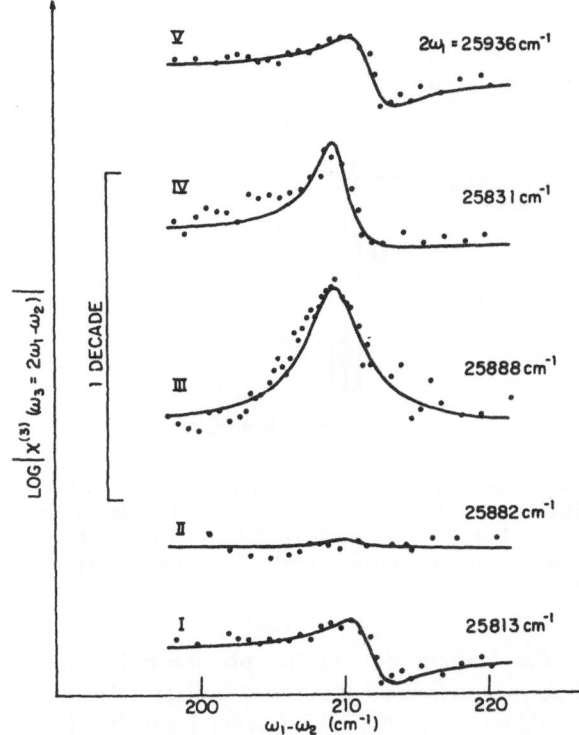

Fig. 4.11. Double resonance structure in the CARS spectrum of CuCl at 15.1 K. $\omega_1 - \omega_2$ is scanned through the Raman line of the phonon-polariton at 208 cm^{-1}, while $2\omega_1$ is varied in steps across the TPA line of the Z_3 exciton-polariton at 25,880 cm^{-1}. The solid curves represent a best fit to the experimental points based on a decomposition of $\chi^{(3)}$ into a background, TPA, and Raman part. $\lambda_1 = 773$ nm [4.55, 75]

constant Γ_{fi}, the TPA term may be written in analogy to (4.30) as [4.29, 73, 74]

$$\chi^{TPA}_{\alpha\beta\beta\alpha}(-\omega_s, \omega_p, \omega_1, -\omega_2) \approx \chi^{TPA}_{\alpha\beta\beta\alpha}(-\omega_1, \omega_1, \omega_1, -\omega_1)$$
$$= \frac{LN}{24\hbar} \frac{[T^{fi}_{\alpha\alpha}(\omega_1)]^* T^{fi}_{\beta\beta}(\omega_1)}{\omega_{fi} - 2\omega_1 - i\Gamma_{fi}}. \tag{4.58}$$

We have assumed the differences between $\omega_s, \omega_p, \omega_1, \omega_2$ to be negligibly small compared to the electronic transition frequencies ω_{ni}, so that the TPA contributions to CARS and D4WM become identical. $T^{fi}(\omega_1)$ represents the TPA transition polarizability referred to a single molecule or unit cell. It is given by

$$T^{fi}_{\alpha\alpha}(\omega_1) = \frac{2}{\hbar} \sum_n \frac{\langle i|M_\alpha|n\rangle \langle n|M_\alpha|f\rangle}{\omega_{ni} - \omega_1}, \tag{4.59}$$

where damping has been omitted.

Due to χ^{TPA}, the nonraman background χ^{NR} can become a complex quantity and additional cross terms have to be considered in the interpretation of CARS spectra [4.47]. A variety of possible distortions is shown in Fig. 4.11 presenting the results of a double resonance study on CuCl [4.55, 75]. On the one hand,

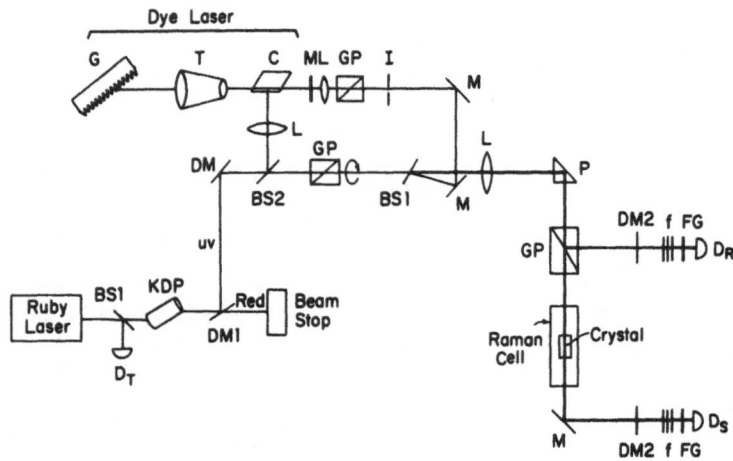

Fig. 4.12. Scheme of the experimental set up for measuring TPA coefficients relative to known Raman cross sections. M – mirror; DM – dichroic mirror; BS – beam splitter; GP – Glan prism; C – dye cell; T – telescope; G – grating; L – lens; P – prism; f – filter; FG – frosted glass; D_R, D_S – photodiode; D_T triggering diode; KDP – frequency doubling crystal. $\lambda_1 = 347.1$ nm [4.34]

$\omega_1 - \omega_2$ is scanned through the Raman line of the phonon-polariton at 208 cm^{-1}. On the other hand, $2\omega_1$ is tuned in steps across the sharp TPA line of the Z_3 exciton-polariton at 25,880 cm^{-1} (3.2087 eV). Curves I and V refer to values of $2\omega_1$ far from resonance and thus reveal the CARS spectrum undisturbed by TPA. In case III, however, χ^{TPA} adds a considerable imaginary amount to χ^{NR} which, according to (4.50, 52), enhances the resonance-type intensity portion I_2.

The interference of χ^R and χ^{TPA} in a double resonance experiment can be utilized to determine the ratio of $|T^{fi}(\omega_1)|$ and $|A^{fi}(\omega_1)|$ and to calibrate one of these quantities by comparison with the other [4.55, 76]. If the line shape of the TPA resonance is completely unknown, it seems to be more convenient to perform a double resonance experiment in the SRS instead of the CARS configuration. Then only the imaginary part of χ^{TPA} is measured and the Raman gain due to the imaginary part of χ^R is linearly superimposed by a TPA background. Hence, the relative magnitude of both can be easily deduced.

In Fig. 4.12 we show an experimental arrangement designed to measure TPA coefficients in the near uv relative to known Raman cross sections [4.34, 73, 74]. A typical result is reproduced in Fig. 4.13 [4.34]. Part (a) presents the profile of the 3062 cm^{-1} Raman line of liquid benzene. The absolute *spectral* differential cross section obtained from spontaneous Raman scattering has been plotted as a function of the Raman shift ω_R. Part (b) shows the relative change of the intensity $I_2 = I(\omega_2)$ due to the interaction with $I(\omega_1)$. As the frequency sum $(\omega_1 + \omega_2)$ already lies in the TPA region of benzene, the Raman gain is shifted down by a TPA background. The TPA coefficient follows from the known Raman cross section and from the ratio of Raman and TPA

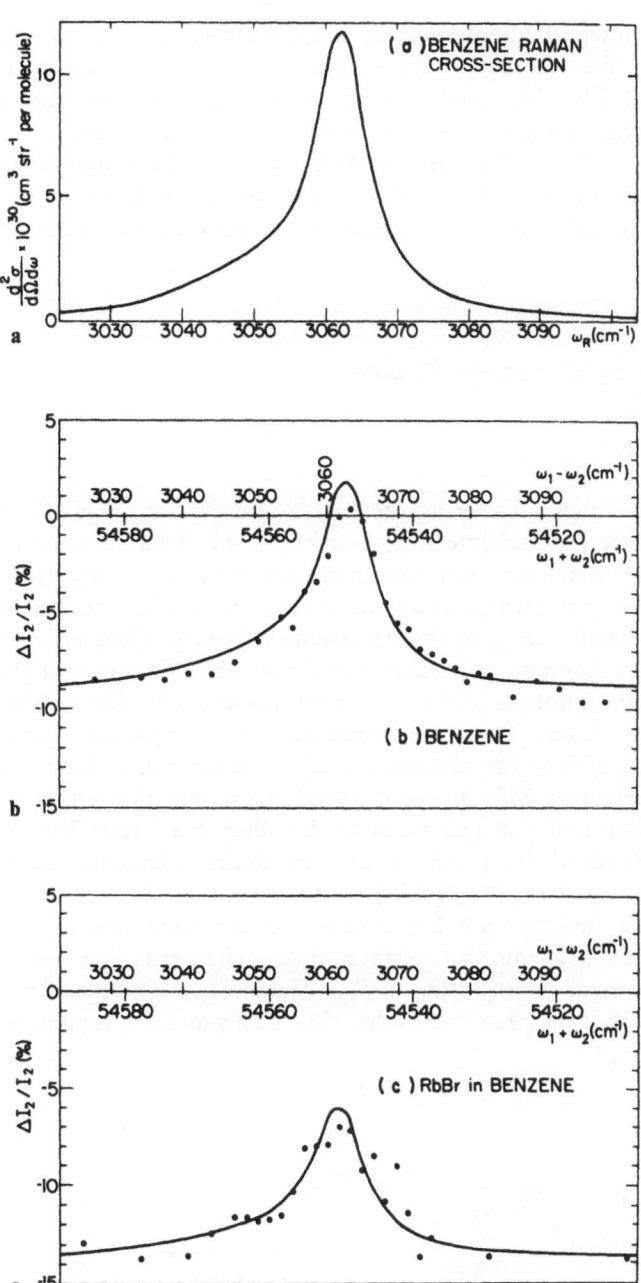

Fig. 4.13. Calibration of TPA coefficients by comparison with known Raman cross sections. (a) Spectral differential cross section of benzene for excitation at 347.1 nm. ω_R = Raman shift. (b) Superposition of Raman gain and TPA loss in benzene. The lineshape of the Raman gain (solid line) is taken from (a). (c) Superposition of Raman gain and TPA loss in benzene as well as TPA loss in RbBr. λ_1 = 347.1 nm [4.34]

contributions to the integral attenuation or amplification, respectively. As illustrated by part (c), the Raman profile is shifted still more if a crystal is immersed in the liquid. The TPA coefficient of this additional sample can be derived from the increase of the background relative to the Raman gain.

Such calibrations of TPA coefficients seem to be more accurate than direct measurements. They are not sensitively influenced by the beam structure and overlap because Raman gain and TPA attenuation depend on these factors in exactly the same way.

4.5 Hyper-Raman Spectroscopy (HRS)

4.5.1 Selection Rules

The process of hyper-Raman scattering (here also abbreviated as HRS) is usually illustrated by the level scheme shown in Fig. 4.14a. If the elementary excitation is a phonon, we can extract the electron-lattice interaction $\mathcal{H}_{EL}$ from the Hamiltonian of the unperturbed system and combine it with the dipole electron-radiation interaction $\mathcal{H}_{ER}$ to the perturbation energy. Then HRS is described by four-vertex diagrams as presented in Fig. 4.14b. The wavy and the broken lines indicate the photons and the phonon, respectively. The bubble denotes the intermediate states. These are virtual electron-hole pairs coupling to the photons via $\mathcal{H}_{ER}$ and to the phonon via $\mathcal{H}_{EL}$. We note that the input photons do not generate two different electron-hole pairs, but that either of them acts on the electron-hole pair generated by the other. The hyper-Raman efficiency S_{HR} is determined by products of four matrix elements, three involving $\mathcal{H}_{ER}$ and one involving $\mathcal{H}_{EL}$ [4.77].

If the crystal under investigation has a centre of inversion, the matrix elements of both $\mathcal{H}_{ER}$ and $\mathcal{H}_{EL}$ must relate states of opposite parity in order to yield nonvanishing products contributing to S_{HR}. Therefore, the parity of the phonon linearly involved in $\mathcal{H}_{EL}$ has to be odd. This is only another version of

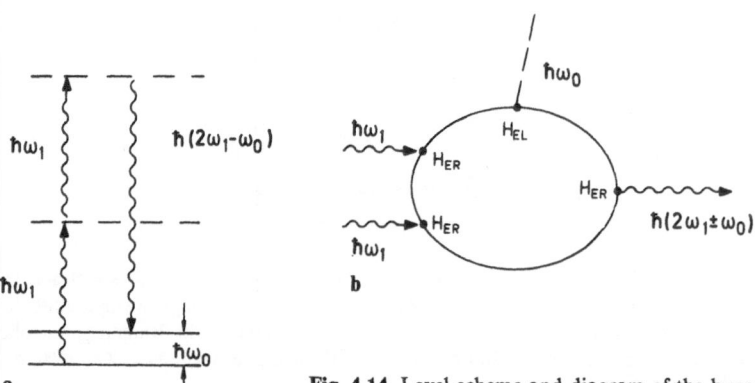

Fig. 4.14. Level scheme and diagram of the hyper-Raman effect

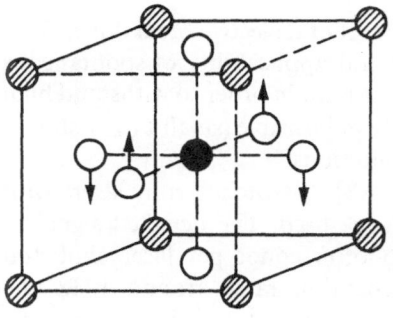

Fig. 4.15. The "silent" mode of $BaTiO_3$ or $SrTiO_3$. Full circle: titanium; empty circles: oxygen; hatched circles: barium or strontium

our earlier statement, that the parity selection rules of HRS are identical with those of one-photon infrared absorption.

As far as phonons are concerned, detailed calculations of S_{HR} have not yet come to the knowledge of the present author. *Jha* and *Woo* [4.77] have deduced general expressions for nonpolar phonons without applying them to a specific material. *Zavorotnev* and *Ovander* [4.78] have developed a theoretical description interpreting HRS as a four-polariton process by which two input polaritons are transformed into *two* scattered ones. As will be illustrated in Sect. 4.5.4, a similar concept has been successfully used for describing HRS via biexcitons in copper halides [4.79].

The symmetry selection rules of HRS were first published by *Cyvin* et al. [4.80]. They extended Placzek's polarizability theory of the Raman effect and considered the transformation properties of the hyper-Raman polarizability $B^{fi}_{\alpha\beta\gamma}(\omega_1)$ we introduced in Sect. 4.1.3. Their tables confirm the two attractive features of HRS:
1) all infrared-active modes are also hyper-Raman-active and thus can be studied by the light scattering technique;
2) HRS allows the observation of "silent" modes which do not carry a dipole moment but only an octupole one.

An example of a "silent" mode is given in Fig. 4.15. The oscillation pattern presented there refers to the F_{2u} mode of cubic perovskites like $BaTiO_3$ or $SrTiO_3$. The oxygen atoms move in such a manner that no dipole moment can arise and only an octupole moment is left [4.45].

4.5.2 Experimental Problems

The order of magnitude of S_{HR} may be roughly estimated by combining (4.10, 43, 45). We find

$$\frac{S_{HR}}{S_R} \approx \frac{E^2 \chi^{(5)}}{\chi^{(3)}} \approx \frac{P^{(5)}}{P^{(3)}} \approx \left(\frac{E}{E_M}\right)^2, \tag{4.60}$$

where we have adopted the notation used in (4.10). Thus, S_{HR} is expected to be several orders of magnitude smaller than a typical value of S_R, even if the incident laser intensity approaches the threshold of dielectric breakdown. This unfavorable situation almost excludes a general applicability of spontaneous HRS. The samples to be studied must be transparent in order to withstand high laser powers. Preferably, they should be highly polarizable (small E_M) or should allow the study of HRS under resonance conditions.

In the pioneering work on HRS [4.4, 81, 82], Q-switched ruby lasers with peak powers in the range of several MW were used. The detected signal in materials like fused quartz was about 1 photon count per laser shot and spectral resolution element. With pulse repetition rates around 1 Hz, the measurement of a hyper-Raman spectrum became a rather time consuming effort, in particular, if done in a single channel scan. Conditions are improved when the ruby laser is replaced by an acousto-optically Q-switched Nd-YAG laser [4.26, 45]. The peak power of this source is lower by a factor of 30. However, the pulse repetition rate up to 10 kHz overcompensates this lack of power and allows a better averaging over laser fluctuations.

Since the hyper-Raman lines are observed in the spectral neighborhood of the second harmonic at 347 or 532 nm, the scattered photons can be detected by the photon counting devices used in standard Raman spectroscopy. The upper limit of the dynamical range is given by the pulse repetition rate of the laser because only a few photon counts can be clearly resolved within a laser pulse width between 5 and 100 ns. The lower limit is determined by the dark noise of the photomultiplier. For a pulse repetition rate of 10 kHz, the effective dark noise can be reduced to about 0.01 counts per second by properly gating the counting electronics synchronously to the laser Q-switch.

A major breakthrough in HRS is expected when optical multichannel systems become generally available and the photon counts of all spectral elements can be simultaneously accumulated (see Chap. 3 of this volume). Concerning quantum efficiency, dark noise and gating facilities, these systems should be comparable with the single-channel photon counting devices. *Savage* and *Maker* [4.83], *French* and *Long* [4.84], as well as *Denisov* et al. [4.85], have already described hyper-Raman spectrometers which allow multichannel detection. The spectral resolution of their instruments, however, seems to be rather low with spectral slitwidths around 20 cm^{-1} and more.

4.5.3 Spontaneous HRS of Phonons and Polaritons

Due to the extremely small cross section, hyper-Raman studies are still rare and concentrate mostly on the effect as such. The following solids have been under investigation: CsBr, CsI, RbI [4.45], CdS [4.87], TiO_2 [4.85], $SrTiO_3$ [4.45, 85, 86], $LiNbO_3$ [4.85], $CaCO_3$ [4.87], $NaNO_3$ [4.20, 88], NH_4Cl [4.19, 20, 83], NH_4Br [4.20], fused quartz [4.4, 85], TiO_2-SiO_2 glass [4.89].

Figure 4.16 shows the rather simple hyper-Raman spectrum of CsI at room temperature [4.45]. The scattering intensity in counts per second has been plotted as a function of the wave number shift from the second harmonic. The

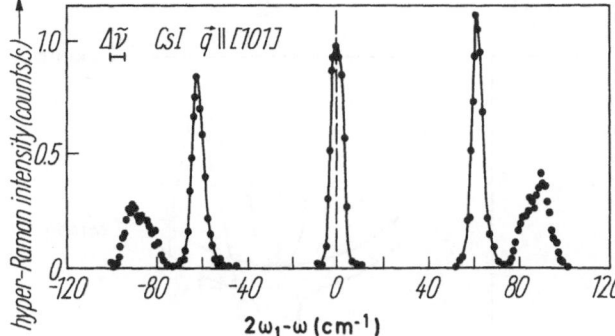

Fig. 4.16. Stokes and anti-Stokes hyper-Raman spectrum of CsI at room temperature. $\lambda_1 = 1.06\,\mu\text{m}$ [4.45]

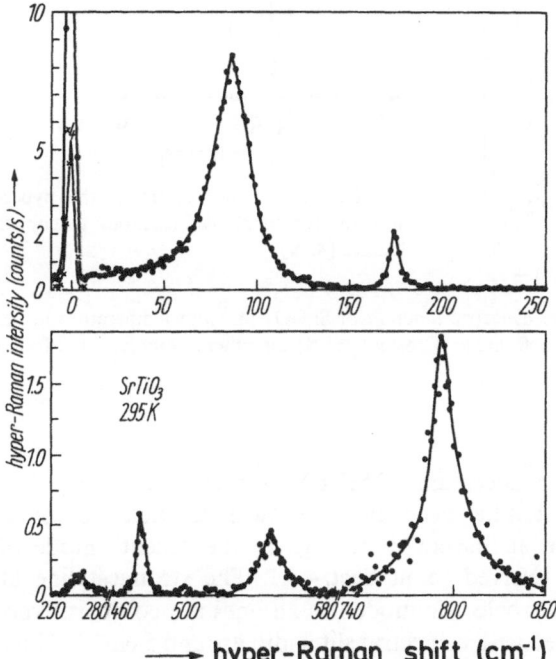

Fig. 4.17. Hyper-Raman spectrum of SrTiO$_3$ at room temperature displaying all Γ-phonons. The line at 175 cm^{-1} is actually a superposition of two lines. $\lambda_1 = 1.06\,\mu\text{m}$ [4.45]

two observed hyper-Raman lines correspond to the TO and LO-phonon near the Γ-point of the Brillouin zone. The ratio of antistokes and Stokes intensity is given by the Boltzmann factor at room temperature and thus indicates that the sample is not being heated up although peak and average laser power are about 20 kW and 3 W, respectively.

Since the sample is centrosymmetric, the hyper-Rayleigh line at $2\omega_1$ is very weak, in particular, if compared with the Rayleigh line encountered in normal Raman spectroscopy. For a perfect crystal, the intensity at $2\omega_1$ should actually vanish completely, because quasi-elastic scattering by acoustic phonons (hyper-Brillouin scattering) or entropy fluctuations is forbidden by symmetry.

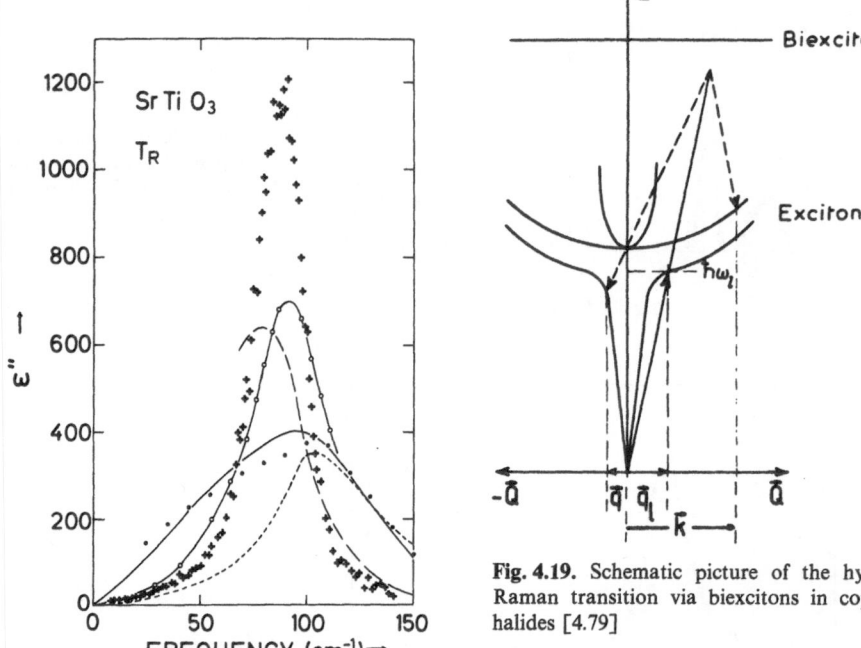

Fig. 4.19. Schematic picture of the hyper-Raman transition via biexcitons in copper halides [4.79]

Fig. 4.18. The imaginary part of the dielectric function of SrTiO₃ at room temperature in the spectral region of the ferroelectric soft mode. Crosses: HRS; all other symbols: far infrared reflection spectroscopy [4.90]

The slightly more complex spectrum of SrTiO$_3$ is presented in Fig. 4.17 [4.45]. A scattering configuration has been chosen allowing the detection of all Γ-phonons. The weakest line at 266 cm^{-1} belongs to the "silent" mode of symmetry type F_{2u} already referred to in Sect. 4.5.1. The strongest line at 88 cm^{-1} corresponds to the ferroelectric mode. As all lines are comparatively broad, they can be well resolved with a spectral slit width around 5 cm^{-1}. Thus, HRS yields mode frequencies and damping constants of SrTiO$_3$ with a precision superior to that of all spectroscopic methods used so far (neutron scattering, infrared reflection, electric field induced Raman scattering).

By application of the fluctuation-dissipation theorem, it is possible to deduce accurate values of the far infrared dielectric function $\varepsilon = \varepsilon' + \varepsilon''$ from the hyper-Raman spectrum. Figure 4.18 compares the imaginary part ε'' as obtained by HRS with the results of various far-infrared reflectivity studies [4.90]. Because of the unusually high reflectivity and small penetration depth of SrTiO$_3$ in the spectral region of the ferroelectric mode, the accuracy of infrared spectroscopy is rather poor, so that considerable discrepancies exist between the data of different authors. As demonstrated by Fig. 4.18, HRS can help in eliminating such uncertainties.

Most of the hyper-Raman works quoted above deal with Raman forbidden or "silent" Γ-phonons. Recently, however, HRS has also been applied to observe the polariton dispersion in $SrTiO_3$ [4.85, 86], TiO_2 and fused quartz [4.85]. The enormous difference between the frequencies of the exciting and the scattered light allows the use of a scattering angle of exactly $0°$. On the other hand, the relatively large dispersion $[n(2\omega_1) - n(\omega_1)]$ confines the observation mainly to the upper or photon-like polariton branches.

Two-phonon structures in a hyper-Raman spectrum were first reported by *Denisov* et al. [4.85]. *Polivanov* and *Sayakhov* [4.87] have found a resonance enhancement of HRS from CdS when tuning the band gap thermally.

4.5.4 Resonant and Stimulated HRS

The first observations of *stimulated* hyper-Raman transitions refer to alkali vapours [4.91]. Here, the process can be strongly enhanced by a twofold resonance characterized by $\omega_1 \approx \omega_{ni}$ and $2\omega_1 \approx \omega_{n'i}$. In sodium, for instance, the $4d$ levels have almost twice the energy of $3p$ levels, so that $3p$ and $4d$ can be utilized as intermediate states $|n\rangle$ and $|n'\rangle$. Then strong resonant and stimulated HRS is observed [4.91]. In contrast to HRS by phonons, however, the final state ($4p$) is far away from the initial one ($3s$), so that the scattered light has frequencies in the near infrared, whereas ω_1 is in the yellow region of the visible spectrum.

Comparably fortunate conditions are provided by the exciton and biexciton energy levels in copper halides. A schematic picture of the doubly resonance enhanced HRS in these materials is given in Fig. 4.19 [4.79]. Inside the crystal, the incident laser photons become polaritons of frequency ω_l and wave vector q_l. Two of them virtually, but almost resonantly excite a biexciton. The biexciton decays into two polaritons. One of them (wave vector q) belongs to the lower branch and thus can escape from the sample as a photon. The other (wave vector k) is on the upper branch and cannot be directly detected.

This type of hyper-Raman emission has been extensively studied by *Grun* and his associates (see [4.79] and references therein). The experimental information can be used for constructing the exciton and polariton branches in copper halides.

References

4.1   N. Bloembergen: *Nonlinear Optics* (W. A. Benjamin, Reading/Massachusetts 1965)
4.2   N. Bloembergen: Am. J. Phys. **35**, 989 (1967)
4.3   P. D. Maker, R. W. Terhune: Phys. Rev. **137**, A801 (1965)
4.4   R. W. Terhune, P. D. Maker, C. M. Savage: Phys. Rev. Lett. **14**, 681 (1965)
4.5   A. Laubereau, W. Kaiser: Rev. Mod. Phys. **50**, 607 (1978); "Coherent Picosecond Interactions", In *Coherent Nonlinear Optics*, ed. by M. S. Feld, V. S. Letokhov, Topics in Current Physics, Vol. 21 (Springer, Berlin, Heidelberg, New York 1980) p. 271
4.6   J. A. Giordmaine, W. Kaiser: Phys. Rev. **144**, 676 (1966)

4.7 A. Yariv: *Quantum Electronics* (Wiley, New York 1975) p. 485
4.8 R. F. Begley, A. B. Harvey, R. C. Byer: Appl. Phys. Lett. **25**, 387 (1974)
4.9 D. A. Long, L. Stanton: Proc. Roy. Soc. Lond. A**318**, 441 (1970)
4.10 C. Flytzanis: In *Quantum Electronics*, Vol. 1A, ed. by H. Rabin, C. L. Tang (Academic Press, New York 1975) p. 9
4.11 R. W. Hellwarth: Progr. Quant. Electr. **5**, 1 (1977)
4.12 M. Maier: Appl. Phys. **11**, 209 (1976)
4.13 V. R. Shen: Rev. Mod. Phys. **48**, 1 (1976)
4.14 S. A. Akhmanov, N. I. Koroteev: Sov. Phys. Usp. **20**, 899 (1977)
4.15 W. M. Tolles, J. W. Nibler, J. R. McDonald, A. B. Harvey: Appl. Spectr. **31**, 253 (1977)
4.16 S. Druet, J. P. Taran: In *Chemical and Biochemical Applications of Lasers*, Vol. IV, ed. by C. B. Moore (Academic Press, New York 1979)
4.17 J. W. Nibler, G. V. Knighten: In *Raman Spectroscopy of Gases and Liquids*, ed. by A. Weber, Topics in Current Physics, Vol. 11 (Springer, Berlin, Heidelberg, New York 1979) p. 253
4.18 M. D. Levenson, J. J. Song: In *Coherent Nonlinear Optics*, ed. by M. S. Feld, V. S. Letokhov, Topics in Current Physics, Vol. 24 (Springer, Berlin, Heidelberg, New York 1980) p. 293
4.19 T. J. Dines, M. J. French, R. J. B. Hall, D. A. Long: *Proc. 5th Intern. Conf. on Raman Spectroscopy*, Freiburg 1976 (H. F. Schulz Verlag, Freiburg 1976) p. 707
4.20 D. A. Long: *Proc. 31st Intern. Meeting of the Societé de Chimie Physique*, Abbaye de Fontevraud 1978 (Elsevier, Amsterdam 1979) p. 153
4.21 M. J. French: In *Chemical Applications of Nonlinear Raman Spectroscopy*, ed. by A. B. Harvey (Academic Press, New York) in press
4.22 L. Ortmann, H. Vogt: Opt. Commun. **16**, 234 (1976); P. Y. Yu and M. Cardona: Solid State Commun. **9**, 1421 (1971)
4.23 S. A. Akhmanov: In *Nonlinear Spectroscopy*, ed. by N. Bloembergen (North-Holland, Amsterdam 1977) p. 239
4.24 H. Mahr: In *Quantum Electronics*, Vol. 1a, ed. by H. Rabin, C. L. Tang (Academic Press, New York 1975) p. 290
4.25 E. Yablonovitch, C. Flytzanis, N. Bloembergen: Phys. Rev. Lett. **29**, 865 (1972)
4.26 H. Vogt, G. Neumann: Opt. Commun. **19**, 108 (1976)
4.27 D. A. Kleinman: Phys. Rev. **126**, 1977 (1962)
4.28 J. A. Armstrong, N. Bloembergen, J. Ducuing, P. S. Pershan: Phys. Rev. **127**, 1918 (1962)
4.29 H. Lotem, R. T. Lynch, Jr., N. Bloembergen: Phys. Rev. A**14**, 1748 (1976)
4.30 N. Bloembergen, H. Lotem, R. T. Lynch, Jr.: Indian J. Pure Appl. Phys. **16**, 151 (1978)
4.31 Y. Prior, A. R. Bogdan, M. Dagenais, N. Bloembergen: Phys. Rev. Lett. **46**, 111 (1981)
4.32 M. Born, K. Huang: *Dynamical Theory of Crystal Lattices* (Clarendon Press, Oxford 1962) p. 203
4.33 Chen-Show Wang: In *Quantum Electronics*, Vol. 1a, ed. by H. Rabin, C. L. Tang (Academic Press, New York 1975) p. 447
4.34 Y. Prior, H. Vogt: Phys. Rev. B**19**, 5388 (1979)
4.35 B. E. Kincaid, J. R. Fontana: Appl. Phys. Lett. **28**, 12 (1976)
4.36 P. Lallemand, P. Simova, G. Bret: Phys. Rev. Lett. **17**, 1239 (1966)
     N. Bloembergen, G. Bret, P. Lallemand, A. Pine, P. Simova: IEEE J. QE-3, 197 (1967)
4.37 S. R. J. Brueck, A. Mooradian: Opt. Commun. **8**, 263 (1973)
4.38 A. Owyoung, C. W. Patterson, R. S. McDowell: Chem. Phys. Lett. **59**, 156 (1978)
4.39 M. D. Levenson, N. Bloembergen: Phys. Rev. B**10**, 4447 (1974)
4.40 D. Heiman, R. W. Hellwarth, M. D. Levenson, G. Martin: Phys. Rev. Lett. **36**, 189 (1976)
4.41 M. D. Levenson, J. J. Song: J. Opt. Soc. Am. **66**, 641 (1976)
     J. J. Song, M. D. Levenson: J. Appl. Phys. **48**, 3496 (1977)
4.42 A. Owyoung, P. S. Peercy: J. Appl. Phys. **48**, 674 (1977)
4.43 S. Kielich: Acta Phys. Polon. **26**, 135 (1964)
4.44 K. Altmann, G. Strey: Z. Naturforsch. **32a**, 307 (1977)
4.45 H. Vogt, G. Neumann: Phys. Stat. Sol. (b) **92**, 57 (1979)
4.46 J. J. Wynne: Comm. Sol. State Phys. **6**, 31 (1974)
4.47 J. W. Fleming, C. S. Johnson, Jr.: J. Raman Spectr. **8**, 284 (1979)

4.48   I.Chabay, G.K.Klauminzer, B.S.Hudson: Appl. Phys. Lett. **28**, 27 (1975)
4.49   G.Laufer, R.B.Miles: Opt. Commun. **28**, 250 (1979)
4.50   A.Compaan, S.Chandra: Opt. Lett. **4**, 170 (1979)
4.51   A.C.Eckbreth: Appl. Phys. Lett. **32**, 421 (1978)
4.52   J.A.Shirley, R.J.Hall, A.C.Eckbreth: Opt. Lett. **5**, 380 (1980)
4.53   Y.Prior: Appl. Opt. **19**, 1741 (1980)
4.54   L.A.Carreira, L.P.Goss, T.B.Malloy, Jr.: J. Chem. Phys. **66**, 2762 (1977)
4.55   R.T.Lynch, Jr., S.D.Kramer, H.Lotem, N.Bloembergen: Opt. Commun. **16**, 372 (1976)
4.56   J.J.Song, G.L.Eesley, M.D.Levenson: Appl. Phys. Lett. **29**, 567 (1976)
4.57   E.Yablonovitch, N.Bloembergen, J.J.Wynne: Phys. Rev. B**3**, 2060 (1971)
4.58   J.J.Wynne: Phys. Rev. B**6**, 534 (1972)
4.59   Y.R.Shen: Appl. Phys. Lett. **23**, 516 (1973)
4.60   M.D.Levenson, C.Flytzanis, N.Bloembergen: Phys. Rev. B**6**, 3962 (1972)
4.61   M.D.Levenson: IEEE J. QE-**10**, 110 (1974)
4.62   S.A.Akhmanov, N.I.Koroteev: Sov. Phys. JETP **40**, 650 (1975)
4.63   S.A.Akhmanov, R.V.Khokhlov, A.P.Sukhorukov: In *Laser Handbook*, Vol. II, ed. by
       F.T.Arecchi, E.O.Schulz-Dubois (North-Holland, Amsterdam 1972) p. 1151
4.64   J.J.Wynne: Phys. Rev. Lett. **29**, 650 (1972); Comm. Sol. State Phys. **7**, 7 (1975)
4.65   Yu.N.Polivanov, A.T.Sukhodol'skii: Sov. J. Quantum Electron. **8**, 962 (1978)
4.66   R.W.Hellwarth: J. Opt. Soc. Am. **67**, 1 (1977);
       A.Tomita: Appl. Phys. Lett. **34**, 463 (1979)
4.67   A.Maruani, J.L.Oudar, E.Batifol, D.S.Chemla: Phys. Rev. Lett. **41**, 1372 (1978)
4.68   D.S.Chemla, A.Maruani, E.Batifol: Phys. Rev. Lett. **42**, 1075 (1979)
4.69   M.A.F.Scarparo, J.H.Lee, J.J.Song: Opt. Lett. **6**, 193 (1981)
4.70   A.Koster, S.Biraud-Laval, R.Reinisch: Sol. State Commun. **27**, 1167 (1978)
4.71   G.L.Eeseley, M.D.Levenson: Opt. Lett. **3**, 178 (1979)
4.72   M.A.Washington, H.Z.Cummins: Phys. Rev. B**15**, 5843 (1977);
       R.Tubino, J.L.Birman: Phys. Rev. B**15**, 5843 (1977)
4.73   H.Lotem, C.B.de Araujo: Phys. Rev. B**16**, 1711 (1977)
4.74   C.B.de Araujo, H.Lotem: Phys. Rev. B**18**, 30 (1978)
4.75   S.D.Kramer, N.Bloembergen: Phys. Rev. B**14**, 4654 (1976)
4.76   H.Lotem, R.T.Lynch, Jr.: Phys. Rev. Lett. **37**, 334 (1976)
4.77   S.S.Jha, J.W.F.Woo: Nuovo Cimento 2B, 167 (1971)
4.78   Yu.D.Zavorotnev, L.N.Ovander: Sov. Phys. Solid State **16**, 1550, 2413 (1975); Sov. J. Quant.
       Electron. **5**, 644 (1975)
4.79   B.Hönerlage, U.Rössler, Vu Duy Phach, A.Bivas, J.B.Grun: Phys. Rev. B**22**, 797 (1980)
4.80   S.J.Cyvin, J.E.Rauch, J.C.Decius: J. Chem. Phys. **43**, 4083 (1965)
4.81   J.F.Verdieck, S.H.Peterson, C.M.Savage, P.D.Maker: Chem. Phys. Lett. **7**, 219 (1970)
4.82   P.D.Maker: Phys. Rev. A**1**, 923 (1970)
4.83   C.M.Savage, P.D.Maker: Appl. Opt. **10**, 965 (1971)
4.84   M.J.French, D.A.Long: J. Raman Spectr. **3**, 391 (1975)
4.85   V.N.Denisov, B.N.Mavrin, V.B.Podobedov, Kh.E.Sterin: Sov. Phys. JETP **75**, 684 (1978);
       Opt. Commun. **26**, 372 (1978) (LiNbO$_3$); Sov. Phys. JETP Lett. **31**, 101 (1980) (SrTiO$_3$); Sov.
       Phys. JETP Lett. **32**, 316 (1980) (quartz); Opt. Commun. **34**, 357 (1980) (TiO$_2$)
4.86   K.Inoue, T.Sameshima: J. Phys. Soc. Jpn. **47**, 2037 (1979); K.Inoue, N.Asai, T.Sameshima: J.
       Phys. Soc. Jpn. **48**, 1787 (1980)
4.87   Yu.N.Polivanov, R.Sh.Sayakhov: Phys. Stat. Sol. (b) **103**, 89 (1981) (CaCO$_3$); Sov. Phys.
       JETP Lett. **29**, 10 (1979) (CdS)
4.88   H.Vogt, G.Neumann: Phys. Stat. Sol. (b) **86**, 615 (1978)
4.89   B.G.Varshal, V.N.Denisov, B.N.Mavrin, G.A.Parlova, V.B.Podobedov, Kh.E.Sterin: Opt.
       Spectrosc. (USSR) **47**, 344 (1980)
4.90   H.Vogt, G.Rossbroich: Phys. Rev. B**24**, 3086 (1981)
4.91   D.Cotter, D.C.Hanna, W.H.W.Tuttlebee, M.A.Yuratich: Opt. Commun. **22**, 190 (1977)

Additional References

1. R. Baumert, I. Broser: A tunable CdS polariton laser. Solid State Commun. **38**, 31 (1981)
2. V. N. Denisov, B. N. Mavrin, V. B. Podobedov, Kh. E. Sterin: Polariton Fermi resonance in hyper-Raman spectra of a calcite crystal. Solid State Commun. **40**, 793 (1981)
3. G. L. Eesley: *Coherent Raman Spectroscopy* (Pergamon, Oxford 1981)
4. C. R. Giuliano: Applications of optical phase conjugation. Phys. Today **34**, 27 (1981)

Subject Index

Springer Series in Solid-State Sciences

Editors: M. Cardona, P. Fulde, H.-J. Queisser

Springer-Verlag
Berlin Heidelberg New York

A monthly journal

Applied Physics A
Solids and Surfaces

Applied Physics A "Solids and Surfaces" is devoted to concise accounts of experimental and theoretical investigations that contribute new knowledge or understanding of phenomena, principles or methods of applied research.
Emphasis is placed on the following fields (giving the names of the responsible co-editors in parentheses):

Solid-State Physics
Semiconductor Physics (**H.J.Queisser,** MPI Stuttgart)
Amorphous Semiconductors (**M.H.Brodsky,** IBM Yorktown Heights)
Magnetism (Materials, Phenomena) (**H.P.J.Wijn,** Philips Eindhoven)
Metals and Alloys, Solid-State Electron Microscopy (**S.Amelinckx,** Mol)
Positron Annihilation (**P.Hautojärvi,** Espoo)
Solid-State Ionics (**W.Weppner,** MPI Stuttgart)

Surface Science
Surface Analysis (**H.Ibach,** KFA Jülich)
Surface Physics (**D.Mills,** UC Irvine)
Chemisorption (**R.Gomer,** U.Chicago)

Surface Engineering
Ion Implantation and Sputtering (**H.H.Andersen,** U.Aarhus)
Laser Annealing (**G.Eckhardt,** Hughes Malibu)
Integrated Optics, Fiber Optics, Acoustic Surface Waves (**R.Ulrich,** TU Hamburg)

Special Features:
Rapid publication (3–4 months)
No page charges for concise reports
50 complimentary offprints
Microform edition available

Articles:
Original reports and short communications.
Review and/or tutorial papers
To be submitted to:
Dr.H.K.V.Lotsch, Springer-Verlag,
P.O.Box 105280, D-6900 Heidelberg, FRG

Subscription information and/or **sample copies** are available from your bookseller or directly from Springer-Verlag, Journal Promotion Dept., P.O.Box 105280, D-6900 Heidelberg, FRG

**Springer-Verlag
Berlin
Heidelberg
New York**